Somaclonal Variation: Basic and Practical Aspects

Carolina Sánchez-Romero
Editor

Somaclonal Variation: Basic and Practical Aspects

Springer

Editor
Carolina Sánchez-Romero
Department of Botany and Plant Physiology
University of Malaga
Malaga, Spain

ISBN 978-3-031-51625-2 ISBN 978-3-031-51626-9 (eBook)
https://doi.org/10.1007/978-3-031-51626-9

This Springer imprint is published by the registered company Springer Nature Switzerland AG
The registered company address is: Gewerbestrasse 11, 6330 Cham, Switzerland

Preface

In vitro culture includes a wide variety of different techniques such as micropropagation, somatic hybridization, haploid and polyploid production, embryo rescue, in vitro mutagenesis, variant selection, or plant regeneration through organogenesis or somatic embryogenesis. These procedures can be applied with different purposes such as clonal mass propagation, virus eradication, germplasm conservation, production of secondary metabolites, basic investigations, or plant breeding by using conventional means or biotechnological tools, such as genetic transformation or gene editing.

Cell and tissue culture only involves mitotic divisions of cells, and therefore, one of the main goals of in vitro culture is the production of genetically uniform plants. However, from the beginnings of in vitro culture, different abnormalities have been observed in the obtained plants. In 1981, Larkin and Scowcroft stated that in vitro culture itself is a source of variation and proposed the term "somaclonal variation" to refer to variations induced by in vitro culture in cells, cultures, regenerated plants, and their progenies. Occurrence of this spontaneous and uncontrolled variation can represent a serious problem in most systems. However, it constitutes an alternative source of variability that can be exploited in plant breeding.

In this book, basic and practical aspects of somaclonal variation are included. Specifically, the first chapters are focused on the genetic and epigenetic basis of somaclonal variation, as well as on the main factors affecting its occurrence. The second part addresses different approaches for somaclonal variation evaluation. The last chapters analyze somaclonal variation in important groups of plants or crop species, examining different aspects of somaclonal variation in these specific cases.

Malaga, Spain — Carolina Sánchez-Romero

Contents

Contributors

Mehdi Alizadeh Department of Horticultural Science, Faculty of Plant Production, Gorgan University of Agricultural Sciences and Natural Resources, Basij, SQ, Gorgan, Golestan, Iran

Isabel Ballesteros Departamento de Genética, Facultad de Ciencias Biológicas, Universidad Complutense de Madrid, Madrid, Spain

Fredy Carrera Departamento de Ciencias de la Vida y la Agricultura, Carrera de Agropecuaria, Universidad de las Fuerzas Armadas ESPE, Santo Domingo, Ecuador

Zahra Rezaei Ghaleh Department of Horticultural Science, Faculty of Plant Production, Gorgan University of Agricultural Sciences and Natural Resources, Basij, SQ, Gorgan, Golestan, Iran

M. Elena González-Benito Department of Biotechnology-Plant Biology, Universidad Politécnica de Madrid, Madrid, Spain

Jennifer Ann Harikrishna Centre for Research in Biotechnology for Agriculture (CEBAR), Universiti Malaya, Kuala Lumpur, Malaysia

Ana M. Henao Biotechnology Group, Faculty of Exact and Natural Sciences, University of Antioquia, Medellín, Colombia

Sandeep Kaushik Department of Environmental Science, Indira Gandhi National Tribal University, Amarkantak, Madhya Pradesh, India

Rosario Linacero Departamento de Genética, Facultad de Ciencias Biológicas, Universidad Complutense de Madrid, Madrid, Spain

Carmen Martín Department of Biotechnology-Plant Biology, Universidad Politécnica de Madrid, Madrid, Spain

Abdul Aziz Mirani Date Palm Research Institute (DPRI), Shah Abdul Latif University, Khairpur, Sindh, Pakistan

Akio Miyao Institute of Crop Science, NARO, Tsukuba, Ibaraki, Japan

Tatiana Osorio Biotechnology Group, Faculty of Exact and Natural Sciences, University of Antioquia, Medellín, Colombia

Manoj K. Rai Department of Environmental Science, Indira Gandhi National Tribal University, Amarkantak, Madhya Pradesh, India

Roshni Rathour Department of Environmental Science, Indira Gandhi National Tribal University, Amarkantak, Madhya Pradesh, India

Carolina Sánchez-Romero Departamento de Botánica y Fisiología Vegetal, Universidad de Málaga, Málaga, Spain

Mostafa Khoshhal Sarmast Department of Horticultural Science, Faculty of Plant Production, Gorgan University of Agricultural Sciences and Natural Resources, Basij, SQ, Gorgan, Golestan, Iran

Amit Singh Department of Environmental Science, Indira Gandhi National Tribal University, Amarkantak, Madhya Pradesh, India

Najamuddin Solangi Date Palm Research Institute (DPRI), Shah Abdul Latif University, Khairpur, Sindh, Pakistan

Chee How Teo Centre for Research in Biotechnology for Agriculture (CEBAR), Universiti Malaya, Kuala Lumpur, Malaysia

Aura I. Urrea Biotechnology Group, Faculty of Exact and Natural Sciences, University of Antioquia, Medellín, Colombia

Srishti Yadav Department of Environmental Science, Indira Gandhi National Tribal University, Amarkantak, Madhya Pradesh, India

Genetic Basis of Somaclonal Variation

Rosario Linacero and Isabel Ballesteros

Abstract Plant tissue culture is a source of genetic changes that are referred to as somaclonal variation. Somaclonal variation as a cause of variability has advantages and disadvantages. It is undesirable when in vitro culture is used as a microprogation tool to clone genetically identical plants. However, as a source of new intraspecific variation, it can be exploited by breeders to obtain new varieties. The molecular basis of somaclonal variation is not known, but both genetic and epigenetic mechanisms appear to be involved. The genetic basis of somaclonal variation, as well as the possible causes and genetic tools used to detect somaclonal variants are reviewed in this chapter.

Keywords Somaclonal variation · Tissue culture · Genetics · Molecular marker

1 Definitions and Background

Plant tissue culture is a source of genetic changes referred to as somaclonal variation. Somaclonal variation, first described by Larkin and Scowcroft (1981), refers to variation arising in the cell cultures, regenerated plants and their progenies. This has been observed in regenerated plants (Kaeppler et al. 2000) and in their progeny (Linacero and Vázquez 1993). Somaclonal variation as a source of genetic variation has advantages and disadvantages. Phenotypic variation is undesirable when in vitro culture is used as a micropropagation tool to clone genetically identical plants, as it causes major economic losses. It can also be a serious setback in plant breeding and commercialization, germplasm conservation and obtainment of transgenic and double haploid plants. However, it can be considered a source of new intraspecific variation that can be exploited by breeders to obtain new varieties with

R. Linacero (✉) · I. Ballesteros
Departamento de Genética, Facultad de Ciencias Biológicas, Universidad Complutense de Madrid, 28040 Madrid, Spain
e-mail: charolin@bio.ucm.es

I. Ballesteros
e-mail: maball09@ucm.es

C. Sánchez-Romero (ed.), *Somaclonal Variation: Basic and Practical Aspects*,
https://doi.org/10.1007/978-3-031-51626-9_1

traits of interest. Some agronomic traits obtained from somaclonal variation have been used in crop breeding (Krishna et al. 2016). The frequency of somaclonal variation is influenced by the genotype of the plant species, the tissue used as explant and the regeneration protocol, among other factors. Detection and quantification of somaclonal variation depends on the method and molecular marker used for analysis. The molecular basis of somaclonal variation is not known, but both genetic and epigenetic mechanisms have been proposed (Kaeppler et al. 2000; Tanurdzic et al. 2008).

In vitro plant regeneration is a complex developmental process, which is only partially understood (Long et al. 2022). Explants taken from different parts of the plant are grown in vitro, in a culture medium with different nutrients and plant hormones. The first stress to which plant cells are exposed is wounding. Thereafter, cells leave their normal position in the plant and are placed in a culture medium to induce, directly or indirectly, a morphogenetic response. Cultured cells undergo a complex process that can trigger the expression of pluripotency or totipotency. They acquire a new developmental program leading to organ or embryo differentiation. The processes associated with dedifferentiation and redifferentiation induce stress in plant cells, resulting in genetic and epigenetic instability (Grafi and Avivi 2004; Grafi et al. 2011; Wang et al. 2013). Normally, cells that are no longer dividing re-enter the cell cycle and, in most cases, undergo a high rate of division and callus formation; this behavior has often been compared to that of tumor cells. The culture medium contains various compounds at different concentrations, including phytohormones. In the case of 2,4-D (2,4-dichlorophenoxyacetic acid), an essential hormone in the induction of somatic embryogenesis, its role in generating stress in cells cultured in vitro has been demonstrated. In addition, the induction of proliferation involves a high rate of cell division, and in the S phase, during DNA replication, systems for detecting and repairing genetic damage must be maintained. Errors in these systems can cause changes in the genetic material of these cells. These mutations can have no apparent effect. However, some of these cells have acquired totipotency or pluripotency and, under the right conditions, are capable of forming a complete plant in which all cells, including those of the germ line, would carry the modification. In this way, plants that should be genetically identical to the plant from which the initial explant was taken may carry changes that are not present in the genome of the donor plant.

2 Sources of Variations in Plant Tissue Culture

Before discussing the origin of somaclonal variation, some of the characteristics and peculiarities of this type of variation will be discussed. Although in many cases the underlying mechanisms are not well understood, there are certain factors that are known to influence the occurrence of somaclonal variation. These factors include:

- Medium composition

The composition of the culture medium, especially the type and concentration of growth regulators. Auxins, in particular the auxin 2,4-D, are considered a factor that may be related to mutation induction. Hormones present in the culture medium have been shown to have a significant effect on the frequency of point mutations and, in the case of 2,4-D, on ploidy changes (Silva and Carvalho 2014).

- Age of culture

The age of the culture seems to influence genetic instability, partly due to the accumulation of changes over time. Some authors suggest that the changes occur in the early stages of the culture, so that the mutation rate does not increase over time, but rather the mutations produced are maintained and accumulate.

- Explant type and regeneration pathway

In terms of explant type, more variation is generally obtained from more differentiated tissues, such as roots, leaves, and stems, than from explants with pre-existing meristems, such as axillary buds. There may be a relationship between the degree of dedifferentiation and disorganization that cultures undergo and the frequency of variation. When shoots are produced directly from pre-existing meristems in the explant, plants are usually unchanged. However, when regeneration involves a prolonged dedifferentiation step, such as the formation of callus or protoplasts, the regenerated plants may show genetic variation, possibly due to a response of the genome to the stress that in vitro conditions impose on plant cells. According to this reasoning, meristem culture would be the most stable, followed, in order, direct formation of organs or embryos, regeneration from callus, cell suspensions and finally protoplast cultures. Regarding the pathway of regeneration, some authors suggest that there is more variation in plants regenerated by organogenesis than in plants regenerated by somatic embryogenesis.

- Genotype and genome size

Not all genotypes show the same rate of variation in in vitro cultures. In some species, the presence of somaclonal variants has not been described, while in others it is relatively common to obtain plants with phenotypes or genotypes different from those of the mother plant. These differences are found not only between species, but also between varieties or cultivars of the same species. The frequency and spectrum of variation observed in regenerated plants seems to be influenced by the ploidy level. Haploids tend to be quite stable while polyploids show a greater number of chromosomal changes. In diploid species, there is more variability; there are species that are very stable in cultivation, such as barley, and others that are very variable, such as rye. Among all the factors that influence somaclonal variation, genotype is perhaps the most important (Shen et al. 2007; Tican et al. 2008; Nwauzoma and Jaja 2013).

Regarding the characteristics of this type of variation, it should be noted that it exhibits several peculiarities that do not occur when other systems are used to induce mutations.

- The spectrum of variation is enormous and different from that obtained with mutagenic treatments. Studies have been carried out comparing the frequency of mutations induced during in vitro culture with those obtained after mutagenic treatments with chemical or physical agents. The results of these studies indicate that both the spectrum of variation and the nature of the phenotypic changes observed are different (Linacero and Vázquez 1993).
- Several alterations per individual. One of the most striking aspects of somaclonal variation is that in vitro cultured plants can accumulate more than one mutation generated during in vitro culture and pass it on to their progeny.
- The appearance of dominant mutations. Mutagenic treatments usually induce changes in which the affected gene loses its function and therefore has a recessive effect, with the altered phenotype being detected in the progeny. In the case of somaclonal variation, dominant mutations are detected, which may represent a significant percentage of the total variation detected.
- There are homozygous variations. This is other of the "peculiar" features of somaclonal variation, the detection of regenerated plants with mutations that are homozygous.
- Some of the mutations do not show Mendelian inheritance, and others are unstable. One example is the transmission of the dwarfism trait found in regenerated rice plants. Oono (1985) obtained dwarf regenerated plants of rice. The progeny resulting by self-fertilization of dwarf plants (homozygous recessive) was uniform, all the plants were dwarf. When crossing a normal height plant (homozygous dominant) with a dwarf plant, the offspring was normal and heterozygous. It would be expected that in the offspring, obtained by self-fertilization of these plants, dwarf plants would appear again, however this does not occur, all the offspring were normal, showing a paramutation-like behavior.

3 Genetic Basis of Somaclonal Variation

Modifications induced during in vitro culture can be genetic or epigenetic, and both are closely related. In this chapter, we will focus on genetic changes. Genetic changes include those that affect ploidy, chromosome number or structure, DNA mutations, polymorphic bands, or sequence divergences (Krishna et al. 2016).

Several mechanisms have been proposed for the genetic basis of somaclonal variation, including pre-existing chimeric tissue, errors during mitosis, somatic overcrossing, point mutations, or activation of mobile elements, which will be discussed.

Are there pre-existing changes in the explant cells cultured in vitro? Several studies have attempted to determine how much of the variation observed in regenerated plants is pre-existing and how much is de novo. Among the thousands of cells that make up an explant, there may be changes prior to in vitro culture that are passed on to daughter cells. If any of these cells are involved in the formation of organs or somatic embryos, the plant regenerated from them would carry the mutation. In this case, the changes detected in the regenerated plants would not have been induced in vitro but would be pre-existing changes in the tissues revealed by in vitro culture. These changes could affect both the number of chromosomes and DNA sequence. During the process of cytodifferentiation in plant development, some cell types increase their ploidy. If these cells are involved in plant regeneration during tissue culture, they may have a different chromosome number than expected, in which case the origin of the change in the regenerated plant would not be the in vitro culture per se. Similarly, if explant cells have nucleotide sequence changes, regenerated plants with these mutations could be obtained, but the mutation would not have occurred during in vitro culture. Sato et al. (2011) analyzed changes detected in regenerated *Saintpaulia ionanta* plants. They found that 87–95% of the fixed mutations originated in vitro, but only 5–13% of the mutations were due to pre-existing changes.

3.1 Cytogenetic Variation

Chromosome studies of plant tissue cultures are used to distinguish chromosome number, relative size, centromere position, and changes in chromosome structure.

In vitro conditions affect mitotic stability, so changes in chromosome number and structure are more common in in vitro regenerated plants than in those obtained by other propagation methods. Cytogenetic abnormalities, including ploidy, changes in chromosome number as well as in chromosome architecture (duplications, translocation, deletion and inversions) have been observed in both in vitro cultured cells and regenerated plants. In some cases, regenerated plants show a chromosomal mosaic, which is often not heritable. In contrast, some regenerated plants are solid mutants, and the cytogenetic change can be transmitted to subsequent generations (Linacero and Vázquez 1992). Although cytogenetic alterations are detected in all regeneration systems, they are more frequent in callus and plants derived from them, since their formation involves "uncontrolled" cell division (Duta-Cornescu et al. 2023).

3.2 Somatic Recombination

Mitotic crossovers have been detected in cells in culture and regenerated plants, including both symmetric and asymmetric recombination. Somatic recombination can explain the occurrence of homozygous mutants in regenerated plants and can also result in deletions or duplications. Auxin has been shown to increase the frequency of somatic recombination (Filkowski et al. 2003).

3.3 Point Mutations

Phenotypic changes in regenerated plants and/or their progeny can result from point mutations (substitutions, additions, or deletions) affecting a coding region. Similarly, point mutations could be involved in many of the modifications found in the banding pattern that appear in DNA fingerprinting and other molecular techniques used to assess variability. Until recently, the presence of point mutations in somaclones has only been demonstrated experimentally in a few cases. The use of high-throughput sequencing techniques such as whole-genome sequencing (WGS) has made it possible to sequence the genomes of regenerated plants with stable phenotypic changes. The study of the whole genome of somaclonal variants in species such as *Arabidopsis thaliana*, rice, cucumber, *Brassica rapa*, orchid, *Pyrus*, allows the analysis of the genetic basis of the changes induced during in vitro culture (Jiang et al. 2011; Zhang et al. 2014; Park et al. 2020; Wang et al. 2019; Skarzyńska et al. 2020; Liu et al. 2022).

Whole-genome sequencing of regenerating Arabidopsis lineages with heritable phenotypic variation showed that the most frequent changes found were single nucleotide substitutions in protein-coding genes. In addition, the frequencies of both base substitutions and small insertions/deletions (InDels) were significantly higher than those observed in spontaneous mutations in seed-derived plants. In addition, the distribution of polymorphisms in the chromosomes was not uniform. It was suggested that alteration in the mechanisms of genetic damage repair and maintenance of sequence integrity would be one of the major causes of this genetic variability (Jiang et al. 2011). The results found in genome-wide studies of regenerated plants of other species are similar in two aspects: most of the changes detected are Single Nucleotide Polymorphisms (SNPs) or InDels, and the estimated rate of change in regenerated plants is much higher than the rate of spontaneous mutation.

3.4 Mobile Elements

Activation of both class I and class II mobile elements is one possible cause of the mutations found in regenerated plants. Mobile elements constitute the majority of the plant genome, although most are inactive because the host genome has mechanisms to silence them to prevent their replication (Smith et al. 2012). However, the plasticity of the plant genome to adapt to adverse conditions allows different types of stress, including that induced by in vitro culture, to activate some of these elements, either through changes in methylation patterns or by inducing the expression of those elements that are not methylated.

The role of mobile elements in somaclonal variation has been demonstrated in different species, such as maize, rice, barley, potato, tobacco, carrot, etc. The activity of different retrotransposons has been verified by detecting new copies of the retrotransposon inserted into the genome of both cell culture lines and regenerants because of in vitro induced activity. In rice, three retrotransposons (*Tos10*, *Tos17*, and *Tos19*) were activated under tissue culture conditions (Hirochika et al. 1996; Hirochika 1997). The most studied, *Tos17*, is only activated by in vitro culture.

The increase in copy number with prolongation of culture time has been demonstrated (Miyao et al. 2003). In addition, the insertion of new copies of *Tos17* was shown to occur preferentially in single-copy sequences, up to 84% of new insertions. Phenotypic analysis of *Tos17* insertion lines showed that half of the mutant lines had at least one phenotypic change (Miyao et al. 2007). However, the contribution of *Tos17* insertions to this phenotypic change was estimated to be < 10%. *Tos17* copies are inactivated in the regenerated plant and remain in their inserted position. These features allow the use of this system as an endogenous mutagen to obtain insertion mutants of great utility in functional genomics studies of this species. The activity of *Karma*, a Long-INterspersed Element (LINE) retrotransposon (Komatsu et al. 2003), and *mPing*, a Miniature Inverted-repeat Transposable Element a class II transposon (Kikuchi et al. 2003), has also been reported in rice. Not all mobile element transcriptional activity leads to new insertions, as in the case of *Tos17*. In some cases, RNA is detected but no new insertions are found, in others new copies are detected in offspring. For example, the *Karma* transcript is activated in cultured cells, but no increase in copy number is detected in cultured cells or regenerated plants; the increase is detected in progeny plants (Komatsu et al. 2003).

In regenerated maize plants, in addition to retrotransposon activity, activation of transposons (class II), specifically activator (*Ac*), suppressor mutator (*Spm*), and mutator transposon, has been detected. In rye, the *RYS1* element was activated in vitro and its position in the genome was changed (Alves et al. 2005).

The genomes of different plant species have different populations of mobile elements and probably different mechanisms of response to stress, different efficiencies to inactivate and/or eliminate these elements, and certainly different capacities to tolerate the action of these elements. All this could be reflected in their response to in vitro conditions, which could explain the different mutation rates found among genotypes. Several works based on the analysis of the genomes of regenerated plants by WGS have evaluated the role of mobile elements in somaclonal variation. In Arabidopsis, the contribution of transposable elements is lower than expected, and in the case of rice, up to 43 mobile elements were analyzed and only the mobilization of *Tos17* was detected (Jiang et al. 2011; Miyao et al. 2012). In wheat, transposable elements were found to be the most likely factors for somaclonal variation at the genomic level (Baránek et al. 2016).

4 Analysis of Somaclonal Variation

Depending on the method used to assess variation, different types of changes have been described, ranging from phenotypic to chromosomal and molecular changes. Changes found in in vitro cultured cells and regenerated plants include those related to genetic information, changes in (1) DNA content, (2) chromosome structure or number, (3) DNA sequence, (4) length of repetitive sequences (both increase and decrease), (5) activation of mobile elements. Detection of changes can be done by analyzing different types of morphological, biochemical, molecular DNA and cytogenetic markers. Detection of changes is related to the type of analysis performed. Genetic changes are detected through their products when morphological or biochemical markers are analyzed, or directly when molecular or cytogenetic markers are analyzed.

4.1 Morphological Markers

Phenotypic analysis involves making crosses and analyzing the progeny to determine, first, whether the trait is inherited and therefore it is caused by a genetic change and, second, by studying segregations, how it is transmitted. It must be considered that if the change that occurs during the in vitro culture is a loss of function, it will not be reflected in the phenotype of the regenerated plant, since it has a recessive behavior. Phenotypic changes observed in regenerated plants affect both quantitative and qualitative traits. In many cases, phenotypic variation in regenerated plants is transient and not transmitted to progeny. In the case where the variation is transmitted to the progeny by Mendelian inheritance, it is assumed to be caused by a mutation. To detect recessive mutations, it is necessary to analyze progeny to determine whether regenerated plants carry such mutations or not. Variations in the phenotype due to changes affecting both nuclear genes, with monogenic or polygenic inheritance, and organelle genes have been detected by analyzing the progeny of regenerated plants. The design of the crosses and their interpretation are described in the section Detection of Somaclonal Variation: Progeny Analysis.

4.2 Cytogenetic Markers

The most used method has been metaphase karyotypic analysis, which makes possible to detect rearrangements (inversions, duplications, translocations) and changes in chromosome number (haploid, polyploid, aneuploid) in both in vitro cultured cells and regenerated plants (Ochatt et al. 2011).

In some cases, the regenerated plants are a chromosomal mosaic and therefore, if the change does not affect the germ line, it will not be transmitted to the progeny. In cases where the regenerated plant is a solid mutant, the cytogenetic change will be transmitted to the next generation (Linacero and Vázquez 1992).

Another technique for analysis of DNA content and ploidy changes is flow cytometry, which is simple and reproducible. This technique has been used to detect changes in ploidy and chromosome number in regenerated plants of several species (Currais et al. 2013).

4.3 *DNA Molecular Markers*

The use of DNA markers makes it possible to detect the presence of genetic alterations in regenerated plants, even if their phenotype is identical to that of the wild type. Most of the markers used are based on DNA analysis by polymerase chain reaction (PCR) and/or sequencing and are the same as those used to analyze genetic variability, from multilocus markers to SNP analysis or WGS. Moreover, one of the advantages of using these markers over the analysis of morphological markers is that they allow the presence of genetic alterations to be detected in regenerated plants without the need for these plants to complete their development. Advantages and disadvantages of using molecular markers versus morphological markers are listed below:

- Molecular markers are neutral and unaffected by environmental changes. In the case of morphological markers, environmental conditions can influence the development of individuals and thus their phenotype.
- An unlimited number of molecular markers can be generated, covering a large part of the genome. Morphological markers are few and have low genome coverage.
- Molecular markers can detect a high degree of polymorphism in DNA. Polymorphism is lower for morphological markers.
- As mentioned above, DNA analysis can be performed at early stages of development, whereas morphological markers of maturity require plants to have completed their development.
- The interpretation of molecular markers is fast, simple and objective. In the case of morphological markers, it is necessary to know the phenotypic characteristics of the species, it is necessary to have some training and in some cases the interpretation can be subjective.

5 Detection of Somaclonal Variation: DNA Molecular Markers

The analysis of regenerated plants after in vitro culture is very important to assess the genetic fidelity.

As mentioned above, the assessment of genetic variation based on phenotypic analysis alone can be misleading; plants may not show phenotypic changes but could be carriers of recessive mutations which can appear in the progeny.

The term molecular marker refers to a DNA-based marker: a polymorphism in the DNA sequence of the species under study that allows individuals to be distinguished. However, in many cases it is not necessary to know all the sequence differences between two or more genomes. It is sufficient to analyze the variations using an appropriate specific DNA polymorphism detection method.

Different types of molecular markers are available and new methods are being developed. To choose the most appropriate marker for a given analysis, it is necessary to consider the advantages and disadvantages of using different types of markers, since they differ in their technical requirements, level of polymorphism detected, reproducibility and cost (Adhikari et al. 2017). However, they all detect genetic variability, are not influenced by the environment, and can be analyzed in the early stages of plant development. These advantages over traditional methods do not mean exclusion but complementarity. Many researchers prefer to use a combination of different markers (Campbell et al. 2011; Noormohammadi et al. 2020).

Here we describe some of the DNA molecular markers commonly used for the analysis of somaclonal variation in tissue culture, excluding techniques based on hybridization as Restriction Fragment Polymorphism (RFLP) or methods with reproducibility problems, such as RAPD (Random Amplified Polymorphic). These markers have been used in many studies, but their reproducibility problems mean that in many cases the scientific community does not accept the results obtained. Table 1 shows some examples of molecular markers used in the analysis of somaclonal variation. The RAPD technique is included in cases where it is used in combination with other markers.

Molecular markers can be classified according to their dominant or recessive nature or according to the primers used for amplification: those that require prior knowledge of the genomic sequence and those that do not.

5.1 Dominant Markers

Dominant markers do not allow homozygotes to be distinguished from heterozygotes for each locus. Most of markers obtained by conventional PCR are dominant, such as ISSR (Inter-Simple Sequence Repeat), AFLP (Amplified Fragment Length Polymorphsim), RAPD, ScoT (Start Codon Targeted Polymorphism) and Retrotransposon-Based Markers.

5.1.1 Inter-Simple Sequence Repeat (ISSR)

ISSR involves the amplification of DNA segments between simple sequence repeats in opposite directions. Prime target multiple genomic loci in a single PCR reaction, amplifying approximately 20–40 fragments of different sizes. ISSR markers are highly polymorphic, simple, rapid, and highly reproducible, with the major advantage that no prior information about the genomic sequence is required. This technique has been used successfully on several occasions to screen for genetic variation in plant tissue culture across a range of species (Ramírez-Mosqueda and Iglesias-Andreu 2015; Tikendra et al. 2019).

Table 1 Examples of somaclonal variation analysis using DNA markers

Species	Type of in vitro culture	Results	Marker	Reference
Artemisia absinthium L.	Indirect (from different explants) and direct (nodal explants) organogenesis	Both markers were able to detect the somaclonal variations in callus regenerated plants, while no variation was detected in the plants regenerated from the nodal explants	ISSR, SSAP	Kour et al. (2014)
Barley (*Hordeum vulgare* L.)	Indirect organogenesis from immature embryos	Level of polymorphism and genetic variability detected was comparable between IRAP and ISSR markers	ISSR, IRAP	Campbell et al. (2011)
Cannabis sativa L.	Indirect organogenesis from stem explants	No somaclonal variation was detected	SSR	Ioannidis et al. (2022)
Date palm (*Phoenix dactylifera* L.)	Somatic embryogenesis from immature inflorescences	Transposable elements activation	IRAP, MITE	Mirani et al. (2020)
Linum usitatissimum L.	Direct organogenesis from hypocotyl	Genetic changes were detected with all markers and seemed to be genotype-dependent	SSR, IRAP, REMAP	Noormohammadi et al. (2020)
Humulus lupulus L.	Indirect organogenesis from different explants	Tetraploids obtained from medium containing the highest BAP concentration. Influence regarding explant was observed in mutants detected by RAPD analysis	RAPD, flow cytometry	Liberatore et al. (2020)
Mimosa (*Albizia julibrissin* Durazz.)	Indirect organogenesis from hypocotyl	The in vitro-raised plants exhibited 0.129–0.438 genetic distance from the mother plant	IRAP, SCoT	Rahmani et al. (2015)
Olea europaea L.	Callus-induced and meristem-induced shoots	DNA amount varied among CIS and MIS. Marker analysis showed that genotype was the major factor in genetic stability	AFLP, flow cytometry	Mirzaei et al. (2021)
Olea europaea L.	Indirect embryogenesis from radicle of mature zygotic embryos	Variation was slightly higher in plants derived from lines maintained in vitro for a longer time	RAPD, SSR	Bradaï et al. (2019)

(continued)

Table 1 (continued)

Species	Type of in vitro culture	Results	Marker	Reference
Orchid (*Dendrobium fimbriatum* Lindl.)	Micropropagation from exalbuminous seeds	Low genetic polymorphism among the regenerants from the hormone enriched medium	RAPD, ISSR ScoT	Tikendra et al. (2021)
Orchid (*Dendrobium chrysotoxum*)	Micropropagation from exalbuminous seeds	Low genetic polymorphism (3.6%) among the in vitro clones and mother plant	RAPD, ISSR	Tikendra et al. (2019)
Pistacia lentiscus L.	Micropropagation from shoot tips	Polymorphism information content (PIC) values lower than 0.5	IRAP	Kılınç et al. (2015)
Plantago major	Callus derived from leaf and root tips	High level of somaclonal variation. Genetic similarity values ranged between 0.16 and 0.71	ISSR	Ghorbanpour and Khadivi-Khub (2015)
Prunus salicina	In vitro propagation using axillary buds	No polymorphism was found	RAPD, ISSR, ScoT	Thakur et al. (2021)
Saffron (*Crocus sativus*)	Micropropagation from corms	Clonal variability of 7.81%	IRAP, SCoT, DAMD	Gautam and Bhattacharya (2021)
Vanilla planifolia Jacks	Indirect organogenesis from immature seeds	71.66% genetic polymorphism detected in regenerated plantlets	ISSR	Ramírez-Mosqueda and Iglesias-Andreu (2015)
Viola uliginosa Besser	Indirect and direct organogenesis	Lower marker polymorphism in indirect organogenesis. Tetrapolidy in indirect organogenesis	AFLP, flow cytometry	Slazak et al. (2015)
Vitis vinifera	Somatic embryogenesis from floral explants	Autotetraploid plants regenerated via somatic embryogenesis without detectable changes in the DNA profile	RAPD, ISSR SSR, flow cytometry	Catalano et al. (2021)

(continued)

Table 1 (continued)

Species	Type of in vitro culture	Results	Marker	Reference
Arabidopsis	Indirect organogenesis	Regenerant mutations included SNPs and InDels. No detected transposed mobile genetic elements activation	WGS	Jiang et al. (2011)
Rice	Organogenesis from callus cultured for 5 months	SNPs, and InDels were detected in regenerated plants. Transposition of *Tos17*	WGS	Miyao et al. (2012)
Rice	Anther culture	SNPs (88.7% DNA polymorphisms), insertions (4.5%) and deletions (6.6%) were detected in the regenerated plants	WGS	Jeong et al. (2013)
Rice	Embryogenic calli from seed	Average density of DNA polymorphisms (for each 100 Kb) was 10 SNPs, 1.93 insertions and 2.6 deletions. Seven new insertions of *Tos17*	WGS	Zhang et al. (2014)
Rice	Indirect organogenesis from seed	The average mutation rate (SNPs and InDels) in regenerants plants ranged from 9.14×10^{-5} to 1.37×10^{-4}	WGS	Qin et al. (2018)
Rice	Indirect organogenesis (agrobacterium-mediated transgenic plants)	7% of the observed variations were located within coding regions	WGS	Park et al. (2019)
Brassica rapa ssp.	Indirect organogenesis	SNPs and InDels were identified in regenerated plants. Five of the SNP occurred in exons, two altered the amino acid compositions	WGS	Park et al. (2020)
Cucumber (*Cucumis sativus* L.)	Direct leaf regeneration, leaf callus regeneration, cytokinin-dependent embryogenic suspension culture	Mainly SNPs polymorphisms detected, most frequently in non-coding regions	WGS	Skarzyńska et al, (2020)

ISSR = inter simple sequence repeats, RAPD = random amplified polymorphic DNA, AFLP = amplified fragment length polymorphism, SSR = simple sequence repeat, SCoT = start codon targeted polymorphism, SSAP = sequence-specific amplification polymorphism; IRAP = inter-retrotransposon amplified polymorphism; REMAP = retrotransposon-microsatellite amplified polymorphism; MITE = miniature inverted-repeat transposable elements; DAMD = amplification of minisatellite DNA; WGS = whole-genome sequencing

5.1.2 Amplified Fragment Length Polymorphsim (AFLP)

The AFLP technique combines the features of RFLP with the flexibility of PCR using primers that recognize the adapters attached to the digested DNA. AFLPs allow amplification of sequences randomly distributed throughout the genome. No prior knowledge of the sequence is required to generate AFLP markers from any organism. AFLP analysis begins by digesting genomic DNA with a combination of restriction enzymes to which fragments are bound by adapters to provide sequences for PCR amplification. AFLP can be visualized on automated DNA sequencers and generates large numbers of marker fragments. This technique has a number of advantages, as it is highly reproducible, reliable and informative. The AFLP technique has been used by a number of authors for the assessment of the genetic instability and somaclonal variation produced by in vitro culture (Slazak et al. 2015; Mirzaei et al. 2021).

5.1.3 Retrotransposon-Based Markers

The use of retrotransposon-based DNA markers in the analysis of genetic variability and diversity has become widespread in recent years (Kalendar et al. 2021). They are used for both genetic mapping and polymorphism identification (Khapilina et al. 2021). These molecular genetic markers have also been used to detect polymorphisms in plants obtained from in vitro culture (Kılınç et al. 2015; Mirani et al. 2020).

In addition, Long Terminal Repeat (LTR) retrotransposons are associated with key genes involved in potential applications such as functional genomics, suggesting their critical role as markers in molecular breeding. Different types of markers have been generated:

- IRAP (Inter-Retrotransposon Amplified Polymorphism) allows amplification between retrotransposons
- REMAP (Retransposon-Microsatellite Amplified Polymorphism)
- SSAP (Sequence-Specific Amplified Polymorphism)
- iPBS (inter-Priming Binding Site)

5.2 Codominat Markers

Codominant markers allow heterozygotes to be distinguished from homozygotes for each locus. PCR amplification of microsatellites are the most used codominant markers.

5.2.1 Simple Sequence Repeat (SSR)

SSR or microsatellite markers consist of short DNA sequence motifs of few bases which are a tandem repeat widely distributed throughout the genome, especially in the euchromatin. Its mutation rate, 10^3–10^6 per cell generation, is up to 10 orders of magnitude greater than point mutations because they are closely associated with failures in DNA mismatch repair systems (Gemayel et al. 2012). Each locus may differ in having a characteristic number of repeats that can be amplified by PCR using primers from the specific flanking sequences. Repeat polymorphisms usually result from the addition or deletion of the entire repeat. The use of SSR markers has a number of advantages because they are codominant, highly informative, abundant in the genome, multiallelic genetic markers, and reproducible and transferable among related species. They present wide applicability, ease to interpret in genotyping, and ease to automate and multiplex in PCR (Chen et al. 2000). Moreover, microsatellites do not require high-throughput technologies and computational resources for their development and analysis. This makes them ideal, fast and simple genotyping assays, but the main drawback is the need to know the genome flanking sequence.

SSR markers have been used to screen for somaclonal variants in different plant species such as olive (Bradaï et al. 2019), *Linum* (Noormohammadi et al. 2020), and cannabis (Ioannidis et al. 2022).

5.2.2 Single Nucleotide Polymosphism (SNP)

SNPs are DNA sequence variations that occur when only one nucleotide base (A, T, G, or C) varies between individuals. SNPs are common markers in the genome with an estimated frequency of one per 500–1000 bp. They are codominant like microsatellites, but unlike microsatellites, their power does not come from the number of alleles, but from the large number of loci that can be scored. SNPs are high-throughput markers, allowing rapid and efficient genotyping of large numbers of samples.

Next-generation sequencing technology (NGS) has enabled whole genome sequencing of individual plants. Recent improvements in next-generation sequencing methods have led to a rapid increase in the availability of reference genomes. The development of next-generation sequencing (NGS) technology allows genome-wide analysis of DNA polymorphisms, both SNPs and InDels, and the cost of analysis and marker development has been reduced.

WGS analyses have been performed on regenerated plants of Arabidopsis (Jiang et al. 2011), rice (Miyao et al. 2012; Zhang et al. 2014; Qin et al. 2018; Park et al. 2019), cucumber (Skarzyńska et al. 2020) and *Brasica rapa* (Park et al. 2020) for variation generated during in vitro culture. A high mutation rate was observed in regenerated Arabidopsis seedlings, with base substitutions being the most common change (Jiang et al. 2011). In rice, analysis of the genomes of three independent regenerated plants revealed a significant number of SNPs and InDels. The rate of change was estimated at 1.74×10^{-6} base substitutions per site per regeneration, 248 times higher than the spontaneous mutation rate in *A. thaliana* (Miyao et al. 2012). In cucumber, the total number of polymorphisms detected varied between somaclones. The polymorphisms, most of them SNPs, occur most frequently in non-coding regions. Between 1 and 3% affect coding sequences and in most cases cause an open reading frame shift (Skarzyńska et al. 2020). In pear, no phenotypic changes were detected. The analysis showed that only 0.065% of the SNPs and 1.429% of the InDels were located in coding regions, leading to changes in the reading frame and the appearance of stop codons (Liu et al. 2022).

6 Detection of Somaclonal Variation: Progeny Analysis

The nomenclature used to refer to the regenerated plants and their descendants follows the same pattern as that used in the genetic analysis of the offspring. Regenerated plants obtained from in vitro cultures are called R0, the letter R comes from the term regenerated and 0 refers to the fact that they are the plants taken from the culture. The progeny of this initial plants, obtained by self-fertilization or by crossing with other plants, form the R1 generation. The progeny of the R1 generation plants form the R2 generation, and so on.

Regenerated plants (R0) may show phenotypic changes compared to plants obtained from seeds. In other cases, the plants do not show phenotypic changes but could be carriers of recessive mutations. In both cases, in order to know if the regenerated plant have any genetic alterations, it is necessary to make crosses and to analyze the phenotype of the plants obtained. The results will show if the plants have a mutation that will be transmitted to the offspring, if the transmission is Mendelian and, through the analysis of the segregations, to determine the type of inheritance. The following are some basic examples of how the presence of dominant or recessive mutations in monogenic traits would be identified:

- Dominant Mutation Inheritance

In the case of cross-pollinated species, if the R0 plant is crossed with a wild phenotype plant, the expected result would be a 1:1 segregation, i.e., half of the individuals would have the mutant phenotype and the other 50% would have the wild phenotype. The analysis of the individuals of the next generation R2 would allow us to confirm the results observed in R1.

- Recessive Mutation Inheritance

In R1 obtained after self-fertilization, segregation for the modified phenotype is expected, so that 75% of the individuals (3/4) would have the wild-type phenotype and the remaining 25% (1/4) would have the modified phenotype. These plants would be homozygous and therefore only plants with the mutant phenotype would be obtained in their progeny.

In the case of cross-pollinated species, if the R0 plant, carrying the recessive mutation, is crossed with a plant with wild-type phenotype, the expected result in R1 would be that all individuals would have the same phenotype. The R1 plants can be homozygous for the wild allele or heterozygous, and to know if the regenerated plant is a carrier of a recessive mutation, it is necessary to go to the next generation by performing a family analysis. The individuals of the R1 generation that come from a regenerated R0 plant form a family and, to obtain the next generation R2, crosses are made between R1 individuals of the same family. The crosses between R1 plants are made randomly and, depending on the genotypes of the crossed plants, it is possible to have uniform families, where all individuals are phenotypically the same, and families where individuals with mutant phenotype appear, with a 3:1 segregation.

Somaclonal variation allows plants with new phenotypes to be obtained in a significantly shorter time than in a classical breeding program and is less costly than other methods. Despite its advantages, somaclonal variation has been underused as a tool for breeders, mainly because it is random, and the phenotype of regenerated plants is unpredictable and, in some cases, unstable.

On the other hand, cultures maintained by clonal propagation provide unique material to select somaclonal variants and analyze their phenotypic consequences.

References

Adhikari S, Saha S, Biswas A et al (2017) Application of molecular markers in plant genome analysis: a review. Nucleus 60:283–297. https://doi.org/10.1007/s13237-017-0214-7

Alves E, Ballesteros I, Linacero R, Vázquez AM (2005) RYS1, a foldback transposon, is activated by tissue culture and shows preferential insertion points into the rye genome. Theor Appl Genet 111:431–436. https://doi.org/10.1007/S00122-005-2013-9

Baránek M, Čechová J, Kovacs T et al (2016) Use of combined MSAP and NGS techniques to identify differentially methylated regions in somaclones: a case study of two stable somatic wheat mutants. PLoS One 11. https://doi.org/10.1371/JOURNAL.PONE.0165749

Bradaï F, Sánchez-Romero C, Martín C (2019) Somaclonal variation in olive (*Olea europaea* L.) plants regenerated via somatic embryogenesis: influence of genotype and culture age on genetic stability. Sci Hortic (Amsterdam) 251:260–266. https://doi.org/10.1016/j.scienta.2019.03.010

Campbell BC, LeMare S, Piperidis G, Godwin ID (2011) IRAP, a retrotransposon-based marker system for the detection of somaclonal variation in barley. Mol Breed 27:193–206. https://doi.org/10.1007/s11032-010-9422-4

Catalano C, Abbate L, Motisi A et al (2021) Autotetraploid emergence via somatic embryogenesis in vitis vinifera induces marked morphological changes in shoots, mature leaves, and stomata. Cells 10. https://doi.org/10.3390/cells10061336

Chen J, Iannone MA, Li MS et al (2000) A microsphere-based assay for multiplexed single nucleotide polymorphism analysis using single base chain extension. Genome Res 10:549–557. https://doi.org/10.1101/GR.10.4.549

Currais L, Loureiro J, Santos C, Canhoto JM (2013) Ploidy stability in embryogenic cultures and regenerated plantlets of tamarillo. Plant Cell Tissue Organ Cult 114:149–159. https://doi.org/10.1007/s11240-013-0311-5

Duta-Cornescu G, Constantin N, Pojoga DM et al (2023) Somaclonal variation—advantage or disadvantage in micropropagation of the medicinal plants. Int J Mol Sci 24:838. https://doi.org/10.3390/ijms24010838

Filkowski J, Besplug J, Burke P et al (2003) Genotoxicity of 2,4-D and dicamba revealed by transgenic *Arabidopsis thaliana* plants harboring recombination and point mutation markers. Mutat Res Toxicol Environ Mutagen 542:23–32. https://doi.org/10.1016/J.MRGENTOX.2003.07.008

Gautam N, Bhattacharya A (2021) Molecular marker based assessment of genetic homogeneity within the in vitro regenerated plants of *Crocus sativus* L.—a globally important high value spice crop. S Afr J Bot 140:461–467. https://doi.org/10.1016/j.sajb.2021.03.038

Gemayel R, Cho J, Boeynaems S, Verstrepen KJ (2012) Beyond junk-variable tandem repeats as facilitators of rapid evolution of regulatory and coding sequences. Genes (Basel) 3:461–480. https://doi.org/10.3390/genes3030461

Ghorbanpour M, Khadivi-Khub A (2015) Somaclonal variation in callus samples of *Plantago major* using inter-simple sequence repeat marker. Caryologia 68:19–24. https://doi.org/10.1080/00087114.2014.998128

Grafi G, Avivi Y (2004) Stem cells: a lesson from dedifferentiation. Trends Biotechnol 22:388–389. https://doi.org/10.1016/J.TIBTECH.2004.06.001

Grafi G, Florentin A, Ransbotyn V, Morgenstern Y (2011) The stem cell state in plant development and in response to stress. Front Plant Sci 2. https://doi.org/10.3389/fpls.2011.00053

Hirochika H (1997) Retrotransposons of rice: their regulation and use for genome analysis. Plant Mol Biol 35:231–240. https://doi.org/10.1007/978-94-011-5794-0_22

Hirochika H, Sugimoto K, Otsuki Y et al (1996) Retrotransposons of rice involved in mutations induced by tissue culture. Proc Natl Acad Sci U S A 93:7783. https://doi.org/10.1073/PNAS.93.15.7783

Ioannidis K, Tomprou I, Mitsis V, Koropouli P (2022) Genetic evaluation of in vitro micropropagated and regenerated plants of *Cannabis sativa* L. using SSR molecular markers. Plants 11:1–17. https://doi.org/10.3390/plants11192569

Jeong IS, Yoon UH, Lee GS et al (2013) SNP-based analysis of genetic diversity in anther-derived rice by whole genome sequencing. Rice 6:1–12. https://doi.org/10.1186/1939-8433-6-1

Jiang C, Mithani A, Gan X et al (2011) Regenerant Arabidopsis lineages display a distinct genome-wide spectrum of mutations conferring variant phenotypes. Curr Biol 21:1385–1390. https://doi.org/10.1016/j.cub.2011.07.002

Kaeppler SM, Kaeppler HF, Rhee Y (2000) Epigenetic aspects of somaclonal variation in plants. Plant Mol Biol 43:179–188. https://doi.org/10.1023/A:1006423110134

Kalendar R, Muterko A, Boronnikova S (2021) Retrotransposable elements: DNA fingerprinting and the assessment of genetic diversity. Methods Mol Biol 2222:263–286. https://doi.org/10.1007/978-1-0716-0997-2_15

Khapilina O, Turzhanova A, Danilova A et al (2021) Primer binding site (PBS) profiling of genetic diversity of natural populations of endemic species *Allium ledebourianum* Schult. Biotech 10:23. https://doi.org/10.3390/biotech10040023

Kikuchi K, Terauchit K, Wada M, Hirano HY (2003) The plant MITE mPing is mobilized in anther culture. Nature 421(6919):167–170. https://doi.org/10.1038/nature01218

Kılınç FM, Süzerer V, Çiftçi YÖ et al (2015) Clonal micropropagation of *Pistacia lentiscus* L. and assessment of genetic stability using IRAP markers. Plant Growth Regul 75:75–88. https://doi.org/10.1007/s10725-014-9933-9

Komatsu M, Shimamoto K, Kyozuka J (2003) Two-step regulation and continuous retrotransposition of the rice LINE-type retrotransposon Karma. Plant Cell 15:1934–1944. https://doi.org/10.1105/TPC.011809

Kour B, Kour G, Kaul S, Dhar MK (2014) In vitro mass multiplication and assessment of genetic stability of in vitro raised *Artemisia absinthium* L. plants using ISSR and SSAP molecular markers. Adv Bot 2014:1–7. https://doi.org/10.1155/2014/727020

Krishna H, lizadeh M, Singh D et al (2016) Somaclonal variations and their applications in horticultural crops improvement. 3 Biotech 6:1–18. https://doi.org/10.1007/s13205-016-0389-7

Larkin PJ, Scowcroft WR (1981) Somaclonal variation—a novel source of variability from cell cultures for plant improvement. Theor Appl Genet 60:197–214. https://doi.org/10.1007/BF02342540

Liberatore CM, Rodolfi M, Beghè D et al (2020) In vitro leaf-derived organogenesis and somaclonal variant detection in *Humulus lupulus* L. In Vitro Cell Dev Biol Plant 56:865–874. https://doi.org/10.1007/s11627-020-10088-7

Linacero R, Vázquez AM (1992) Cytogenetic variation in rye regenerated plants and their progeny. Genome 35:428–430. https://doi.org/10.1139/g92-063

Linacero R, Vázquez AM (1993) Somaclonal variation in rye. Mutat Res Lett 302:201–205. https://doi.org/10.1016/0165-7992(93)90105-5

Liu Q, Yang Y, Liu J et al (2022) Establishment of regeneration system of *Pyrus* and the genetic stability analysis of regenerated population. Plant Cell Tissue Organ Cult 152(1):215–228. https://doi.org/10.1007/S11240-022-02378-2

Long Y, Yang Y, Pan G, Shen Y (2022) New insights into tissue culture plant-regeneration mechanisms. Front Plant Sci 13:2112. https://doi.org/10.3389/FPLS.2022.926752

Mirani AA, Teo CH, Markhand GS et al (2020) Detection of somaclonal variations in tissue cultured date palm (*Phoenix dactylif*era L.) using transposable element-based markers. Plant Cell Tissue Organ Cult 141:119–130. https://doi.org/10.1007/s11240-020-01772-y

Mirzaei L, Yadollahi A, Kermani MJ et al (2021) Evaluation of genetic stability in olive callus-induced and meristem-induced shoots using flow cytometry and amplified fragment length polymorphism techniques. Plant Methods 17:1–15. https://doi.org/10.1186/s13007-021-007 24-7

Miyao A, Tanaka K, Murata K et al (2003) Target site specificity of the *Tos17* Retrotransposon shows a preference for insertion within genes and against insertion in retrotransposon-rich regions of the genome. Plant Cell 15:1771. https://doi.org/10.1105/TPC.012559

Miyao A, Iwasaki Y, Kitano H et al (2007) A large-scale collection of phenotypic data describing an insertional mutant population to facilitate functional analysis of rice genes. Plant Mol Biol 63:625. https://doi.org/10.1007/S11103-006-9118-7

Miyao A, Nakagome M, Ohnuma T et al (2012) Molecular spectrum of somaclonal variation in regenerated rice revealed by whole-genome sequencing. Plant Cell Physiol 53:256–264. https://doi.org/10.1093/pcp/pcr172

Noormohammadi Z, Habibollahi H, Farahani F, Sheidai M (2020) Molecular analysis of somaclonal variation in *Linum usitatissimum* and *L. album*. Plant Gene 24:100261. https://doi.org/10.1016/j.plgene.2020.100261

Nwauzoma A, Jaja E (2013) A review of somaclonal variation in plantain (*Musa* spp): mechanisms and applications. J Appl Biosci 67:5252–5260. https://doi.org/10.4314/jab.v67i0.95046

Ochatt SJ, Patat-Ochatt EM, Moessner A (2011) Ploidy level determination within the context of in vitro breeding. Plant Cell Tissue Organ Cult 104:329–341. https://doi.org/10.1007/s11240-011-9918-6

Oono K (1985) Putative homozygous mutations in regenerated plants of rice. Mol Gen Genet MGG 198:377–384. https://doi.org/10.1007/BF00332926

Park D, Park SH, Kim YS et al (2019) NGS sequencing reveals that many of the genetic variations in transgenic rice plants match the variations found in natural rice population. Genes Genomics 41:213–222. https://doi.org/10.1007/s13258-018-0754-5

Park JS, Park JH, Kim SJ, Park YD (2020) Genome analysis of tissue culture-derived variations in regenerated *Brassica rapa* ssp. *pekinensis* plants using next-generation sequencing. Hortic Environ Biotechnol 61:549–558. https://doi.org/10.1007/s13580-020-00237-7

Qin Y, Shin KS, Woo HJ, Lim MH (2018) Genomic variations of rice regenerants from tissue culture revealed by whole genome re-sequencing. Plant Breed Biotechnol 6:426–433. https://doi.org/10.9787/PBB.2018.6.4.426

Rahmani MS, Pijut PM, Shabanian N, Nasri M (2015) Genetic fidelity assessment of in vitro-regenerated plants of *Albizia julibrissin* using SCoT and IRAP fingerprinting. In Vitro Cell Dev Biol Plant 51:407–419. https://doi.org/10.1007/s11627-015-9692-y

Ramírez-Mosqueda MA, Iglesias-Andreu LG (2015) Indirect organogenesis and assessment of somaclonal variation in plantlets of *Vanilla planifolia* Jacks. Plant Cell Tissue Organ Cult 123:657–664. https://doi.org/10.1007/s11240-015-0868-2

Sato M, Hosokawa M, Doi M (2011) Somaclonal variation is induced de novo via the tissue culture process: a study quantifying mutated cells in *Saintpaulia*. PLoS ONE 6:e23541. https://doi.org/10.1371/JOURNAL.PONE.0023541

Shen X, Chen J, Kane ME, Henny RJ (2007) Assessment of somaclonal variation in *Dieffenbachia* plants regenerated through indirect shoot organogenesis. Plant Cell Tissue Organ Cult 91:21–27. https://doi.org/10.1007/s11240-007-9273-9

Silva TCR, Carvalho CR (2014) Vertical heterogeneity of DNA ploidy level assessed by flow cytometry in calli of *Passiflora Cincinnata*. In Vitro Cell Dev Biol Plant 50:158–165. https://doi.org/10.1007/s11627-013-9582-0

Skarzyńska A, Pawełkowicz M, Pląder W (2020) Genome-wide discovery of DNA variants in cucumber somaclonal lines. Gene 736. https://doi.org/10.1016/j.gene.2020.144412

Slazak B, Sliwinska E, Saługa M et al (2015) Micropropagation of *Viola uliginosa* (Violaceae) for endangered species conservation and for somaclonal variation-enhanced cyclotide biosynthesis. Plant Cell Tissue Organ Cult 120:179–190. https://doi.org/10.1007/s11240-014-0592-3

Smith AM, Hansey CN, Kaeppler SM (2012) TCUP: a novel hAT transposon active in maize tissue culture. Front Plant Sci 3:6. https://doi.org/10.3389/fpls.2012.00006

Tanurdzic M, Vaughn MW, Jiang H et al (2008) Epigenomic consequences of immortalized plant cell suspension culture. PLOS Biol 6:e302. https://doi.org/10.1371/JOURNAL.PBIO.0060302

Thakur M, Rakshandha SV, Chauhan A (2021) Genetic fidelity assessment of long term in vitro shoot cultures and regenerated plants in Japanese plum cvs Santa Rosa and Frontier through RAPD, ISSR and SCoT markers. S Afr J Bot 140:428–433. https://doi.org/10.1016/j.sajb.2020.11.005

Tican A, Câmpeanu G, Chiru N, Ivanovici D (2008) Using of unconventional methods for obtaining somaclonal variations, having as goal making of new potato varieties with resistance at diseases and pests. Roum Biotechnol Lett 13:3791–3798

Tikendra L, Koijam AS, Nongdam P (2019) Molecular markers based genetic fidelity assessment of micropropagated *Dendrobium chrysotoxum* Lindl. Meta Gene 20:100562. https://doi.org/10.1016/j.mgene.2019.100562

Tikendra L, Potshangbam AM, Dey A et al (2021) RAPD, ISSR, and SCoT markers based genetic stability assessment of micropropagated *Dendrobium fimbriatum* Lindl. var. oculatum Hk. f.- an important endangered orchid. Physiol Mol Biol Plants 27:341–357. https://doi.org/10.1007/s12298-021-00939-x

Wang X, Wu R, Lin X et al (2013) Tissue culture-induced genetic and epigenetic alterations in rice pure-lines, F1 hybrids and polyploids. BMC Plant Biol 13:1–12. https://doi.org/10.1186/1471-2229-13-77

Wang C, Tian M, Zhang Y, Ouyang T (2019) Molecular spectrum of somaclonal variation in PLB-regenerated *Oncidium* revealed by SLAF-seq. Plant Cell Tissue Organ Cult 137:541–552. https://doi.org/10.1007/s11240-019-01589-4

Zhang D, Wang Z, Wang N et al (2014) Tissue culture-induced heritable genomic variation in rice, and their phenotypic implications. PLoS ONE 9:5–7. https://doi.org/10.1371/journal.pone.0096879

Epigenetic Basis of Somaclonal Variation

Rosario Linacero and Isabel Ballesteros

Abstract Epigenetic changes induced in in vitro culture affect gene expression in a more or less permanent way, and in some cases are transmitted to the progeny. These changes affect chromatin structure and include changes in DNA methylation, histone modifications and the expression of small RNAs involved in DNA methylation (RNA-directed DNA methylation; RdDM). DNA methylation is one of the best studied epigenetic marks in regenerated plants, and changes in the methylation pattern have been reported to be the cause of somaclonal variation in many species. This chapter reviews the epigenetic basis of somaclonal variation and the tools used to detect it.

Keywords Somaclonal variation · Tissue culture · Epigenetics · DNA methylation · Epiallele

1 Background

Tissue culture is used for clonal propagation, plant breeding, germplasm conservation, and in biotechnology programs. Cells present in tissues cultured in vitro are clones of the starting material (explant), thus they are genetically identical. Therefore, the plants regenerated from them should, at least in theory, be phenotypically and genetically identical. However, there are numerous examples of somaclonal variation in tissue-cultured plants (Kaeppler et al. 2000; Miguel and Marum 2011; Neelakandan and Wang 2012). During in vitro culture, changes occur that can lead to phenotypic and genotypic variation in the regenerated plants, which could be due to chromosomal alterations, point mutations, activation of transposons, or changes in epigenetic marks (Bednarek and Orłowska 2020).

R. Linacero (✉) · I. Ballesteros
Departamento de Genética, Facultad de Ciencias Biológicas, Universidad Complutense de Madrid, 28040 Madrid, Spain
e-mail: charolin@bio.ucm.es

I. Ballesteros
e-mail: maball09@ucm.es

C. Sánchez-Romero (ed.), *Somaclonal Variation: Basic and Practical Aspects*,
https://doi.org/10.1007/978-3-031-51626-9_2

It has been proposed that epigenetic modifications act as promoters of the variation observed in tissue culture (Kaeppler et al. 2000; Valledor et al. 2007). This can include both heritable and non-heritable changes. Sometimes, some regenerated plants with an altered phenotype could recover the wild type phenotype, or their progeny did not inherit the abnormal trait. This can be explained if the affected gene has been hypermethylated and, therefore, its expression has been repressed. Conversely, hypomethylation can explain the reversion to the normal phenotype. Though methylation changes that are passed on to offspring are increasingly reported (Ashapkin et al. 2020), in most cases the methylation does not persist into the next generation and the altered phenotype is not observed in the offspring. Even if this type of variation is not heritable, it is important in cases where regenerated plants are the end product, such as in micropropagation of forest trees (Valledor et al. 2007). In addition, it should be considered that the coexistence of genetic and epigenetic changes has been described in regenerated plants of barley, rye or wheat (Bednarek et al. 2007; Linacero et al. 2011; Baránek et al. 2016).

2 Epigenetics Aspect of Morphogenesis

Morphogenesis requires dramatic changes in both cellular identities and cell growth patterns. These changes are accompanied by the activation or repression of numerous genes throughout the genome. It seems unlikely that these changes in gene expression are achieved only by regulation. Epigenetic pathways are also involved in the regulation of gene expression in in vitro cultures.

Plants have a profound ability to regenerate organs from differentiated somatic tissues. *In planta*, regeneration capacity is only required when damage occurs. Several epigenetic mechanisms have been shown to actively repress regenerative potential during development; POLYCOMB REPRESSIVE COMPLEX 2 (PRC2) is a chromatin modifier that maintains transcriptional repression through lysine 27 trimethylation of histone H3, H3K27me3 (Bouyer et al. 2011). Loss of function PRC2 mutants developed roots that initially were indistinguishable from those of the wild type, but fail to maintain their differentiated state, developing callus and embryo-like structures. Some authors proposed that highly differentiated cells retain the ability to undergo embryogenesis and that this ability must be tightly regulated, in this case epigenetically repressed by PRC2. Some of the genes that regulate regeneration, such as *LEAFY COTYLEDON* 2 (*LEC2*), *WOUND-INDUCED DEDIFFERENTIATION* 3 (*WIND3*), *WUSCHEL* (*WUS*) and *SHOOT MERISTEMLESS* (*STM*), are under PRC2-mediated repression (Horstman et al. 2017). Histone deacetylation, which is also involved in transcriptional repression, may serve as another safeguard to prevent premature initiation of somatic embryogenesis. Arabidopsis plants treated with a histone deacetylase inhibitor produced structures similar to leaf embryos. Similarly, loss-of-function mutants of two histone deacetylases produced embryo-like structures in shoots. These phenotypes were associated with the ectopic expression of several embryonic regulators, such as *LEC1* and *LEC2*, and could be suppressed by the *lec1* mutation. Treatment with histone deacetylase inhibitors combined with heat treatment increased the efficiency of somatic embryogenesis in *Brassica napus* microspores. It is possible that heat stress and histone deacetylation converge in the positive modulation of embryonic regulators to initiate the embryonic program. Another key epigenetic factor in the repression of ectopic embryonic traits is the PICKLE protein (PKL), a member of the chromatin ATPase remodeling family (Long et al. 2022). Genetic studies in Arabidopsis also suggested the involvement of other epigenetic mechanisms in the control of organ regeneration. Mutations in *METHYLTRANSFERASE 1* (*MET1*) enhanced shoot regeneration in shoot induction medium, and this phenotype was accompanied by increased expression of several *MET1* targets, including *WUS*. Like many other regulators of regeneration, the *WUS* locus contained several epigenetic marks that were modified when WUS expression is increased during shoot regeneration (Bouyer et al. 2011).

3 Epigenetics Basis of Somaclonal Variation

Epigenetic mechanisms contribute to the maintenance of gene expression patterns during plant development. Chromatin modifications are considered epigenetic marks and include DNA methylation changes, histone modifications and small RNAs (sRNAs) involved in DNA methylation (RNA-directed DNA methylation; RdDM).

DNA methylation in plants has two functions: to protect the genome by inactivating transposable elements (TE), known as silencing, and to regulate gene expression during development. DNA methylation is the major epigenetic mark that protects the genome from TE disruption (Zakrzewski et al. 2017).

Plants and mammals differ in the reprogramming of DNA methylation. In plants, symmetric DNA methylation is maintained in the germline, allowing for epigenetic inheritance. Besides, asymmetric DNA methylation is erased in male gametes of plants and subsequently re-established after fertilization. Nonetheless, imprinted genes retain RdDM and escape germline reprogramming.

DNA methylation involves the addition of a methyl group to a cytosine, but while in mammals the cytosines that are methylated are almost exclusively those that are present in CpG islands; in plants, methylation can occur in three different contexts: CG, CHG, or CHH (where H is A, T, or C). Among these three types of cytosine methylation in plants, the frequency of methylation of CG sequences is the highest, followed by CHG and CHH sequences.

This cytosine modification is abundant in the genomes of most eukaryotes and can be symmetric or asymmetric. Cytosine modification in plants is mediated by several enzymes that catalyze the transfer of the methyl group and are called DNA methyltransferases. In Arabidopsis, DRM2 (DOMAIN-REARRANGING METHYLTRANSFERASE 2) is the enzyme responsible for de novo methylation. In plants, much of the de novo DNA methylation is mediated by the RdDM pathway, in which DRM2 catalyzes DNA methylation, while MET1 (METHYLTRANSFERASE

1) and CMT3 (CHROMOMETHYLASE 3) are involved in methylation maintenance, identifying sites to be methylated on the newly synthesized strand during DNA replication (Zhang et al. 2018).

DNA methylation can also be removed through a DNA cleavage and repair mechanism by DNA glycosylases, such as DEMETER (DME) or SILENCING REPRESSOR 1 (ROS1). Proteins involved in chromatin remodeling, such as DECREASED DNA METHYLATION 1 (DDM1), are also involved in GC methylation. The RdDM pathway is very important. It is involved in several biological processes such as paramutation, repression of transposon activity, biotic and abiotic stress responses, and methylation patterning during reproduction and is mediated by Small Interferfering RNAs (siRNA). The siRNAs are commonly associated with transcriptional silencing and target CG or CHH sites for methylation. Borges et al. (2021) proposed that loss of small RNA-directed DNA methylation is associated with germline reprogramming and somaclonal variation. In both cases, 24 nt siRNAs and non-CG methylation decreased in actively dividing cells.

In plants, as in most eukaryotes, histones can undergo post-translational modifications that affect the chromatin state and thus the regulation of gene expression. Acetylation of histone H3 or H4 is often associated with transcription. Acetylation involves the transfer of an acetyl group from acetyl-coenzyme A (acetyl-CoA) to the α-amino group of lysine. This process is the result of a balance between the activity of two antagonistic families of enzymes: histone acetyltransferases (HATs) and histone deacetylases (HDACs). Like other histone modifications, histone acetylation is reversible and histone deacetylases play a key role in gene repression. It is currently known that there are more than 150 different histone modifications that require an extensive enzymatic machinery (Pikaard and Scheid 2014).

The appearance in regenerated plants of theoretical mutants with high reversion rates, paramutation-like behavior, or modifications in the developmental pattern may indicate the presence of epigenetic modifications in which DNA methylation could be involved. This modification may alter gene expression and may be more or less permanent and maintained throughout the life of the regenerated plant. There is evidence that DNA methylation patterns are altered by tissue culture. Genome-wide changes in DNA methylation following tissue culture have been reported in Arabidopsis, rice, or maize (Tanurdzic et al. 2008; Stroud et al. 2013; Stelpflug et al. 2014). In general, a global genome-wide decrease in methylation is observed during callus formation, although there is significant variation between and within species. In addition, other active epigenetic marks, such as H3 and H4 acetylation, were increased in callus compared to somatic cells, whereas repressive marks were globally reduced in pluripotent cells (Lee and Seo 2018).

These epigenetic modifications can affect the phenotype of regenerated plants, the changes can be transient and not transmitted to the progeny. However, in plants, variations in DNA methylation can be associated with heritable differences in gene expression in the absence of changes in the DNA sequence. On the other hand, hotspots of DNA methylation instability, as a subset of hypomethylated regions, were found in several independent regenerants or cell lines (Han et al. 2018). Heritable changes, such as the Karma epiallele in oil palm, were associated with loss of CHG

DNA methylation in a gene nested transposable element, resulting in the production of aberrant transcripts (Ong-Abdullah et al. 2015). Other epialleles resulting from tissue culture that are transmitted to the progeny have been characterized (Krizova et al. 2009; Rhee et al. 2010; Han et al. 2018). Borges et al. (2021) suggested that epialleles arising in in vitro cultures are stably maintained by sRNAs and H3K9me2 independently of RdDM. Analysis of methylomes from different tissues and cell types suggests that loss of methylation during replication promotes germline reprogramming and epigenetic variation in regenerated plants. Heritable changes would be maintained while transient changes would be eliminated in the progeny of plants regenerated from tissue culture. Wibowo et al. (2018), proposed that the type of tissue used for regeneration may influence the degree of reprogramming achieved, as somaclonal variation may also be due to incomplete programming of the epigenome of the tissue of origin.

DNA methylation is one of the best studied epigenetic marks, and changes in methylation pattern have been reported to be the cause of somaclonal variation in many species (González et al. 2013; Machczyńska et al. 2015). The tissue culture environment induces abiotic stress during regeneration and is responsible for global genome-wide methylation. A long callus phase induces the methylation event, leading to variation during plant regeneration.

Multiple studies have examined methylation in regenerated plants to analyze whether their methylation patterns differ from those of the donor plants (Xu et al. 2004; Li et al. 2007; Smýkal et al. 2007), or between plants obtained in vitro (Peraza-Echeverria et al. 2001). However, limited work has been done to analyze whether epigenetic changes are associated with genetic changes. Guo et al. (2007) described a significant correlation between genetic variations detected by ISSR (Inter Simple Sequence Repeats) and epigenetic changes obtained by MSAP (Methylation-Sensitive Amplification Polymorphism) in *Codonopsis Lanceolata.* In addition, a significant correlation between MSAP and SSAP (Sequence-Specific Amplification Polymorphism) was observed in *Hordeum brevisubulatum*, whereas changes detected by MSAP and AFLP (Amplified Fragment Length Polymorphism) markers were not statistically correlated (Li et al. 2007). In rye, Linacero et al. (2011) proposed that the genetic variation was related to the presence of CCGG targets, suggesting that there was a common mechanism connecting genetic and epigenetic phenomena. Likewise, in garlic (*Allium sativum*), it was observed that while genetic changes accumulated with the age of the in vitro culture, epigenetic changes stabilized after 6 months (Gimenez et al. 2016).

4 Genetics Versus Epigenetics

Some authors postulate that it is difficult to distinguish between genetic and epigenetic changes that occur in tissue cultures (Vázquez 2001; Smulders and de Klerk 2011).

Changes in the methylation pattern may be the starting point of a series of events leading to the occurrence of mutations. It is known that methylation can have mutagenic effects and that altered DNA methylation can affect heterochromatinization and the duration of DNA replication. In turn, this alteration in DNA replication time can also affect the accuracy of the genetic damage repair machinery, leading to variations in the form of nucleotide substitutions and small insertions or deletions. Similarly, hypermethylated cytosines are known to undergo deamination resulting in C-T or G-A transitions.

The clearest example of the relationship between the two mechanisms is the activation of mobile elements. As mentioned previously (see chapter on Genetic basis of somaclonal variation), activation of class I and II mobile elements can be induced during in vitro culture. Several studies have linked mobile element activation in in vitro cultures to changes in DNA methylation. Methylation in the CHG and CHH context is highly enriched in transposons; this methylation is mediated by siRNAs in the RdDM pathway. The changes in methylation can lead to the activation of mobile elements that cause mutations by disrupting the sequences of the genes in which they are inserted. In the case of transposable element mobilization, changes in the expression of genes adjacent to the insertion site have also been described, both by induction and repression of transcription, since mobile elements can carry sequences with a regulatory function in gene expression.

The use of methylation mutants, which show a decrease in methylation levels, has confirmed that many of these elements are inactivated by methylation and activated by demethylation. In vitro culture is considered a source of stress and stress causes, among other epigenetic changes, the modification of DNA methylation levels. As a consequence, some of the mobile elements are transcribed and even mobilized, increasing their copy number in the genome. The retrotransposon *Tos17* is normally inactive in the rice genome and is activated in vitro, and the use of methylation mutants has shown that its activation is due to a decrease in methylation (Cheng et al. 2006). Activation of *Tos17* leads to an increase in its copy number, which, when inserted at different positions in the rice genome, generates stable mutations (Miyao et al. 2007). It is therefore likely that epigenetic modifications induced by tissue culture are the cause of some of the mutations detected in the regenerated plants; mutations caused by new insertions or by deaminations are stable and transmitted by Mendelian inheritance.

In rice, *Karma*, a LINE element, is transcribed in cultured cells, but new copies are not detected in regenerated plants. However, an increase in copy number was detected in the offspring of the regenerated plants. The number of copies increased over several generations without the retrotransposon recovering the methylation level it had before its passage through in vitro culture (Komatsu et al. 2003). Activation of other mobile elements, such as *mPing*, a MITE-type transposon, *CACTA*, a class II transposon, or the *MERE1* retrotransposon, in in vitro culture has also been associated with changes in methylation (Ngezahayo et al. 2009; Rakocevic et al. 2009).

In the oil palm (*Elaeis guineensis* Jacq.), a "mantled" phenotype is observed in regenerated plants. The phenotypic change affects the shape of the flower and the oil content of the fruit, which is the reason why this species is used for commercial purposes. Hypomethylation of a LINE retrotransposon, similar to rice *Karma*, located in an intron of the homeotic gene *DEFICIENS*, was detected in all clones with the modified phenotype. The phenotypic modification was associated with alternative splicing leading to premature termination of the protein. Hypermethylated alleles/epialleles produced normal fruits and hypomethylated alleles caused homeotic transformation and loss of oil accumulation in the fruit. Analysis of the epigenomes of plants with the mantled phenotype, by EWAS (Epigenome-Wide Association Study), showed that this phenotype was a consequence of epigenetic modifications of the *Karma* transposon located in the *EgDEF1* gene due to small RNA loss and methylation at CHG and CHH sites (Ong-Abdullah et al. 2015).

5 Detection of Epigenetic Changes

There are several methods to detect changes in the methylation pattern in regenerated plants. There are two main types of approaches: (1) Methods based on the use of restriction enzymes with different methylation sensitivity, such as the combination of MspI and HpaII isoschizomers, including MSAP and Methylation Sensitive Inter-Simple Sequence Repeat (ISSR), and (2) methods to detect changes in the genome-wide methylation pattern, such as bisulfite genome sequencing (Cokus et al. 2008) or chromatin immunoprecipitation (ChIP) (Kaufmann et al. 2010). Some examples of the detection of DNA methylation changes in a variety of crops and plant species are listed in Table 1.

5.1 Methylation Sensitive Amplified Polymorphism (MSAP)

The MSAP technique is used to assess total methylation. MSAP is a modification of the AFLP technique in which methylation-sensitive restriction endonucleases are used to digest DNA. These restriction enzymes, as the isoschizomers HpaII and MspI, differentially cleave their target sequence CCGG based on methylation sensitivity differences of cytosine residues after digestion and amplification of the digested fragments, which allows the identification of methylation sensitive amplification polymorphism (Yaish et al. 2014). A related approach, methylation amplified fragment length polymorphism (metAFLP), allows quantitative analysis of both sequence changes and DNA methylation changes affecting the restriction sites of the Acc65I-KpnI endonucleases (Orłowska and Bednarek 2020).

Both MSAP and metAFLP methods do not require genome sequencing and have been efficiently used to detect somaclonal variants in different plants (Coronel et al. 2018; Gimenez et al. 2016; Francischini et al. 2017; Goyali et al. 2018).

Table 1 Examples of epigenetic variation analysis

Species	Type of in vitro culture	Results	Technique(s)	Reference
Allium sativum L.	Micropropagation from meristem culture	Epigenetic polymorhism reached the major variation after 6 months and then stabilized, being demethylation and CG methylation the principal conversions	AFLP, MSAP	Gimenez et al. (2016)
Coffea arabica	Somatic embryogenesis (embriogenic suspensions)	Low polymorphism between mother and regenerated plants	AFLP, MSAP	Bobadilla Landey et al. (2013)
Coffea Arabica	Indirect embryogenesis	Methylation polymorphism was low and ranged between 0.087 and 0.149%	AFLP, MSAP, SSAP	Bobadilla Landey et al. (2015)
Sugarcane	Micropropagation	Higher rate of loss of methylation (hypomethylation) than of gain of methylation (hypermethylation) was observed in the somaclonal variants	MSAP	Francischini et al. (2017)
Blueberry (*Vaccinium angustifolium* Ait.)	Micropropagation through node culture	Methylation polymorphism was detected in in vitro propagated plants, but not in vivo propagated plants	MSAP	Goyali et al. (2018)
Arabidopsis	Indirect organogenesis from root	Higher average methylation values and higher number of non-fully methylated sites flanking transposable elements in regenerated and transgenic plants. than in control plants	MetAFLP, TMD	Coronel et al. (2018)
Barley (*Hordeum vulgare* L.)	Androgenesis or somatic embryogenesis	RP-HPLC showed induction of DNA methylation in regenerants and conservation of the methylation level in the progeny. The SSAP and MSTD approach revealed that donor plants and regenerants were more polymorphic than respective progeny	RP-HPLC, SSAP, MSTD	Orłowska et al. (2016)

(continued)

Table 1 (continued)

Species	Type of in vitro culture	Results	Technique(s)	Reference
Arabidopsis	Plant cell suspension cultures	A prolonged cultivation period was related with increase in heterochromatin-specific epigenetic markers, i.e., global DNA methylation, H3K9	HPLC	Kwiatkowska et al. (2014)
Rye *(Secale cereale)*	Embryogenic calli from inmature embryo	73 and 30% of rye plants showed at least one genetic change, and 50 and 73% carried at least one methylation change	ISSR, msISSR	Linacero et al. (2011)
Arabidopsis	Embryogenic callus	Several differentially expressed miRNAs. Overexpression of miR167 inhibited somatic embryo formation	MiRNA microarray	Su et al. (2016)
Rice	Embryogenic calli from seed	Widespread changes in DNA methylation, small RNA accumulation and gene expression in regenerated plants	sRNA-Seq, Microarray MSAP	Wang et al. (2022)
Cucumber (*Cucumis sativus* L.)	Direct leaf regeneration, leaf callus regeneration, cytokinin-dependent embryogenic suspension culture	Differentially expressed miRNAs	sRNA-seq	Pawełkowicz et al. (2022)
Oil palm (*Elaeis guineensis*)	Micropropagation through cell culture of immature apex leaf tissue	Loss of Karma retrotransposon methylation in tissue culture contributes to the origin of mantled somaclonal variant	WGBS	Ong-Abdullah et al. (2015)
Apple	In vitro shoot cultures, and acclimatized in vitro plants	Differentially methylated genes (DMGs). Higher level of methylation in in vitro cultures than in acclimatized in vitro plants	WGBS	Gulyás et al. (2019)
Pinapple	Indirect organogenesis	mCG and mCHG was hypermethylated, whereas mCHH was hypomethylated in the genome of regenerated plants compared to donor explant	WGBS	Lin et al. (2019)

MetAFLP = methylation sensitive amplified fragment length polymorphism; MSAP = methylation-sensitive amplified polymorphism; TMD = transposon methylation display; SSAP = sequence-specific amplification polymorphism; RP-HPLC = reversed phase high pressure liquid chromatography; MSTD = methyl-sensitive transposon display; ms-ISSR = methyl-sensitive inter simple sequence repeats; sRNA-seq = small RNA sequencing; WGBS = whole-genome bisulphite Sequencing

5.2 *Methylation-Sensitive Inter-Simple Sequence Repeat (ISSR)*

Methylation-sensitive ISSR technique is used to analyze the presence of DNA methylation status changes between clonal plants obtained from the same cell line. As in case of MSAP, DNA is digested with methylation-sensitive restriction endonucleases. After digestion, the digested DNA is amplified using ISSR primers. These primers target multiple genomic loci in a single PCR reaction and amplify multiple amplicons of different sizes.

The use of ISSR fingerprinting on HpaII/MspI digested DNA represents a simpler method to identify DNA segments with differential C-methylation at the CCGG sites than the MSAP technique. The methylation-sensitive ISSR method does not require sequencing and has been used to detect somaclonal variants in several species such as rye (Linacero et al. 2011).

5.3 *High Performance Liquid Chromatography (HPLC)*

Quantification of global DNA methylation can be performed using separation techniques such as HPLC. These techniques are based on the digestion of genomic DNA to quantify the 5-methyl cytosines (Jaligot et al. 2000). The technique is specific and sensitive, making it useful for the rapid quantification of global DNA methylation. In some species, such as oil palm (*Elaeis guineensis*), *Cedrus* spp., or barley (*Hordeum vulgare*), HPLC has been used to detect DNA methylation in regenerated plants (Jaligot et al. 2000; Renau-Morata et al. 2005; Orłowska et al. 2016).

5.4 *Methylome Analysis Based on Next-Generation Sequencing (NGS)*

5.4.1 Whole Genome Bisulfite Sequencing (WGBS)

Bisulfite sequencing is one of the most widely used techniques to analyze DNA methylation. It is a qualitative and quantitative, efficient approach to identify 5-methylcytosine and provide results at base pair resolution. Genomic DNA is treated with sodium bisulfite to cause deamination of unmethylated cytosines, resulting in their conversion to uracil, while methylated cytosines remain stable. The bisulfite-treated DNA is amplified by PCR using specific primers, the amplified fragments are sequenced and the methylated cytosines can be identified.

This technique is also suitable for whole genome analysis using NGS techniques. In this case, the bisulfite-treated DNA is used directly for sequencing, providing high resolution results. Although WGBS is the most direct approach as it allows the identification and detection of the genome-wide pattern of methylated cytosines, it is expensive, which limits its application.

This technique has been successfully used to detect methylation changes in plant tissue culture, e.g., in long-term in vitro cultures of shoots and regenerated apple plants, regenerated pineapple and oil palm plants (Ong-Abdullah et al. 2015; Gulyás et al. 2019; Lin et al. 2019). In strawberry (*Fragaria nilgerrensis*), WGBS revealed that CG sites had the highest methylation level at all micropropagation stages, with an average of 49.5%, followed by CHG (33.2%) and CHH (12.4%). Global methylation levels alternately decreased and increased throughout the tissue culture process, and the distribution of DNA methylation was heterogeneous among different genetic regions (Cao et al. 2021).

5.4.2 Immunoprecipitation

Protein-DNA interactions can be analyzed by chromatin immunoprecipitation (ChIP), in which specific chromatin is immunoprecipitated by interaction with an antibody that detects the epigenetic mark of interest. This method, coupled with high-throughput sequencing techniques (ChIP-SEQ), allows the detection of the sequence associated with the epigenetic mark, such as histone modifications. It has been used in Arabidopsis to detect the DNA transcription factor binding site during the DNA methylation process (Wang et al. 2010). Chromatin immunoprecipitation in combination with NGS has made it possible to analyze genome-wide cytosine methylation changes during the cell cycle, showing that epialleles that arise spontaneously in plant cell cultures were stably maintained by siRNA and lysine demethylation of histone H3 (Borges et al. 2021).

The immunoprecipitation method also allows the analysis of DNA methylation using 5-methylcytidine antibodies (Mohn et al. 2009). This method is called methylated DNA immunoprecipitation (MeDIP) and can be combined with other techniques as NGS for the detection of methylated DNA.

5.4.3 Micro-RNA (miRNA)

In in vitro culture, changes in the transcription of miRNAs have been analyzed during the processes of dedifferentiation and pluripotency, rather than in relation to their role in somaclonal variation. The molecular pathway of shoot regeneration mainly involves phytohormones and genes related to the shoot apical meristem (Meng et al. 2010). The expression of these genes appears to be regulated at the transcriptional and post-transcriptional levels, and many of these pathways are regulated by microRNAs (miRNAs), which act as signals that restrict gene expression in specific plant tissues (Vashisht and Nodine 2014; Axtell and Meyers 2018; Singh et al. 2018). The regulatory roles of miRNAs have been highlighted during somatic embryogenesis of many plants and are essential as regulators of metabolism in long-term cultures (Nodine and Bartel 2010; Zhang et al. 2015; Su et al. 2016), as well as their involvement in epigenetic processes such as RNA-directed DNA methylation. However, the role of miRNAs in the context of somaclonal variation remains uncertain. It has been proposed that miRNAs may play an important role in different aspects of epigenetic regulation leading to somaclonal variation (Pawełkowicz et al. 2022).

Chromosomal changes, mutations in DNA, altered DNA sequence and, transposable elements are the most common genetic alterations detected in regenerated plants, while DNA methylation and chromatin remodeling are considered the main epigenetic changes underlying the phenotypic changes observed in the regenerated plants compared to their original clones.

References

Ashapkin VV, Kutueva LI, Aleksandrushkina NI, Vanyushin BF (2020) Epigenetic mechanisms of plant adaptation to biotic and abiotic stresses. Int J Mol Sci 21:7457. https://doi.org/10.3390/ijms21207457

Axtell MJ, Meyers BC (2018) Revisiting criteria for plant microRNA annotation in the era of big data. Plant Cell 30:272–284. https://doi.org/10.1105/tpc.17.00851

Baránek M, Čechová J, Kovacs T et al (2016) Use of combined MSAP and NGS techniques to identify differentially methylated regions in somaclones: a case study of two stable somatic wheat mutants. PLoS ONE 11:e0165749. https://doi.org/10.1371/JOURNAL.PONE.0165749

Bednarek PT, Orłowska R (2020) Plant tissue culture environment as a switch-key of (epi)genetic changes. Plant Cell Tissue Organ Cult 140:245–257. https://doi.org/10.1007/s11240-019-01724-1

Bednarek PT, Orłowska R, Koebner RMD, Zimny J (2007) Quantification of the tissue-culture induced variation in barley (*Hordeum vulgare* L.). BMC Plant Biol 7:1–9. https://doi.org/10.1186/1471-2229-7-10

Bobadilla Landey R, Cenci A, Georget F et al (2013) High genetic and epigenetic stability in *Coffea arabica* plants derived from embryogenic suspensions and secondary embryogenesis as revealed by AFLP, MSAP and the phenotypic variation rate. PLoS ONE 8:e56372. https://doi.org/10.1371/journal.pone.0056372

Bobadilla Landey R, Cenci A, Guyot R et al (2015) Assessment of genetic and epigenetic changes during cell culture ageing and relations with somaclonal variation in *Coffea arabica*. Plant Cell Tissue Organ Cult 122:517–531. https://doi.org/10.1007/s11240-015-0772-9

Borges F, Donoghue MTA, LeBlanc C et al (2021) Loss of small-RNA-directed DNA methylation in the plant cell cycle promotes germline reprogramming and somaclonal variation. Curr Biol 31:591-600.e4. https://doi.org/10.1016/j.cub.2020.10.098

Bouyer D, Roudier F, Heese M et al (2011) Polycomb repressive complex 2 controls the embryo-to-seedling phase transition. PLoS Genet 7:1002014. https://doi.org/10.1371/journal.pgen.1002014

Cao Q, Feng Y, Dai X et al (2021) Dynamic changes of DNA methylation during wild strawberry (*Fragaria nilgerrensis*) tissue culture. Front Plant Sci 12:1–13. https://doi.org/10.3389/fpls.2021.765383

Cheng C, Daigen M, Hirochika H (2006) Epigenetic regulation of the rice retrotransposon *Tos17*. Mol Genet Genomics 276:378–390. https://doi.org/10.1007/s00438-006-0141-9

Cokus SJ, Feng S, Zhang X et al (2008) Shotgun bisulphite sequencing of the *Arabidopsis* genome reveals DNA methylation patterning. Nature 452:215–219. https://doi.org/10.1038/nature06745

Coronel CJ, González AI, Ruiz ML, Polanco C (2018) Analysis of somaclonal variation in transgenic and regenerated plants of *Arabidopsis thaliana* using methylation related metAFLP and TMD markers. Plant Cell Rep 37:137–152. https://doi.org/10.1007/s00299-017-2217-x

Francischini JHMB, Kemper EL, Costa JB et al (2017) DNA methylation in sugarcane somaclonal variants assessed through methylation-sensitive amplified polymorphism. Genet Mol Res 16:16029585. https://doi.org/10.4238/gmr16029585

Gimenez MD, Yañez-Santos AM, Paz RC et al (2016) Assessment of genetic and epi-genetic changes in virus-free garlic (*Allium sativum* L.) plants obtained by meristem culture followed by in vitro propagation. Plant Cell Rep 35:129–141. https://doi.org/10.1007/s00299-015-1874-x

González AI, Sáiz A, Acedo A et al (2013) Analysis of genomic DNA methylation patterns in regenerated and control plants of rye (*Secale cereale* L.). Plant Growth Regul 70:227–236. https://doi.org/10.1007/s10725-013-9794-7

Goyali JC, Igamberdiev AU, Debnath SC (2018) DNA methylation in lowbush blueberry (*Vaccinium angustifolium* Ait.) propagated by softwood cutting and tissue culture. Can J Plant Sci 98:1035–1044. https://doi.org/10.1139/cjps-2017-0297

Gulyás A, Dobránszki J, Kiss E et al (2019) Changes in DNA methylation pattern of apple long-term in vitro shoot culture and acclimatized plants. J Plant Physiol 239:18–27. https://doi.org/10.1016/j.jplph.2019.05.007

Guo WL, Wu R, Zhang YF et al (2007) Tissue culture-induced locus-specific alteration in DNA methylation and its correlation with genetic variation in *Codonopsis lanceolata* Benth. et Hook. f. Plant Cell Rep 26:1297–1307. https://doi.org/10.1007/s00299-007-0320-0

Han Z, Crisp PA, Stelpflug S et al (2018) Heritable epigenomic changes to the maize methylome resulting from tissue culture. Genetics 209:983–995. https://doi.org/10.1534/genetics.118.300987

Horstman A, Bemer M, Boutilier K (2017) A transcriptional view on somatic embryogenesis. Regeneration 4:201. https://doi.org/10.1002/REG2.91

Jaligot E, Rival A, Beulé T et al (2000) Somaclonal variation in oil palm (*Elaeis guineesis* Jacq.): the DNA methylation hypothesis. Plant Cell Rep 19:684–690. https://doi.org/10.1007/s00299 9900177

Kaeppler SM, Kaeppler HF, Rhee Y (2000) Epigenetic aspects of somaclonal variation in plants. Plant Mol Biol 43:179–188. https://doi.org/10.1023/A:1006423110134

Kaufmann K, Muiño JM, Østerås M et al (2010) Chromatin immunoprecipitation (ChIP) of plant transcription factors followed by sequencing (ChIP-SEQ) or hybridization to whole genome arrays (ChIP-CHIP). Nat Protoc 53(5):457–472. https://doi.org/10.1038/nprot.2009.244

Komatsu M, Shimamoto K, Kyozuka J (2003) Two-step regulation and continuous retrotransposition of the rice LINE-type retrotransposon Karma. Plant Cell 15:1934–1944. https://doi.org/10.1105/tpc.011809

Krizova K, Fojtova M, Depicker A, Kovarik A (2009) Cell culture-induced gradual and frequent epigenetic reprogramming of invertedly repeated tobacco transgene epialleles. Plant Physiol 149:1493. https://doi.org/10.1104/PP.108.133165

Kwiatkowska A, Zebrowski J, Oklejewicz B et al (2014) The age-dependent epigenetic and physiological changes in an *Arabidopsis* T87 cell suspension culture during long-term cultivation. Biochem Biophys Res Commun 447:285–291. https://doi.org/10.1016/j.bbrc.2014.03.141

Lee K, Seo PJ (2018) Dynamic epigenetic changes during plant regeneration. Trends Plant Sci 23:235–247. https://doi.org/10.1016/J.TPLANTS.2017.11.009

Li X, Xiaoming AE, Ae Y et al (2007) Genetic and epigenetic instabilities induced by tissue culture in wild barley (*Hordeum brevisubulatum* (Trin.) Link). Plant Cell Tissue Organ Cult 90:153–168. https://doi.org/10.1007/s11240-007-9224-5

Lin W, Xiao X, Zhang H et al (2019) Whole-genome bisulfite sequencing reveals a role for DNA methylation in variants from callus culture of pineapple (*Ananas comosus* l.). Genes (Basel) 10:877. https://doi.org/10.3390/genes10110877

Linacero R, Rueda J, Esquivel E et al (2011) Genetic and epigenetic relationship in rye, *Secale cereale* L., somaclonal variation within somatic embryo-derived plants. In Vitro Cell Dev Biol Plant 47:618–628. https://doi.org/10.1007/s11627-011-9407-y

Long Y, Yang Y, Pan G, Shen Y (2022) New insights into tissue culture plant-regeneration mechanisms. Front Plant Sci 13:926752. https://doi.org/10.3389/fpls.2022.926752

Machczyńska J, Zimny J, Bednarek PT (2015) Tissue culture-induced genetic and epigenetic variation in triticale (× *Triticosecale* spp. Wittmack ex A. Camus 1927) regenerants. Plant Mol Biol 89:279–292. https://doi.org/10.1007/s11103-015-0368-0

Meng L, Zhang S, Lemaux PG (2010) Toward molecular understanding of in vitro and in planta shoot organogenesis. CRC Crit Rev Plant Sci 29:108–122. https://doi.org/10.1080/07352681003617327

Miguel C, Marum L (2011) An epigenetic view of plant cells cultured in vitro: somaclonal variation and beyond. J Exp Bot 62:3713–3725. https://doi.org/10.1093/jxb/err155

Miyao A, Iwasaki Y, Kitano H et al (2007) A large-scale collection of phenotypic data describing an insertional mutant population to facilitate functional analysis of rice genes. Plant Mol Biol 63:625. https://doi.org/10.1007/S11103-006-9118-7

Mohn F, Weber M, Schübeler D, Roloff T-C (2009) Methylated DNA Immunoprecipitation (MeDIP). In: Tost J (ed) DNA methylation: methods and protocols. Humana Press, Totowa, NJ, pp 55–64. https://doi.org/10.1007/978-1-59745-522-0_5

Neelakandan AK, Wang K (2012) Recent progress in the understanding of tissue culture-induced genome level changes in plants and potential applications. Plant Cell Rep 31:597–620. https://doi.org/10.1007/s00299-011-1202-z

Ngezahayo F, Xu C, Wang H et al (2009) Tissue culture-induced transpositional activity of *mPing* is correlated with cytosine methylation in rice. BMC Plant Biol 9:91. https://doi.org/10.1186/1471-2229-9-91

Nodine MD, Bartel DP (2010) MicroRNAs prevent precocious gene expression and enable pattern formation during plant embryogenesis. Genes Dev 24:2678–2692. https://doi.org/10.1101/gad.1986710

Ong-Abdullah M, Ordway JM, Jiang N et al (2015) Loss of *Karma* transposon methylation underlies the mantled somaclonal variant of oil palm. Nature 525:533–537. https://doi.org/10.1038/nature15365

Orłowska R, Bednarek PT (2020) Precise evaluation of tissue culture-induced variation during optimisation of in vitro regeneration regime in barley. Plant Mol Biol 103:33–50. https://doi.org/10.1007/s11103-020-00973-5

Orłowska R, Machczyńska J, Oleszczuk S et al (2016) DNA methylation changes and TE activity induced in tissue cultures of barley (*Hordeum vulgare* L.). J Biol Res 23:1–12. https://doi.org/10.1186/s40709-016-0056-5

Pawełkowicz ME, Skarzyńska A, Koter MD et al (2022) miRNA profiling and its role in multi-omics regulatory networks connected with somaclonal variation in cucumber (*Cucumis sativus* L.). Int J Mol Sci 23:4317. https://doi.org/10.3390/ijms23084317

Peraza-Echeverria S, Herrera-Valencia VA, Kay AJ (2001) Detection of DNA methylation changes in micropropagated banana plants using methylation-sensitive amplification polymorphism (MSAP). Plant Sci 161:359–367. https://doi.org/10.1016/S0168-9452(01)00421-6

Pikaard CS, Scheid OM (2014) Epigenetic regulation in plants. Cold Spring Harb Perspect Biol 6:a019315. https://doi.org/10.1101/CSHPERSPECT.A019315

Rakocevic A, Mondy S, Tirichine L et al (2009) *MERE1*, a low-copy-number copia-type retroelement in *Medicago truncatula* active during tissue culture. Plant Physiol 151:1250–1263. https://doi.org/10.1104/pp.109.138024

Renau-Morata B, Nebauer SG, Arrillaga I, Segura J (2005) Assessments of somaclonal variation in micropropagated shoots of *Cedrus*: consequences of axillary bud breaking. Tree Genet Genomes 1:3–10. https://doi.org/10.1007/S11295-004-0001-X

Rhee Y, Sekhon RS, Chopra S, Kaeppler S (2010) Tissue culture-induced novel epialleles of a Myb transcription factor encoded by pericarp color1 in maize. Genetics 186:843–855. https://doi.org/10.1534/GENETICS.110.117929

Singh A, Gautam V, Singh S et al (2018) Plant small RNAs: advancement in the understanding of biogenesis and role in plant development. Planta 248:545–558. https://doi.org/10.1007/s00425-018-2927-5

Smulders MJM, de Klerk GJ (2011) Epigenetics in plant tissue culture. Plant Growth Regul 63:137–146. https://doi.org/10.1007/S10725-010-9531-4/FIGURES/3

Smýkal P, Valledor L, Rodríguez R, Griga M (2007) Assessment of genetic and epigenetic stability in long-term in vitro shoot culture of pea (*Pisum sativum* L.). Plant Cell Rep 26:1985–1998. https://doi.org/10.1007/S00299-007-0413-9

Stelpflug SC, Eichten SR, Hermanson PJ et al (2014) Consistent and heritable alterations of DNA methylation are induced by tissue culture in maize. Genetics 198:209–218. https://doi.org/10.1534/genetics.114.165480

Stroud H, Ding B, Simon SA et al (2013) Plants regenerated from tissue culture contain stable epigenome changes in rice. Elife 2013:1–14. https://doi.org/10.7554/eLife.00354

Su YH, Liu YB, Zhou C et al (2016) The microRNA167 controls somatic embryogenesis in *Arabidopsis* through regulating its target genes ARF6 and ARF8. Plant Cell Tissue Organ Cult 124:405–417. https://doi.org/10.1007/s11240-015-0903-3

Tanurdzic M, Vaughn MW, Jiang H et al (2008) Epigenomic consequences of immortalized plant cell suspension culture. PLOS Biol 6:e302. https://doi.org/10.1371/JOURNAL.PBIO.0060302

Valledor L, Hasbún R, Meijón M et al (2007) Involvement of DNA methylation in tree development and micropropagation. Plant Cell Tissue Organ Cult 91:75–86. https://doi.org/10.1007/s11240-007-9262-z

Vashisht D, Nodine MD (2014) MicroRNA functions in plant embryos. Biochem Soc Trans 42:352–357. https://doi.org/10.1042/BST20130252

Vázquez AM (2001) Insight into somaclonal variation. Plant Biosyst 135:57–62. https://doi.org/10.1080/11263500112331350650

Wang C, Xu J, Zhang D et al (2010) An effective approach for identification of in vivo protein-DNA binding sites from paired-end ChIP-Seq data. BMC Bioinform 11:1–8. https://doi.org/10.1186/1471-2105-11-81

Wang N, Yu Y, Zhang D et al (2022) Modification of gene expression, DNA methylation and small RNAs expression in rice plants under in vitro culture. Agronomy 12:1675. https://doi.org/10.3390/agronomy12071675

Wibowo A, Becker C, Durr J et al (2018) Partial maintenance of organ-specific epigenetic marks during plant asexual reproduction leads to heritable phenotypic variation. Proc Natl Acad Sci U S A 115:E9145–E9152. https://doi.org/10.1073/pnas.1805371115

Xu Z, Yan X, Maurais S et al (2004) Jittery, a mutator distant relative with a paradoxical mobile behavior: excision without reinsertion. Plant Cell 16:1105–1114. https://doi.org/10.1105/TPC.019802

Yaish M, Peng M, Rothstein SJ (2014) Global DNA methylation analysis using methyl-sensitive amplification polymorphism (MSAP). Methods Mol Biol 1062:285–298. https://doi.org/10.1007/978-1-62703-580-4_16

Zakrzewski F, Schmidt M, Van Lijsebettens M, Schmidt T (2017) DNA methylation of retrotransposons, DNA transposons and genes in sugar beet (*Beta vulgaris* L.). Plant J 90:1156–1175. https://doi.org/10.1111/tpj.13526

Zhang M, Dong Y, Nie L et al (2015) High-throughput sequencing reveals miRNA effects on the primary and secondary production properties in long-term subcultured Taxus cells. Front Plant Sci 6:604. https://doi.org/10.3389/fpls.2015.00604

Zhang H, Lang Z, Zhu J-K (2018) Dynamics and function of DNA methylation in plants. Nat Rev Mol Cell Biol 19:489–506. https://doi.org/10.1038/s41580-018-0016-z

Factors Affecting Tissue Culture-Induced Variations

Carolina Sánchez-Romero

Abstract In vitro culture can induce somaclonal variation, giving rise to the appearance of alterations in the cultured cells and tissues, as well as in the regenerated plants and their progenies. A wide range of alterations can emerge as a consequence of somaclonal variation, which can be detected at different levels, such as phenotypic, cytogenetic, biochemical or molecular. Nature and frequency of changes are uncontrolled due to the unpredictable nature of somaclonal variation. Thus, somaclonal variation is considered an important drawback intrinsically associated to the use of in vitro culture techniques and, in fact, it constitutes a very important problem in culture systems in which true-to-typeness is expected. Nevertheless, it can also be used as a source of genetic variability which can be exploited in plant improvement. Comprehensive knowledge of the effect of different factors affecting somaclonal variation, such as the genotype, explant source, culture conditions, culture age or the in vitro culture method, can help to design strategies in order to reduce or promote somaclonal variation, depending on the objectives that have determined the use of in vitro culture.

Keywords Culture age · Explant source · Genotype · In vitro culture · Plant growth regulator · Somaclonal variation

1 Introduction

In vitro culture is at the base of different techniques with multiple applications in different fields such as commercial plant production, basic research or plant breeding by conventional or biotechnological means. During in vitro culture, cells and tissues are exposed to different stresses, from explant excision to the artificial conditions on which they are incubated, which may induce genetic and epigenetic changes (Neelakandan and Wang 2012).

C. Sánchez-Romero (✉)
Departamento de Botánica y Fisiología Vegetal, Universidad de Málaga, Campus de Teatinos s/n, 29071 Málaga, Spain
e-mail: c.sanchez@uma.es

C. Sánchez-Romero (ed.), *Somaclonal Variation: Basic and Practical Aspects*,
https://doi.org/10.1007/978-3-031-51626-9_3

The term ‘somaclonal variation’ was coined by Larkin and Scowcroft (1981) to refer to alterations induced by in vitro culture in the cells, tissues, regenerated plants and their progenies. Somaclonal variation (SV) occurs in a spontaneous, random and uncontrolled way (Karp 1994) and, therefore, it is unavoidable and unpredictable in nature (Jain 2001). Very variable rates of SV have been reported depending on the culture system, with cases in which the genetic integrity was maintained versus systems in which a high proportion of in vitro culture-derived plants exhibited variations.

The occurrence of SV constitutes an important problem in commercial clonal propagation, germplasm conservation and in genetic transformation or gene editing experiments, in which no alterations due to tissue culture are expected. However, SV can be used as an alternative source of genetic variability in breeding programs. This option is especially interesting when other breeding methods are limited (Vitamvas et al. 2019), as occurs in trees and long-lived perennial species, vegetatively propagated plants and crops with polygenic traits or having a narrow genetic base (DeVerno 1995; Brar and Jain 1998).

Due to the importance of SV and its consequences on multiple systems in which in vitro culture is used as the main technology or as a complementary tool, it is very important to know factors affecting its appearance. The comprehensive understanding of the factors influencing SV can help to design different strategies to increase or reduce its incidence, depending on the objectives of the in vitro culture system.

Although there is no effective method known to control this phenomenon (Bhojwani and Razdan 1996), it is well established that some factors affect the occurrence of SV and the frequency of changes. Somaclonal variations can be generated during in vitro culture or can be due to pre-existing variability in the tissues used as explant (Peschke and Phillips 1992). In this chapter, different factors affecting variations induced by in vitro culture, such the composition of the culture medium, in vitro culture method, starting explant, culture age, or the genotype of the mother plant, will be studied.

2 Culture Conditions

Plant stability is affected by the conditions to which cells and tissues are exposed during in vitro culture.

2.1 Culture Medium

Medium composition is a key factor influencing SV. Although plant growth regulators (PGRs) play a main role inducing SV, other components also contribute to the occurrence of SV.

In *Agave angustifolia*, nutrient limitation altered the phenotype and increased DNA methylation (Duarte-Aké and De-la-Peña 2021). However, in maize, hypomethylation was reported under nitrogen and phosphorus deficiency (Mager and Ludewig 2018). The concentration of ammonium nitrate affected SV in *Senecio cruentus*, with higher frequency of variation found in shoot cultures growing on proliferation medium devoiding this compound (Sivanesan and Jeong 2012).

2.1.1 Plant Growth Regulators

Exogenously applied PGRs play an essential role in tissue culture as they control morphogenesis in vitro. In fact, optimal concentrations and precise combinations of hormones are required for efficient micropropagation and induction of specific regeneration pathways.

Although different investigations reported no effect of PGRs on SV (Bennici et al. 2004; Venkatachalam et al. 2007), they are known to be one of the causes of SV induction. In fact, several frequently used PGRs, such as 2,4-dichlorophenoxyacetic acid (2,4-D), 1-naphthalene acetic acid (NAA), and 6-benzylaminopurine (BAP), are considered as the most prominent factors causing genetic variability (Sales and Butardo 2014).

Different works have investigated the effect of different PGRs and their concentrations on SV. SV especially increases when auxins and cytokinins were added to the culture medium (Liberatore et al. 2020). In embryogenic suspension cultures of cucumber, greater variation was found in auxin-containing medium compared to that observed in cytokinin-containing medium (Ładyżyński et al. 2002). Similarly, in PLBs of Dendrobium Sabin Blue, different SV level was observed depending on the PGR. In this case, kinetin gave rise to the highest variation by both Inter-Simple Sequence Repeat (ISSR) and Directed Amplification of Minisatellite DNA (DAMD) analysis compared to NAA and thidiazuron (TDZ) (Chin et al. 2019).

Hormone combinations can also influence the occurrence of SV. Thus, incidence of mantled flowering in oil palm depended on auxin/cytokinin balance, with a relatively high auxin/cytokinin ratio resulting in the lowest incidence of variation, while a low auxin/cytokinins ratio gave rise to a high incidence of this alteration (Eeuwens et al. 2002).

Important differences have also been reported among hormones of the same group. In rye mature embryo culture, high level of auxins caused a reduction in the genomic template stability (GTS) value using Random Amplified Polymorphic DNA (RAPD) analysis, with 98.4% in calli maintained on MS medium with 2 mg l^{-1} dicamba versus 81.7% in those cultured on MS medium with 12 mg l^{-1} picloram. Moreover, DNA hypermethylation was observed at high concentrations of 2,4-D and picloram, whereas DNA hypomethylation was observed with dicamba supplementation (Aydin et al. 2016). In caladium, leaf-colour variants differed more with auxin type than with different concentrations of a specific auxin. Using equal molar concentration of different auxins, the number of variants was higher in culture media containing

2,4,5-trichlorophenoxyacetic acid (2,4,5-T) or 2,4-D than in those supplemented with NAA or IBA (Ahmed et al. 2004).

Some auxins, as 2,4-D are especially mutagenic. During in vitro culture of oil palm zygotic embryos, 2,4-D supplementation gave rise to significant variations (Giorgetti et al. 2011). Using high 2,4-D concentrations also caused of SV in strawberry (Nehra et al. 1992), soybean (Gesteira et al. 2008) and cotton (Jin et al. 2008).

In relation to NAA, its effects in Caladium (Ahmed et al. 2004) and in PLBs of Dendrobium Sabin Blue orchid (Chin et al. 2019) were relatively low compared to other auxins, such as indole-3-acetic acid (IAA), 2,4-D and 2,4,5-T. However, an increase in SV was observed in *Galanthus transcaucasicus* when high concentrations of NAA were incorporated in the culture medium (Asadi et al. 2021).

In relation to cytokinins, although Reuveni et al. (1993) stated that high levels of cytokinins did not directly affect SV in 'Cavendish' bananas, and Duarte-Aké and de-la-Peña (2021) suggested that cytokinin could help to reduce variation, different effects have been observed when cytokinins were added to the culture medium. Thus, high concentrations of BAP and kinetin induced genetic variability in banana (Trujillo and de García 1996). In PLBs of Dendrobium Sabin Blue, kinetin gave rise to the highest percentage of polymorphism by both ISSR and DAMD band profiling compared to the other PGRs, like NAA and TDZ. In *Ananas comosus*, high concentration of BAP (4.0 mg l^{-1}) alone or combined with indolebutyric acid (IBA) produced dwarf variants (100%) (Halim et al. 2018). Similarly, genetic variability greatly increased in rice callus cultures incubated in high levels of BAP (30 mg l^{-1}), compared to lower concentrations (2 mg l^{-1}) (Oono 1985). Bairu et al. (2006) also reported a high rate of SV in 'Cavendish' banana (Musa AAA, cv. 'Zelig') cultured with high BAP concentration; and in *Solanum melongena* L., the addition of BAP to the culture medium induced a high rate of polymorphism in adventitious shoots (Mançano et al. 2019). TDZ was used in shoot cultures of Dendrobium Second Love hybrid and SV did not appear after six subculture cycles (Ferreira et al. 2006). However, regenerated Indian soybean plants showed morphological and agronomic variations when it was cultured in TDZ-supplemented medium (Radhakrishnan and Kumari 2008).

Different causes may explain the effects of PGRs on SV. Auxins and cytokinins may affect the cell cycle (Halim et al. 2018). Therefore, while speeding up the cell cycle for proliferation, PGRs in the culture medium can disrupt it. Moreover, cell cycle involves duplication of genetic materials. Consequently, application of PGRs increases the risk of genetic alterations and can affect the chromosome number and ploidy levels (Smulders and de Klerk 2011). Modulation of DNA methylation by PGRs has also been reported (Us-Camas et al. 2014; Campos-Rivero et al. 2017). Additionally, as pointed out by Pastelín Solano et al. (2019), an endogenous accumulation of PGRs in tissues probably occurs throughout subcultures, causing the excess regulators to act as a mutagenic agent.

In relation to auxins, genetic and epigenetic modifications have been found due to this PGR. In general, increased auxin concentration, especially 2,4-D and picloram led to a decrease in the GTS value, and epigenetic variations have also been observed as a consequence of auxin treatments (Aydin et al. 2016). However, while DNA hypermethylation occurred at higher concentrations of 2,4-D, picloram, and NAA, DNA hypomethylation was observed in dicamba-supplemented media (Loschiavo et al. 1989; Aydin et al. 2016). The presence of auxins can also cause chromosomic aberrations (Bairu et al. 2011; Mançano et al. 2019).

The synthetic auxin 2,4-D, widely used in tissue culture, has a proven mutagenic effect (Bhojwani and Dantu 2013). Thus, 2,4-D has been associated with occurrence of genetic changes (Ładyżyński et al. 2002; Mohanty et al. 2008) and promotion of DNA methylation, causing epigenetic variations (Baklouti et al. 2022). Additionally, it has been reported that the presence of 2,4-D in the culture media significantly decreased the mitotic index and increased the ploidy level in cells (Saleh et al. 2019). Nevertheless, Aydin et al. (2016) pointed that the reason behind this variation may be the stress inherent in cellular deprogramming induced by 2,4-D. Dicamba has also been considered a mutagenic agent (Aydin et al. 2016). However, substitution of 2,4-D by NAA or IAA reduced chromosomal aberrations (Bhojwani and Dantu 2013).

Respect to cytokinins, their use has been found to increase the occurrence of polyploid cells (Liberatore et al. 2020). Specifically, BAP has been associated with epigenetic aberrations (Smulders and de Klerk 2011), cell-cycle disturbance (Bairu et al. 2011), and increase in chromosome number (Giménez et al. 2001). Moreover, in *Malus xiaojinensis*, DNA methylation decreased with increasing concentrations of BAP in the culture medium (Huang et al. 2012) and genes associated with DNA methylation were negatively correlated with BAP concentration. As reported by Halim et al. (2018), higher histone deacetylase (HDAC) enzyme activity was in general measured in variant plants regenerated from culture media supplemented with high concentration of BAP (4.0 mg/L) alone or combined with IBA. Kinetin also caused extensive DNA hypomethylation in carrot root cultures (Arnholdt-Schmitt 1993).

In relation to gibberellins, in onio, the mitotic index and cell size decreased with increasing concentration of GA_3 (Ud-Deen and Kabir 2009).

2.2 *Physical Factors*

Besides chemical composition of the culture medium, some physical factors may also influence the stability of in vitro cultures.

An effect of the temperature on SV has been repeatedly reported (Sivanesan and Jeong 2012; Krishna et al. 2016; Duarte-Aké and De-la-Peña 2021). In *Lilium longiflorum*, development of albino seedlings from somatic embryos increased as the incubation temperature was raised above 10–15 °C (Bhojwani and Razdan 1996). However, in *Senecio cruentus*, SV was observed when the cultures were maintained at 15°, but not at 25 °C (Sivanesan and Jeong 2012). The temperature is known to affect the cell cycle in plants and an inadequate control of the cell cycle in vitro has been suggested as one of the causes of SV (Karp 1994).

The physical state of the culture medium has also been related to occurrence of SV as it influences the cytological behaviour of the cultured cells (Bhojwani and Razdan 1996). In *Hevea*, higher polyploidy level was observed when cells were cultured in suspension but it decreased when they were recultured as callus on a solid medium (Wilson et al. 2006).

Other physical factors such as light intensity, osmolarity, humidity and agitation rate of the culture medium also affect the incidence of SV in plants (Krishna et al. 2016; Duarte-Aké and De-la-Peña 2021). Gas exchange can also play a role in SV due to its influence on ethylene accumulation, which cause epigenetic changes (Karp 1994).

Differences in a number of physical factors can be found in different in vitro culture systems and, consequently, in the stress imposed to the cultured cells and tissues. In Persian lime, analysis by ISSR markers revealed differences in SV frequency depending on the culture system used: semisolid medium, partial immersion in liquid culture medium, and the temporary immersion RITA® system. Polymorphism analysis by using ISSR markers revealed that the highest percentage of variation (26.5%) was obtained in shoots cultivated on semisolid medium. The lowest SV occurrence was observed in shoots propagated in the RITA® bioreactors, where the polymorphism percentage remained below 11% during all the subcultures tested (Bulbarela-Marini et al. 2023). The RITA® system exhibits important differences compared to other culture procedures since it promotes gas exchange and increases contact of tissues with the culture medium (Bulbarela-Marini et al. 2023).

3 In Vitro Culture Method

The different in vitro culture techniques present important differences at multiple levels. Culture medium requirements significantly vary depending on the propagation method, being especially important the combinations of PGRs, which have a relevant role in SV induction (Armijos-González et al. 2021). The tissue used as starting material and the treatments required for in vitro culture establishment are also factors influencing SV that present important differences depending on the in vitro culture method. Therefore, the type and magnitude of the stress imposed on cultured cells significantly vary depending on the method used (Krishna et al. 2016) and, in consequence, the nature and frequency of SV.

Different studies have investigated the influence of the in vitro culture procedures on SV. For example, in *Oroxylum indicum*, plants directly produced from axillary buds were phenotypically, cytologically and genetically more identical to their mother plants than those indirectly regenerated. While 10% polymorphism by RAPD analysis was found in plants obtained from axillary buds, 24.59% was estimated in plants regenerated via indirect organogenesis (Gokhale and Bansal 2015). In *Cinchona officinalis*, indirect shoot regeneration induced SV detected by ISSR analysis, whereas direct shoot formation gave rise to true-to-type plants (Armijos-González et al. 2021). Multiplication of *Ophiorrhiza mungos* by direct organogenesis from axillary and terminal buds also gave rise to plants genetically identical to the donor plant, as revealed by the absence of detectable genetic variation evaluated by using RAPD markers; however, indirect shoot regeneration induced SV (Kaushik et al. 2015).

In general, axillary shoot proliferation, with direct formation of plant structures from meristems pre-existing in the explant, minimizes the risk of instability (Karp 1994) and it retains clonal fidelity or gives rise to low frequencies of variation (Sivanesan and Jeong 2012; Gokhale and Bansal 2015). In indirect organogenesis, however, SV has been reported in different species, such as *Solanum melongena* (Zayova et al. 2010), *Arracacia xanthorrhiza* (Vitamvas et al. 2019), or *Oroxylum indicum* (Gokhale and Bansal 2015). SV has also been found in somatic embryogenesis systems of different species, such as *Citrus paradisi*, oil palm, rose, potato, grapevine, coffee (Krishna et al. 2016), yacon (Viehmannova et al. 2014), or olive (Leva et al. 2012; Bradaï et al. 2016a, 2019). Nevertheless, many investigations reported genetic uniformity in plants developed using this regeneration method (Fourré et al. 1997; Gallego et al. 1997; Thakur et al. 1999; Harvengt et al. 2001; Jayanthi and Mandal 2001; Pinto et al. 2004; Loureiro et al. 2005; Mallón et al. 2010; Rai et al. 2012). According to Vasil (1987), regeneration via somatic embryogenesis has better chances of obtaining genetically uniform plants than through organogenic differentiation. As proposed by Venkatachalam et al. (2003), it may reduce the frequency of genetic variation in regenerated plants due to developmental constraints during embryogenesis, that select against genetically altered plants.

In general, regeneration methods including a callus phase is likely to provoke more variability than direct pathways, as the callus phase accumulates most of the abnormality generated during in vitro culture (Phillips et al. 1994).

Cellular organization is a critical feature determining SV (Rani and Raina 2000). In fact, SV has been related to disorganized growth, and the longer the duration of the disorganized phase the greater the chances of generating SV (Karp 1994). In the same line, the greater the departure from organized growth (Karp 1994) or the more the organizational structure of the plant is broken down, the greater the chance of occurring variations (Araújo et al. 2001; Cooper et al. 2006).

Important differences related to cellular organization are found in the different in vitro culture techniques, with methods in which the cellular organization of the initial explant is not altered or minimally changed, such as axillary shoot proliferation or meristem tip culture, to systems in which regeneration is achieved after a phase of disorganized callus or cell suspension culture (Bairu et al. 2011).

Different causes could explain the higher variability found in in vitro methods involving a callus phase. Callus formation is considered a culture procedure imposing stress to cells (Smulders and de Klerk 2011), as it involves a dedifferentiation phase, followed by uncontrolled cell divisions, and a redifferentiation process, which can result in genetic and epigenetic changes (Kaeppler et al. 2000; Vázquez 2001; Leva et al. 2012; Grafi and Barak 2015). Moreover, according to Karp (1994), in disorganized growth, the constraints which act to eliminate genetic variations in normal meristems are suppressed, or mechanisms of genetic instability are induced.

The nature of the callus also may affect the variability observed in the regenerated plants (Bhojwani and Razdan 1996). A true callus is a mass of dedifferentiated cells proliferating in an unorganized manner. However, calli composed by a mass of suppressed organs or proliferating embryos, rather than completely unorganized tissue, also develop during in vitro tissue culture. While a true callus phase is likely to introduce considerable variability, in these calli it is more likely to retain a high degree of chromosomal stability (Bhojwani and Razdan 1996). This could explain higher genetic stability found after a callus phase in tissue cultures and regenerants of different species, such as asparagus, barley, guinea grass, maize, or wheat (Bhojwani and Razdan 1996).

A variation in the abnormalities detected with the nature of callus has also been reported in somatic embryogenesis systems. In oil palm, whereas slow growing, nodular and compact calli produced 5% variant palms, plants derived from fast growing calli exhibited up to 100% variation (Bhojwani and Dantu 2013). In the same species, significantly different variability was observed between embryogenic and non-embryogenic calli of the same genetic background. According to Giorgetti et al. (2011), it could imply that specific aberrations are strongly selected against in the acquisition of embryogenic competence.

A strong correlation between proliferation rate and SV frequency has been in general established, with higher instability as the multiplication rate increases (Bairu et al. 2006; Martínez-Estrada et al. 2017). In Persian lime, however, this correlation was not observed, as the higher multiplication rate was achieved in the RITA® bioreactors, the culture system producing less SV across subcultures (Bulbarela-Marini et al. 2023). Similarly, in banana, Khan et al. (2011) found an increased number of somaclonal variants as the multiplication rate of propagules decreased.

An interesting point is the evolution of SV during the application of an in vitro culture procedure as it may be a helpful tool to predict the moment at which genetic instability is initiated (Lopes et al. 2009). Narváez et al. (2019) investigated SV during somatic embryogenesis in olive by using RAPD markers. Similarly to that reported in *Picea glauca* (DeVerno et al. 1999), they found variations in callus but not in the regenerated plants. According to DeVerno et al. (1999) and González-Benito et al. (2014), embryogenic callus would be formed by a mix of stable and unstable cells, but only cells with an unaltered genome or accumulating not important variations would be able to regenerate plants.

4 Type of Explant

Type of tissue used as starting material can affect the frequency and nature of SV (Sahijram et al. 2003; Kawiak and Łojkowska 2004; Chuang et al. 2009; Krishna et al. 2016). In this line, Cao et al. (2016) found differences in the percentage of somaclonal variants in plants regenerated from mature and immature leaves of caladium, with higher rates of SV in plants obtained from immature leaves, independently on the culture medium used. In *Cucumis melo*, a larger frequency of tetraploids was observed in regenerants derived from immature cotyledons compared to those obtained from apical meristems (Adelberg et al. 1994). In rough lemon, plants regenerated from nucellar tissues showed no variation by RAPD analysis, while 20–30% of plants obtained from ovaries showed different banding patterns (Savita et al. 2015).

In general, highly differentiated tissues such as roots, leaves, or stems, produce more variants than explants like axillary buds and shoot tips, which have pre-existing meristems (Duncan 1996) and no dedifferentiation is required. Using older or more specialized tissues as explant source also increases the chances of SV in the regenerated plants (Krishna et al. 2016).

As previously indicated, cellular organization plays a main role in the appearance of SV. While in explants with pre-existing meristems, cellular organization is only slightly affected, in other systems, in which dedifferentiation and redifferentiation are required and regeneration is achieved after a callus phase, it is drastically altered (Karp 1994). Nevertheless, SV has also been found in plants derived from organized meristems (Rani and Raina 2000), vegetative buds (Martínez-Estrada et al. 2017) and nodal cuttings (Soliman et al. 2014). In fact, in bananas, Israeli et al. (1996) reported that shoot-tips caused more variation than somatic embryogenesis. However, this result was attributed to a possible dissociation of chimeras.

Nevertheless, a relationship between the type of explant and the in vitro propagation method cannot be ignored as, in general, plant regeneration from differentiated tissues normally takes place directly, from atypical points of origin, or mostly indirectly, through a callus stage (Pijut et al. 2012). In fact, according to Liberatore et al. (2020), the risk of SV when using differentiated tissues as explant source is especially high when indirect organogenesis or embryogenesis occurs, since callus formation entails a dedifferentiation and redifferentiation process, which can result in cellular disorganization and genetic and epigenetic changes.

Different causes can explain differences due to the type of tissue used as starting material. One of them is the existence of pre-existing variations in the tissue used as explant source (Peschke and Phillips 1992). According to D'Amato (1977), gross changes in the genome including endo-polyploidy, polyteny and amplification or diminution of DNA sequences could occur during somatic differentiation in normal plant growth and development. In fact, in some species like oil palm, the genome is naturally unstable, accumulating abnormalities during their growth in vivo (Giorgetti et al. 2011). The extent of genetic and epigenetic variations accumulated during normal plant development significantly varied depending of different factors such as the explant age and type of tissue and organ (Bhojwani and Dantu 2013). Thus, SV can arise from somatic variations already present in the mother plant (Karp 1994). This variability present in the tissue used as explant can become evident after in vitro culture, probably due to separation of chimeras (Bairu et al. 2011).

5 Length of Culture Period

The length of the culture period is another factor influencing SV. In general, higher variability was found in cultures maintained for longer periods and in their regenerants (Etienne and Bertrand 2003; Bradaï et al. 2016a, 2019; Krishna et al. 2016). Nevertheless, despite the increase in SV normally detected with time in culture, as stated by Tremblay et al. (1999), the effect of culture age is probably underestimated, since a decrease in cultures regeneration ability is usually in parallel observed (Klimaszewska et al. 2009; Park et al. 2011; Bradaï et al. 2016b). Genetic or epigenetic variations accumulated during in vitro culture can affect the integrity of the developmental pathway (Brar and Jain 1998), thus compromising its progression. In fact, Wang and Wang (2012), proposed the existence of a certain 'threshold' for every plant species, above which, no regenerants would be obtained.

Culture time also affects the nature of the alterations (Plader et al. 1998; Ładyżyński et al. 2002) and the accumulation of abnormalities (Bradaï et al. 2016a). In olive, Bradaï et al. (2016a) found that the percentage of regenerated plants showing morphological variations and the accumulation of variant phenotypes per plant were significantly affected by the culture age, although this effect was strongly influenced by the genotype.

Nevertheless, absence of SV after long-term culture has also been reported. In cassava cultures maintained in vitro for 10 years, no polymorphism was found by using Restriction Fragment Length Polymorphism (RFLP) and RAPD markers (Angel et al. 1996). In shoot cultures of *Vanilla planifolia* micropropagated for 10 years, genetic fidelity was also confirmed by using RAPD and ISSR markers (Sreedhar et al. 2007). In these cases, the genetic stability despite long subculture duration have been attributed to their genetic make-up, which could be inherently resistant to alterations due to in vitro culture (Chin et al. 2019).

In some investigations, the influence of culture time has been assessed by examining the effect of the number of subcultures. Accordingly with the above, the frequency of variation normally increases with increasing the number of subculture cycles. Thus, during in vitro propagation of *Vanilla planifolia*, the lowest polymorphism (0.84%) by using ISSR markers was observed in plants obtained from the first subculture. Shoots from 2 to 5 subcultures showed less than 4% polymorphism, and percentages of polymorphism greater than 15% were found after the sixth subcultures. The highest variation (19.49%) was detected in plants from the tenth subculture, the latest one tested (Solano et al. 2019). In *Musa*, Rodrigues et al. (1998) showed that somaclonal variants appeared from the fifth sub-culture (1.3%) onwards and increased to 3.8% after 11 subcultures.

In this line, Côte et al. (2001) proposed a mathematical model, by which the occurrence of SV can be calculated as a function of the number of multiplication cycles:

$$\%V = [1 - (1 - p)^n] \times 100$$

where, %V = percentage of SV, n = the number of multiplication cycles, p = the expected variation probability (with the value '1 − p' being the expected probability to get a true-to-type plant).

Different causes have been proposed to explain this culture age effect on SV. As previously indicated, in vitro culture itself acts as a mutagenic system because cells are subjected to stressful conditions (Jain 2001). Therefore, the longer the culture period, the greater the chances of SV occurrence, with accumulation of new variations as time in culture progress. As proposed by Bradaï et al. (2016a), since almost all the DNA variation arises from DNA replication and cell division in vitro (Wang and Wang 2012), higher variation frequencies are expected with prolonged cell proliferation periods. In this line, Brar and Jain (1998) stated that the age effect on increased frequency of mutation was primarily due to sequential accumulation of mutations over time rather than an increased mutation rate in old cultures. Nevertheless, activation of certain types of retrotransposons and increase in retrotransposon copy number (Hirochika et al. 1996), as well as induction of DNA hypermethylation (Rival et al. 2013) have been found in a time-dependent fashion. Accumulation in tissues of some culture media components with a mutagenic effect, like PGRs, can also occurs throughout subcultures, thus contributing to amplify the influence of culture duration (Bradaï et al. 2016a). Additionally, repeated subcultures promote multiplication of the alterations generated in prior cycles (Chin et al. 2019).

There are also evidences pointing to an effect of the frequency of subculture on SV. Investigations in different species, such as alfalfa, carrot, *Nigella sativa*, tobacco, or *Begonia rex*, revealed that frequent transfers yield more stable cultures than long subculture intervals (Bhojwani and Razdan 1996). Increased SV found in cultures maintained through long subculture cycles could be due to different causes as multiple modifications can occur in culture conditions throughout the subculture cycle. Factors like chemical composition (including PGRs), pH, moisture content, or gel strength, can change over time or due to the growing plant tissues. Additionally, different metabolites, such as phenols, can be secreted to the culture medium. These circumstances could increase the stress level at which cultures are normally subjected during in vitro culture. Nevertheless, Israeli et al. (1995) reported high SV rates when the period between subcultures is short and more subcultures are performed in a given time.

However, some studies reported lower SV frequencies in plants regenerated from older cultures (Franklin et al. 1989; Wang et al. 1992; Rodríguez López et al. 2010) These results and other counterintuitive findings led to Wang and Wang (2012) to propose a novel hypothesis applying the natural selection theory to in vitro cell populations. Based in five premises, they suggested three models for in vitro cell selection: (a) the selection model for V-win and V-callus, (b) the selection model for T-win and T-callus, and (c) the selection model for a draw and D-callus.

6 Genotype

The genotype of the parent plant also affects SV. Investigations carried out in different species, such as banana (Reuveni et al. 1993), grasspea (Tripathy et al. 2016), olive (Bradaï et al. 2016b, 2019), black and white spruce (Tremblay et al. 1999), cabbage (Abd-Elhamed 2018), chickpea (Alghamdi et al. 2021), Phalaenopsis (Tokuhara and Mii 1998), oil palm (Eeuwens et al. 2002), or *Coffea arabic*a (Etienne and Bertrand 2003), support this genotype effect on SV.

Important differences in stability have been found among species and within the same species, among cultivars and varieties. Thus, oil palm is an intrinsically variable system, particularly prone to genome instability. According to Giorgetti et al. (2011), natural unsteadiness of the chromosomal and genomic asset is greatly enhanced under in vitro culture conditions, giving rise to large genetic instability in the regenerating structures. At an intraspecific level, important differences have been reported among banana cultivars, with overall variation rates ranging from 3 to 20%, depending on the cultivar (Hwuang and Ko 1987). However, in *Phalaenopsis*, the percentage of somaclonal variants ranged from 0 to 100%, depending on the variety (Tokuhara and Mii 1998).

The genotype of the explant donor plant influences both the frequency and type of variation (Bradaï et al. 2016a). Phenotypic analysis of olive plants regenerated via somatic embryogenesis revealed a significant influence of the embryogenic line in the different parameters tested, including frequency of the variant phenotypes, percentage of regenerated plants showing morphological alterations, and accumulation of variant phenotypes per plant (Bradaï et al. 2016a). The results obtained also revealed the existence of genotype-dependent abnormal morphologies, with altered phenotypes only observed in plants derived from a specific cell line. An important variation in the similarity coefficients determined by Simple Sequence Repeat (SSR) and RAPD markers was also found in these plants depending on the genotype (Bradaï et al. 2019).

Although in some cases it is has been reported that the frequency of variation can be genotype-independent (Baklouti et al. 2022), the genotype has been pointed out as a fundamental factor in the appearance of SV in cultures and their regenerants, with genotypes with low, moderate or high risk of variation. In fact, in banana (Reuveni et al. 1993), black and white spruces (Tremblay et al. 1999), and olive (Bradaï et al. 2016a), the genotype has been considered the main factor affecting variability during in vitro culture.

Differences in genotype stability are related to differences in the genetic make-up. Some constituents of the genome make it more prone to undergo both genetic and epigenetic changes during in vitro culture (Bairu et al. 2011). Genotypes containing transposons are more susceptible to mutate under these conditions than those lacking these elements (Peschke and Phillips 1991). Ploidy level of the explant source can also affect the tendency of cultures to display SV. Generally, ploidy levels lower than the usual ploidy level of the respective species prove to be more or less unstable (Bhojwani and Razdan 1996). In *Crepis capillaris*, Sacristán (1971) found that diploidization was more common in haploid lines, than the occurrence of tetraploids in diploid lines. Besides, in general, polyploid genotypes tended to show higher variation than diploid genotypes of the same species (Bhojwani and Razdan 1996). Moreover, variations involving changes in chromosome number or structure are more easily recovered in regenerants of polyploids than in those of diploids and haploids, since polyploids can tolerate greater genomic alterations. However, gene mutations are better expressed in haploids and diploids, although show better survival in polyploids, particularly when such changes are deleterious (Brar and Jain 1998). The genetic make-up can also give rise to a differential sensitivity to the stress caused by in vitro culture (Karp 1994), thus resulting in a differential occurrence of SV.

Due to the great influence of the genotype, depending on the objectives of the in vitro culture technology, the identification of especially stable or highly susceptible genotypes can be essential. Thus, in in vitro methods aiming clonal propagation, it would be advisable to identify highly stable genotypes. However, in order to generate new variability to broaden the genetic basis in breeding programs, highly unstable genotypes could constitute a valuable material for producing sublines displaying interesting traits.

7 Concluding Remarks

In vitro culture techniques are increasingly being used due to their multiple applications in different fields, from the industry to basic research. SV is a phenomenon intrinsically linked to cells and tissue culture. However, sufficient attention is normally not paid to SV occurrence and in most cases plant stability is assumed. However, investigations on SV revealed that it is not an irrelevant event but its incidence can be very high depending on the experimental system.

Somaclonal variations occurring during in vitro culture are unpredictable in nature and there is no method to control their occurrence Nevertheless, knowing the factors affecting SV can help to "control" its incidence. In this way, in vitro culture conditions and other important aspects such as culture duration, the genotype, the type of explant, or the in vitro method, can be adapted to promote or reduce SV depending on the objectives of the culture system, focused on maintaining the genetic stability of the plant material or on the generation of new variability for different purposes.

References

Abd-Elhamed RS (2018) Somaclonal variation of developing callus of three cabbage cultivars under salinity. Biosci Res 15:3519–3530

Adelberg JW, Rhodes BB, Skorupska HT, Bridges WC (1994) Explant origin affects the frequency of tetraploid plants from tissue cultures of melon. HortScience 29:689–692. https://doi.org/10.21273/HORTSCI.29.6.689

Ahmed EU, Hayashi T, Yazawa S (2004) Auxins increase the occurrence of leaf-colour variants in caladium regenerated from leaf explants. Sci Hortic (Amsterdam) 100:153–159. https://doi.org/10.1016/j.scienta.2003.08.012

Alghamdi SS, Migdadi HM, Khan MA et al (2021) Assessment of somaclonal variations in embryo-derived axillary shoots of chickpea using molecular markers. Legum Res 44:508–514. https://doi.org/10.18805/LR-580

Angel F, Barney VE, Tohme J, Roca WM (1996) Stability of cassava plants at the DNA level after retrieval from 10 years of in vitro storage. Euphytica 90:307–313. https://doi.org/10.1007/BF00027481

Araújo LG, Prabhu AS, Filippi MC, Chaves LJ (2001) RAPD analysis of blast resistant somaclones from upland rice cultivar IAC 47 for genetic divergence. Plant Cell Tissue Organ Cult 67:165–172. https://doi.org/10.1023/A:1011960225472

Armijos-González R, Espinosa-Delgado L, Cueva-Agila A (2021) Indirect shoot regeneration using 2,4-D induces somaclonal variations in *Cinchona officinalis*. Floresta e Ambient 28:1–8. https://doi.org/10.1590/2179-8087-FLORAM-2021-0017

Arnholdt-Schmitt B (1993) Rapid changes in amplification and methylation pattern of genomic DNA in cultured carrot root explants (*Daucus carota* L.). Theor Appl Genet 85:793–800. https://doi.org/10.1007/BF00225021

Asadi N, Zarei H, Hashemi-Petroudi SH, Mousavizadeh SJ (2021) Micropropagation and assessment of somaclonal variation in *Galanthus transcaucasicus* in vitro plantlets. Ornam Hortic 27:505–515. https://doi.org/10.1590/2447-536X.v27i4.2320

Aydin M, Arslan E, Taspinar MS et al (2016) Analyses of somaclonal variation in endosperm-supported mature embryo culture of rye (*Secale cereale* L.). Biotechnol Biotechnol Equip 30:1082–1089. https://doi.org/10.1080/13102818.2016.1224980

Bairu MW, Aremu AO, van Staden J (2011) Somaclonal variation in plants: causes and detection methods. Plant Growth Regul 63:147–173. https://doi.org/10.1007/s10725-010-9554-x

Bairu MW, Fennell CW, van Staden J (2006) The effect of plant growth regulators on somaclonal variation in Cavendish banana (*Musa* AAA cv. 'Zelig'). Sci Hortic (Amsterdam) 108:347–351. https://doi.org/10.1016/j.scienta.2006.01.039

Baklouti E, Beulé T, Nasri A et al (2022) 2,4-D induction of somaclonal variations in in vitro grown date palm (*Phoenix dactylifera* L. cv Barhee). Plant Cell Tissue Organ Cult 150:191–205. https://doi.org/10.1007/s11240-022-02259-8

Bennici A, Anzidei M, Vendramin GG (2004) Genetic stability and uniformity of *Foeniculum vulgare* Mill. regenerated plants through organogenesis and somatic embryogenesis. Plant Sci 166:221–227. https://doi.org/10.1016/j.plantsci.2003.09.010

Bhojwani SS, Dantu PK (2013) Plant tissue culture: An introductory text. Springer, New Delhi. https://doi.org/10.1007/978-81-322-1026-9

Bhojwani SS, Razdan MK (1996) Plant tissue culture: Theory and practice, a revised edition. Elsevier, North Holland

Bradaï F, Pliego-Alfaro F, Sánchez-Romero C (2016a) Somaclonal variation in olive (*Olea europaea* L.) plants regenerated via somatic embryogenesis: Influence of genotype and culture age on phenotypic stability. Sci Hortic (Amsterdam) 213:208–215. https://doi.org/10.1016/j.scienta.2016.10.031

Bradaï F, Pliego-Alfaro F, Sánchez-Romero C (2016b) Long-term somatic embryogenesis in olive (*Olea europaea* L.): Influence on regeneration capability and quality of regenerated plants. Sci Hortic (Amsterdam) 199:23–31. https://doi.org/10.1016/j.scienta.2015.12.010

Bradaï F, Sánchez-Romero C, Martín C (2019) Somaclonal variation in olive (*Olea europaea* L.) plants regenerated via somatic embryogenesis: Influence of genotype and culture age on genetic stability. Sci Hortic (Amsterdam) 251:260–266. https://doi.org/10.1016/j.scienta.2019.03.010

Brar DS, Jain SM (1998) Somaclonal variation: mechanism and applications in crop improvement. In: Jain SM, Brar DS, Ahloowalia BS (eds) Somaclonal variation and induced mutations in crop improvement. Springer, Dordrecht, pp 15–37

Bulbarela-Marini JE, Gómez-Merino FC, Galindo-Tovar ME et al (2023) Ratio of somaclonal variation and the phytohormonal content of *Citrus* × *latifolia* in three in vitro culture systems. J Plant Growth Regul 42:3356–3364. https://doi.org/10.1007/s00344-022-10796-x

Campos-Rivero G, Osorio-Montalvo P, Sánchez-Borges R et al (2017) Plant hormone signaling in flowering: An epigenetic point of view. J Plant Physiol 214:16–27. https://doi.org/10.1016/j.jplph.2017.03.018

Cao Z, Sui S, Cai X et al (2016) Somaclonal variation in 'Red Flash' caladium: morphological, cytological and molecular characterization. Plant Cell Tissue Organ Cult 126:269–279. https://doi.org/10.1007/s11240-016-0996-3

Chin CK, Lee ZH, Mubbarakh SA et al (2019) Effects of plant growth regulators and activated charcoal on somaclonal variations of protocorm-like bodies (PLBs) of Dendrobium Sabin Blue orchid. Biocatal Agric Biotechnol 22:101426. https://doi.org/10.1016/j.bcab.2019.101426

Chuang SJ, Chen CL, Chen JJ et al (2009) Detection of somaclonal variation in micro-propagated *Echinacea purpurea* using AFLP marker. Sci Hortic (Amsterdam) 120:121–126. https://doi.org/10.1016/j.scienta.2008.09.020

Cooper C, Crowther T, Smith BM et al (2006) Assessment of the response of carrot somaclones to *Pythium violae*, causal agent of cavity spot. Plant Pathol 55:427–432. https://doi.org/10.1111/j.1365-3059.2006.01355.x

Côte FX, Teisson C, Perrier X (2001) Somaclonal variation rate evolution in plant tissue culture: Contribution to understanding through a statistical approach. In Vitro Cell Dev Biol Plant 37:539–542. https://doi.org/10.1007/s11627-001-0093-z

D'Amato F (1977) Cytogenetics of differentiation in tissue and cell culture. In: Reinert J, Bajaj Y (eds) Applied and fundamental aspects of plant cell, tissue and organ culture. Springer, New York, pp 343–464

DeVerno LL (1995) An evaluation of somaclonal variation during somatic embryogenesis. In: Jain SM, Gupta P, Newton R (eds) Somatic embryogenesis in woody plants. Kluwer, Dordrecht, pp 361–377

DeVerno LL, Park YS, Bonga JM et al (1999) Somaclonal variation in cryopreserved embryogenic clones of white spruce [*Picea glauca* (Moench) Voss.]. Plant Cell Rep 18:948–953. https://doi.org/10.1007/s002990050689

Duarte-Aké F, De-la-Peña C (2021) High cytokinin concentration and nutrient starvation trigger DNA methylation changes in somaclonal variants of *Agave angustifolia* Haw. Ind Crops Prod 172. https://doi.org/10.1016/j.indcrop.2021.114046

Duncan RR (1996) Tissue culture-induced variation and crop improvement. Adv Agron 58:201–240

Eeuwens CJ, Lord S, Donough CR et al (2002) Effects of tissue culture conditions during embryoid multiplication on the incidence of "mantled" flowering in clonally propagated oil palm. Plant Cell Tissue Organ Cult 70:311–323. https://doi.org/10.1023/A:1016543921508

Etienne H, Bertrand B (2003) Somaclonal variation in *Coffea arabica*: effects of genotype and embryogenic cell suspension age on frequency and phenotype of variants. Tree Physiol 23:419–426. https://doi.org/10.1093/treephys/23.6.419

Ferreira WDM, Kerbauy GB, Costa APP (2006) Micropropagation and genetic stability of a Dendrobium hybrid (Orchidaceae). In Vitro Cell Dev Biol Plant 42:568–571. https://doi.org/10.1079/IVP2006820

Fourré J-L, Berger P, Niquet L, André P (1997) Somatic embryogenesis and somaclonal variation in Norway spruce: morphogenetic, cytogenetic and molecular approaches. Theor Appl Genet 94:159–169. https://doi.org/10.1007/s001220050395

Franklin CI, Mott RL, Vuke TM (1989) Stable ploidy levels in long-term callus cultures of loblolly pine. Plant Cell Rep 8:101–104. https://doi.org/10.1007/BF00716849

Gallego FJ, Martinez I, Celestino C, Toribio M (1997) Testing somaclonal variation using RAPDs in *Quercus suber* L. somatic embryos. Int J Plant Sci 158:563–567

Gesteira A, Otoni W, Barros E (2008) RAPD-based detection of genomic instability in soybean plants derived from somatic embryogenesis. Plant Breed 121:269–271. https://doi.org/10.1046/j.1439-0523.2002.00708.x

Giménez C, de García E, de Enrech NX, Blanca I (2001) Somaclonal variation in banana: cytogenetic and molecular characterization of the somaclonal variant CIEN BTA-03. In Vitro Cell Dev Biol Plant 37:217–222. https://doi.org/10.1007/s11627-001-0038-6

Giorgetti L, Castiglione MR, Turrini A et al (2011) Cytogenetic and histological approach for early detection of "mantled" somaclonal variants of oil palm regenerated by somatic embryogenesis: first results on the characterization of regeneration system. Caryologia 64:221–232. https://doi.org/10.1080/00087114.2002.10589787

Gokhale M, Bansal Y (2015) Assessment of genetic fidelity in somaclonal variants with cytological and RAPD analysis in *Oroxylum indicum* (L.) Vent.—An endangered medicinal tree. Indian J Biotechnol 14

González-Benito ME, Miñano HS, Ibáñez M, Martin C (2014) Sequential study of the genetic stability of callus and regenerated shoots in chrysanthemum. Propag Ornam Plants 14:57

Grafi G, Barak S (2015) Stress induces cell dedifferentiation in plants. Biochim Biophys Acta Gene Regul Mech 1849:378–384. https://doi.org/10.1016/j.bbagrm.2014.07.015

Halim NAA, Tan BC, Midin MR et al (2018) Abscisic acid and salinity stress induced somaclonal variation and increased histone deacetylase (HDAC) activity in *Ananas comosus* var. MD2. Plant Cell Tissue Organ Cult 133:123–135. https://doi.org/10.1007/s11240-017-1367-4

Harvengt L, Trontin J-F, Reymond I et al (2001) Molecular evidence of true-to-type propagation of a 3-year-old norway spruce through somatic embryogenesis. Planta 213:828–832. https://doi.org/10.1007/s004250100628

Hirochika H, Sugimoto K, Otsuki Y et al (1996) Retrotransposons of rice involved in mutations induced by tissue culture. Proc Natl Acad Sci U S A 93:7783–7788. https://doi.org/10.1073/pnas.93.15.7783

Huang H, Han SS, Wang Y et al (2012) Variations in leaf morphology and DNA methylation following in vitro culture of *Malus xiaojinensis*. Plant Cell Tissue Organ Cult 111:153–161. https://doi.org/10.1007/s11240-012-0179-9

Hwuang SC, Ko W (1987) Somaclonal variation of banana and screening for resistance to Fusarium wilt. In Perseley G, De Langhe E (eds) Bananas and plantains breeding strategies. ACIAR Proceedings Series 21:151–156

Israeli Y, Ben-Bassat D, Reuveni O (1996) Selection of stable banana clones which do not produce dwarf somaclonal variants during in vitro culture. Sci Hortic (Amsterdam) 67:197–205. https://doi.org/10.1016/S0304-4238(96)00955-7

Israeli Y, Lahav E, Reuveni O (1995) In vitro culture of bananas. In Gowen S (ed) Bananas and plantains. Springer, Dordrecht, pp 147–178

Jain SM (2001) Tissue culture-derived variation in crop improvement. Euphytica 118:153–166. https://doi.org/10.1023/A:1004124519479

Jayanthi M, Mandal PK (2001) Plant regeneration through somatic embryogenesis and RAPD analysis of regenerated plants in *Tylophora indica* (Burm. F. Merrill.). In Vitro Cell Dev Biol Plant 37:576–580. https://doi.org/10.1007/s11627-001-0101-3

Jin S, Mushke R, Zhu H et al (2008) Detection of somaclonal variation of cotton (*Gossypium hirsutum*) using cytogenetics, flow cytometry and molecular markers. Plant Cell Rep 27:1303–1316. https://doi.org/10.1007/s00299-008-0557-2

Kaeppler SM, Kaeppler HF, Rhee Y (2000) Epigenetic aspects of somaclonal variation in plants. Plant Mol Biol 43:179–188. https://doi.org/10.1023/A:1006423110134

Karp A (1994) Origins, causes and uses of variation in plant tissue cultures. In: Vasil I, Thorpe T (eds) Plant cell and tissue culture. Springer, Dordrecht, pp 139–151

Kaushik PS, Swamy MK, Balasubramanya S, Anuradha M (2015) Rapid plant regeneration, analysis of genetic fidelity and camptothecin content of micropropagated plants of *Ophiorrhiza mungos* Linn.—a potent anticancer plant. J Crop Sci Biotechnol 18:1–8. https://doi.org/10.1007/s12892-014-0001-9

Kawiak A, Łojkowska E (2004) Application of RAPD in the determination of genetic fidelity in micropropagated *Drosera* plantlets. In Vitro Cell Dev Biol Plant 40:592–595

Khan S, Saeed B, Kauser N (2011) Establishment of genetic fidelity of in-vitro raised banana plantlets. Pak J Bot, 43:233–242
Klimaszewska K, Noceda C, Pelletier G et al (2009) Biological characterization of young and aged embryogenic cultures of *Pinus pinaster* (Ait.). In Vitro Cell Dev Biol Plant 45:20–33. https://doi.org/10.1007/s11627-008-9158-6
Krishna H, Alizadeh M, Singh D et al (2016) Somaclonal variations and their applications in horticultural crops improvement. 3 Biotech 6:1–18. https://doi.org/10.1007/s13205-016-0389-7
Ładyżyński M, Burza W, Malepszy S (2002) Relationship between somaclonal variation and type of culture in cucumber. Euphytica 125:349–356. https://doi.org/10.1023/A:1016017825907
Larkin PJ, Scowcroft WR (1981) Somaclonal variation—a novel source of variability from cell cultures for plant improvement. Theor Appl Genet 60:197–214. https://doi.org/10.1007/BF02342540
Leva AR, Petruccelli R, Rinaldi LMR (2012) Somaclonal variation in tissue culture: A case study with olive. In Leva A, Rinaldi LMR (eds). Recent advances in plant in vitro culture. IntechOpen, pp 123–150
Liberatore CM, Rodolfi M, Beghè D et al (2020) In vitro leaf-derived organogenesis and somaclonal variant detection in *Humulus lupulus* L. In Vitro Cell Dev Biol Plant 56:865–874. https://doi.org/10.1007/s11627-020-10088-7
Lopes T, Capelo A, Brito G et al (2009) Genetic variability analyses of the somatic embryogenesis induction process in *Olea* spp. using nuclear microsatellites. Trees Struct Funct 23:29–36. https://doi.org/10.1007/s00468-008-0251-6
Loschiavo F, Pitto L, Giuliano G et al (1989) DNA methylation of embryogenic carrot cell cultures and its variations as caused by mutation, differentiation, hormones and hypomethylating drugs. Theor Appl Genet 77:325–331. https://doi.org/10.1007/BF00305823
Loureiro J, Pinto G, Lopes T et al (2005) Assessment of ploidy stability of the somatic embryogenesis process in *Quercus suber* L. using flow cytometry. Planta 221:815–822. https://doi.org/10.1007/s00425-005-1492-x
Mager S, Ludewig U (2018) Massive loss of DNA methylation in nitrogen-, but not in phosphorus-deficient *Zea mays* roots is poorly correlated with gene expression differences. Front Plant Sci 9. https://doi.org/10.3389/fpls.2018.00497
Mallón R, Rodríguez-Oubiña J, González ML (2010) In vitro propagation of the endangered plant *Centaurea ultreiae*: Assessment of genetic stability by cytological studies, flow cytometry and RAPD analysis. Plant Cell Tissue Organ Cult 101:31–39. https://doi.org/10.1007/s11240-009-9659-y
Mançano AP, de Oliveira SB, de Oliveira GR et al (2019) Genetic stability among in vitro eggplant clones induced by different plant growth regulators. Biotecnol Veg 19:75–83
Martínez-Estrada E, Caamal-Velázquez JH, Salinas-Ruíz J, Bello-Bello JJ (2017) Assessment of somaclonal variation during sugarcane micropropagation in temporary immersion bioreactors by intersimple sequence repeat (ISSR) markers. In Vitro Cell Dev Biol Plant 53:553–560. https://doi.org/10.1007/s11627-017-9852-3
Mohanty S, Panda MK, Subudhi E, Nayak S (2008) Plant regeneration from callus culture of *Curcuma aromatica* and in vitro detection of somaclonal variation through cytophotometric analysis. Biol Plant 52:783–786. https://doi.org/10.1007/s10535-008-0153-x
Narváez I, Martín C, Jiménez-Díaz RM et al (2019) Plant regeneration via somatic embryogenesis in mature wild olive genotypes resistant to the defoliating pathotype of *Verticillium dahliae*. Front Plant Sci 10:1471. https://doi.org/10.3389/fpls.2019.01471
Neelakandan AK, Wang K (2012) Recent progress in the understanding of tissue culture-induced genome level changes in plants and potential applications. Plant Cell Rep 31:597–620. https://doi.org/10.1007/s00299-011-1202-z
Nehra NS, Kartha KK, Stushnott C, Giles KL (1992) The influence of plant growth regulator concentrations and callus age on somaclonal variation in callus culture regenerants of strawberry. Plant Cell Tissue Organ Cult 29:257–268. https://doi.org/10.1007/BF00034361

Oono K (1985) Putative homozygous mutations in regenerated plants of rice. Mol Gen Genet MGG 198:377–384. https://doi.org/10.1007/BF00332926

Park SY, Cho HM, Moon HK et al (2011) Genotypic variation and aging effects on the embryogenic capability of *Kalopanax septemlobus*. Plant Cell Tissue Organ Cult 105:265–270. https://doi.org/10.1007/s11240-010-9862-x

Pastelín Solano MC, Salinas Ruíz J, González Arnao MT et al (2019) Evaluation of in vitro shoot multiplication and ISSR marker based assessment of somaclonal variants at different subcultures of vanilla (*Vanilla planifolia* Jacks). Physiol Mol Biol Plants Int J Funct Plant Biol 25:561–567. https://doi.org/10.1007/s12298-019-00645-9

Peschke VM, Phillips RL (1991) Activation of the maize transposable element *Suppressor-mutator (Spm)* in tissue culture. Theor Appl Genet 81:90–97. https://doi.org/10.1007/BF00226117

Peschke VM, Phillips RL (1992) Genetic implications of somaclonal variation in plants. Adv Genet 30:41–75

Phillips RL, Kaeppler SM, Olhoft P (1994) Genetic instability of plant tissue cultures: breakdown of normal controls. Proc Natl Acad Sci U S A 91:5222–5226. https://doi.org/10.1073/pnas.91.12.5222

Pijut P, Beasley R, Lawson S et al (2012) In vitro propagation of tropical hardwood tree species—A review (2001–2011). Propag Ornam Plants 12:25–51

Pinto G, Loureiro J, Lopes T, Santos C (2004) Analysis of the genetic stability of *Eucalyptus globulus* Labill. somatic embryos by flow cytometry. Theor Appl Genet 109:580–587. https://doi.org/10.1007/s00122-004-1655-3

Plader W, Malepszy S, Burza W, Rusinowski Z (1998) The relationship between the regeneration system and genetic variability in the cucumber (*Cucumis sativus* L.). Euphytica 103:9–15. https://doi.org/10.1023/A:1018359726626

Radhakrishnan R, Kumari B (2008) Morphological and agronomic evaluation of tissue culture derived Indian soybean plants. Acta Agric Slov Acta Agric Slov 91:391–396

Rai MK, Phulwaria M, Harish et al (2012) Genetic homogeneity of guava plants derived from somatic embryogenesis using SSR and ISSR markers. Plant Cell Tissue Organ Cult 111:259–264. https://doi.org/10.1007/s11240-012-0190-1

Rani V, Raina SN (2000) Genetic fidelity of organized meristem-derived micropropagated plants: A critical reappraisal. In Vitro Cell Dev Biol Plant 36:319–330. https://doi.org/10.1007/s11627-000-0059-6

Reuveni O, Golubowicz S, Israeli Y (1993) Factors influencing the occurrence of somaclonal variations in micropropagated bananas. Acta Hortic 336:357–364

Rival A, Ilbert P, Labeyrie A et al (2013) Variations in genomic DNA methylation during the long-term in vitro proliferation of oil palm embryogenic suspension cultures. Plant Cell Rep 32:359–368. https://doi.org/10.1007/s00299-012-1369-y

Rodrigues PHV, Tulmann Neto A, Cassieri Neto P, Mendes BMJ (1998) Influence of the number of subcultures on somaclonal variation in micropropagated Nanicão (*Musa* spp., AAA group). Acta Hortic 490:469–474

Rodríguez López CM, Wetten AC, Wilkinson MJ (2010) Progressive erosion of genetic and epigenetic variation in callus-derived cocoa (*Theobroma cacao*) plants. New Phytol 186:856–868. https://doi.org/10.1111/j.1469-8137.2010.03242.x

Sacristán MD (1971) Karyotypic changes in callus cultures from haploid and diploid plants of *Crepis capillaris* (L.) Wallr. Chromosoma 33:273–283. https://doi.org/10.1007/BF00284945

Sahijram L, Soneji JR, Bollamma KT (2003) Analyzing somaclonal variation in micropropagated bananas (*Musa* spp.). In Vitro Cell Dev Biol Plant 39:551–556

Saleh A, Taha R, Mahmad N et al (2019) Detection of somaclonal variation in micropropagated and acclimatized plantlets of *Oryza sativa* MRQ 74 from stem explant. Planta Daninha 37:e019171165

Sales EK, Butardo NG (2014) Molecular analysis of somaclonal variation in tissue culture derived bananas using MSAP and SSR markers. World Acad Sci Eng Technol Int J Biol Biomol Agric Food Biotechnol Eng 8:615–622

Savita PPK, Virk GS, Nagpal A (2015) An efficient somatic embryogenesis protocol for *Citrus jambhiri* and assessment of clonal fidelity of plantlets using RAPD markers. J Plant Growth Regul 34:309–319. https://doi.org/10.1007/s00344-014-9465-6

Sivanesan I, Jeong BR (2012) Identification of somaclonal variants in proliferating shoot cultures of *Senecio cruentus* cv. Tokyo Daruma. Plant Cell Tissue Organ Cult 111:247–253. https://doi.org/10.1007/s11240-012-0186-x

Smulders MJM, de Klerk GJ (2011) Epigenetics in plant tissue culture. Plant Growth Regul 63:137–146. https://doi.org/10.1007/s10725-010-9531-4

Solano MCP, Salinas Ruíz J, González Arnao MT et al (2019) Evaluation of in vitro shoot multiplication and ISSR marker based assessment of somaclonal variants at different subcultures of vanilla (*Vanilla planifolia* Jacks). Physiol Mol Biol Plants 25:561–567. https://doi.org/10.1007/s12298-019-00645-9

Soliman H, Metwali E, Almaghrabi OAH (2014) Micropropagation of *Stevia rebaudiana* Betroni and assessment of genetic stability of in vitro regenerated plants using inter simple sequence repeat (ISSR) marker. Plant Biotechnol 31:249–256. https://doi.org/10.5511/plantbiotechnology.14.0707a

Sreedhar RV, Venkatachalam L, Bhagyalakshmi N (2007) Genetic fidelity of long-term micropropagated shoot cultures of vanilla (*Vanilla planifolia* Andrews) as assessed by molecular markers. Biotechnol J 2:1007–1013. https://doi.org/10.1002/biot.200600229

Thakur R, Goto S, Ishii K, Jain SM (1999) Monitoring genetic stability in *Quercus serrata* Thunb. somatic embryogenesis using RAPD markers. J Res 4:157–160

Tokuhara K, Mii M (1998) Somaclonal variations in flower and inflorescence axis in micropropagated plants through flower stalk bud culture of *Phalaenopsis* and *Doritaenopsis*. Plant Biotechnol 15:23–28. https://doi.org/10.5511/plantbiotechnology.15.23

Tremblay L, Levasseur C, Tremblay F (1999) Frequency of somaclonal variation in plants of black spruce (*Picea mariana*, Pinaceae) and white spruce (*P. glauca*, Pinaceae) derived from somatic embryogenesis and identification of some factors involved in genetic instability. Am J Bot 86:1373–1381

Tripathy SK, Panda A, Nayak PK et al (2016) Somaclonal variation for genetic improvement in grasspea (*Lathyrus sativus* L.). Legum Res 39:329–335. https://doi.org/10.18805/lr.v0iOF.6853

Trujillo I, de García E (1996) Estrategias para la obtención de variantes somaclonales resistentes a la Sigatoka amarilla. InfoMusa 5:12–13

Ud-Deen M, Kabir G (2009) Effects of growth regulators on root tip cells of onion. Bangladesh J Bot 38:99–102

Us-Camas R, Rivera-Solís G, Duarte-Aké F, De-la-Peña C (2014) In vitro culture: an epigenetic challenge for plants. Plant Cell Tissue Organ Cult 118:187–201. https://doi.org/10.1007/s11240-014-0482-8

Vasil IK (1987) Developing cell and tissue culture systems for the improvement of cereal and grass crops. J Plant Physiol 128:193–218. https://doi.org/10.1016/S0176-1617(87)80234-1

Vázquez AM (2001) Insight into somaclonal variation. Plant Biosyst Int J Deal Asp Plant Biol 135:57–62. https://doi.org/10.1080/11263500112331350650

Venkatachalam L, Sreedhar RV, Bhagyalakshmi N (2007) Genetic analyses of micropropagated and regenerated plantlets of banana as assessed by RAPD and ISSR markers. In Vitro Cell Dev Biol Plant 43:267–274. https://doi.org/10.1007/s11627-007-9028-7

Venkatachalam P, Geetha N, Priya P et al (2003) Somatic embryogenesis. In: Jaiwal PK, Singh RP (eds) Improvement strategies of leguminosae biotechnology. Springer, Dordrecht, pp 87–132

Viehmannova I, Bortlova Z, Vitamvas J et al (2014) Assessment of somaclonal variation in somatic embryo-derived plants of yacon [*Smallanthus sonchifolius* (Poepp. and Endl.) H. Robinson] using inter simple sequence repeat analysis and flow cytometry. Electron J Biotechnol 17:102–106. https://doi.org/10.1016/j.ejbt.2013.12.011

Vitamvas J, Viehmannová I, Hlásná Cepková P, et al (2019) Assessment of somaclonal variation in indirect morphogenesis-derived plants of *Arracacia xanthorrhiza*. Pesqui Agropecuária Bras 54. https://doi.org/10.1590/s1678-3921.pab2019.v54.00301

Wang QM, Wang L (2012) An evolutionary view of plant tissue culture: somaclonal variation and selection. Plant Cell Rep 31:1535–1547. https://doi.org/10.1007/s00299-012-1281-5

Wang X-H, Lazzeri PA, Lörz H (1992) Chromosomal variation in dividing protoplasts derived from cell suspensions of barley (*Hordeum vulgare* L.). Theor Appl Genet 85:181–185. https://doi.org/10.1007/BF00222857

Wilson H, Eisa M, Irwin S (2006) The effects of agitated liquid medium on in vitro cultures of *Hevea brasiliensis*. Physiol Plant 36:399–402. https://doi.org/10.1111/j.1399-3054.1976.tb02264.x

Zayova E, Vassilevska-Ivanova R, Stoeva D, Kraptchev B (2010) Somaclonal variation through indirect organogenesis in eggplant (*Solanum melongena* L.). Biol Divers Conserv 3:1–5

Molecular Markers for the Detection and Analysis of Somaclonal Variation

Carmen Martín and M. Elena González-Benito

Abstract A diverse number of molecular techniques are nowadays available to approach the analysis of the genetic stability of plant material regenerated from tissue culture. Different molecular markers based on PCR provide useful tools to check the genetic and epigenetic variations associated to in vitro culture. Arbitrary amplifications, such as those obtained through RAPD, ISSR and AFLP markers, allow the assessment of the appearance of somaclonal variation, even in species with an unknown genome. Besides, specific markers as SSR may be employed in species for which genome information is available. Likewise, epigenetic changes are detected through molecular markers such as MSAP and metAFLP, based on AFLP technique. Recent studies are incorporating new sequencing technologies to explore changes in the genome of the regenerated plants capable of detecting genetic and epigenetic changes, allowing to evaluate simultaneously changes in gene expression associated to these variations. Regardless of the detection technique employed, the appropriate selection of samples and controls is important to obtain a reliable analysis of the genetic and epigenetic changes associated to a tissue culture process.

Keywords Genetic stability · Epigenetic changes · DNA markers · Genome assessment · Tissue culture

1 Introduction

Plant tissue culture processes may generate genetic instability, producing genetic mutations and/or epigenetic changes. Those permanent and heritable variations have been referred to as somaclonal variation since Larkin and Scowcroft (1981) publication. The in vitro environment is stressful to the plant cells and imposes a genomic shock which may derive in genetic rearrangements (Ranghoo-Sanmukhiya 2021).

Depending on the purpose for which in vitro culture techniques are used and the considered source of explant, it is possible to distinguish different protocols. Each of

C. Martín (✉) · M. E. González-Benito
Department of Biotechnology-Plant Biology, Universidad Politécnica de Madrid, Madrid, Spain
e-mail: mariacarmen.martin@upm.es

C. Sánchez-Romero (ed.), *Somaclonal Variation: Basic and Practical Aspects*,
https://doi.org/10.1007/978-3-031-51626-9_4

them presents a series of requirements regarding the type of explant, culture medium and culture conditions. These determining factors may impose different levels of stress to the cells and tissues, which may imply differences in the probability of genetic and epigenetic instability appearance in the regenerated plant material. Even considering and carefully choosing the type of explant, the kind and concentration of plant growth regulators (PGR) and the culture conditions, the occurrence of somaclonal variation is unavoidable in most cases and unpredictable.

Although the number of studies to understand the mechanisms and factors behind somaclonal variation occurrence has increased in the last years (Leva and Rinaldi 2017; Manchanda et al. 2018), many aspects of the mechanisms involved remain undefined. For this reason, and in order to check the stability of the cultures, it is unavoidable to explore the genome-wide change through the sequence analysis (complete or fragments) of the genome of the regenerated plants (Krishna et al. 2016). Early detection of possible differences that may have arisen during in vitro culture is also an important factor, which can help in the selection of regenerated material, as well as in the decisions regarding culture maintenance or the following culture phases.

Molecular markers are small fragments of the genome, which may represent characteristics of the whole genome allowing its study. Not all sections of the genome are equally informative and, therefore, molecular markers may vary in their effectiveness for genome study. Molecular markers have been used to assess genetic stability in plant tissue cultures for many decades. The first studies were based on Restriction Fragment Length Polymorphism (RFLP), a non-Polymerase Chain Reaction (PCR) based technique, scarcely used nowadays due to its high complexity and to be time consuming. Since the use of RFLP to analyse the molecular changes produced in protoplast-derived rice plants (Brown et al. 1990), many studies have employed molecular markers to detect genetic changes associated to plant tissue culture procedures. The appearance of PCR technique led to the development of many molecular markers based on it, which have been used to detect somaclonal variation. The most well-known and employed markers belong to this group, in which it is possible to distinguish between specific and arbitrary markers, depending on whether primers designed from known genome sequences or arbitrary primers are used, respectively (Garrido-Cardenas et al. 2018). Most recently, and related to the advances in sequence analyses, sequence-based marker techniques are being incorporated to somaclonal variation studies.

Molecular markers for the analysis of epigenetic changes have also been developed and their use in the assessment of stability in tissue culture derived plants is frequent. Among the different epigenetic mechanisms, DNA methylation, and particularly cytosine methylation, is the most studied one in plants. The most widely used techniques to analyse DNA methylation are those based on methylation-sensitive restriction enzymes or on bisulfite modification (González-Benito et al. 2020). As for genetic stability studies, the first technique used for epigenetic stability analysis was RFLP, in particular a modification of this molecular marker using for the DNA digestion step the isoschizomeric enzymes *Msp*I and *Hpa*II, which have different sensitivity to methylation. The study of oil palm (*Elaeis guineensis* Jacq.) "mantled" somaclonal variants, which display alterations in floral organ structure, obtained from somatic embryogenesis is among the first works applying this technique to evaluate epigenetic changes (Jaligot et al. 2002).

Other procedures that use tissue culture and, therefore, may be affected by genetic and epigenetic instability, are those employed to obtain transgenic plants. Transformation processes usually require tissue culture steps to generate a whole transgenic plant from transformed cells or tissues. The effect of tissue culture conditions on the transgenic material may result in somaclonal variation and can, therefore, contribute to differences among the transgenic regenerated plants. Most of the studies on variability in transgenic plants focus on those traits affected by the transgenes. However, gene expression is a complex mechanism, and if somaclonal variation has occurred during the process, it may be affecting transgene expression. On the other hand, variations of the level expression of a foreign gene may have influence on the whole genome (Bhat and Srinivasan 2002). For this reason, when tissue culture is used in transformation protocols, it is important to check the genetic and epigenetic stability of transgenic regenerated plants in order to discard variations due to somaclonal variation and not directly produced by genetic transformation.

Therefore, the use of in vitro culture, in their different methodologies or with different objectives, may be associated with instability in the genome (genetic or epigenetic). Consequently, it is important to establish a reference control, that ensures that the changes detected are due to the in vitro procedures and not to the physiological/developmental state of the samples; this is especially relevant for the epigenetic studies. Molecular markers allow fast analyses to detect changes. Depending on the applied tissue culture technique and the different processes that the cultivated material goes through, it is convenient to establish one or several moments of analysis, if possible, throughout the process, in addition to the study of the regenerated plants. A summary scheme with the recommended sampling moments of the process to evaluate the genetic and/or epigenetic stability of the plant material is represented in Fig. 1.

2 Molecular Markers for the Detection of Genetic Variation

Although plants regenerated from tissue culture processes are expected to be homogenous and clonal to the mother plant, it is well known that several factors as source of explant, growth regulators, culture conditions and time of culture may affect genetic and epigenetic stability. These changes arise as a response to stressful conditions associated to tissue culture, and some of them have been considered an adaptation attempt to culture environment (Neelakandan and Wang 2012).

Regarding the genetic changes generated by tissue culture, the most frequent ones are those related to DNA sequence variations, such as single base pair changes and small ‘indels’, which may or may not be associated with phenotypic variations. This loss of sequence fidelity is related to the alteration of the efficiency of DNA replication and repair machinery under in vitro conditions, mainly due to a reduction

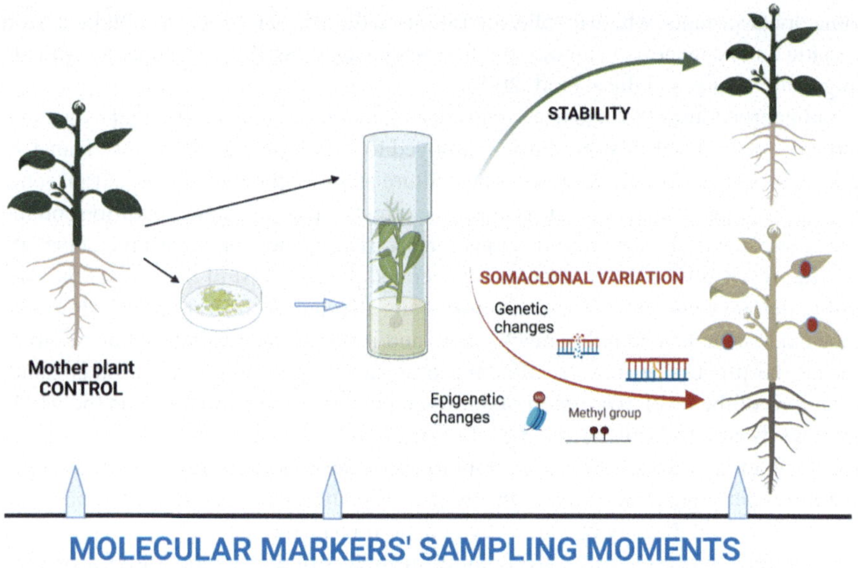

Fig. 1 Scheme of the genetic and/or epigenetic instability generated during tissue culture process and the moments recommended for their detection using molecular markers. Created with BioRender.com

of cellular control (Neelakandan and Wang 2012). Other possible alterations may affect ploidy level or chromosomal structure.

Molecular markers allow for the detection of DNA sequence variations, even without a previous knowledge of the studied genome. Genetic changes observed in in vitro regenerated plants are largely stochastic, unpredictable, and non-reproducible; therefore, molecular markers that enable a wide screening of the whole genome may be a very useful tool to detect this kind of genetic variations. As it was previously stated, the first developed DNA molecular marker was RFLP technique. From this first attempt to nowadays a wide variety of techniques have emerged, mainly based on PCR procedure. Among this group, it is possible to distinguish arbitrary or specific techniques, depending on the specificity of the primers used and prior knowledge of the genome under study.

One of the most extendedly used classification for DNA markers is based on the detection method, and, consequently, three main categories may be distinguished: hybridization, PCR and DNA sequence dependent molecular markers (Amiteye 2021). The techniques mostly used in the analysis of genetic stability of in vitro cultures are briefly explained below.

2.1 *Hybridization Based Techniques*

Restriction Fragment Length Polymorphism (RFLP) Although this technique is scarcely used nowadays, its importance in the first works to detect somaclonal variation by molecular markers makes necessary its description. Restriction enzymes are used to digest DNA samples; thereafter, the digested products are separated by electrophoresis on agarose gel and transferred to a membrane (Southern blot). Labelled DNA probes are used for hybridization of the membrane. This allows the detection of polymorphisms in the length of the restriction fragments which are complementary to the probe. Variations arise from mutations that are mainly due to single nucleotide base-pair deletions or insertions. Therefore, these sequence changes at the enzyme recognition site produce fragments that vary in length.

2.2 *PCR Based Techniques*

PCR-based methods have had a rapid diffusion in the detection of somaclonal variation, as in many other biological areas, mainly due to its simplicity, speed and, in some cases, low cost, compared to other techniques such as RFLP. Examples of the use of the techniques described below can be found in Table 1.

2.2.1 Arbitrary Techniques

The main advantage of this group of techniques is that prior knowledge of the studied genome is not necessary. On the other hand, one of their limitations is that they generate dominant markers, so it is not possible to distinguish homozygous from heterozygous individuals.

Random Amplified Polymorphic DNA (RAPD)

It is one of the simplest techniques to handle and with the lowest cost, and, possibly for these reasons, it is widely used for the detection of somaclonal variation. The method consists in the use of a single arbitrary primer for DNA amplification. The primers are short, usually 10-mer, and their sequences are arbitrary. Amplification products are easily analysed by electrophoretic separation. Variations in the complementary sequence of the primers or deletions/insertions between two recognition sites produce differences in the presence or absence of amplified fragments, or differences in length. When the technique is applied with precision, the results can be very informative and reliable, and have the advantage of being very quickly obtained. This technique has been widely used in many species due to its ease of use and speed of analysis.

Table 1 Techniques used to detect genetic and epigenetic changes in different in vitro culture procedures

Type of regeneration/in vitro technique	Genetic variation detection technique	Epigenetic variation detection technique	Species	Genetic/epigenetic variation results	Reference
Shoot multiplication by axillary buds					
Shoot multiplication from nodal segments	RAPD, ISSR		*Dendrocalamus strictus*	Stability	Goyal et al. (2015)
Shoot propagation, 15 subcultures	ISSR		*Saccharum officinarum*	No evidence of polymorphism	Hsie et al. (2015)
Shoot multiplication from nodal segments	SCoT		*Ansellia africana*	Higher genetic instability with 6-benzyladenine than with meta-topolin	Bhattacharyya et al. (2017)
Shoot multiplication from apex culture	ISSR		*Saccharum* spp.	Subcultures with the highest percentage of polymorphism and genetic distances:1st, 9th, and 10th	Martínez-Estrada et al. (2017)
Shoot multiplication from one-node cuttings	ISSR		*Prunus cerasifera*	No genetic variation	Nasri et al. (2019)
Shoot multiplication from nodal segments for 10 subculture cycles	ISSR		*Vanilla planifolia*	Increased polymorphism percentage after the 5th subculture cycle	Solano et al. (2019)
Shoot multiplication by axillary buds	RAPD, ISSR		*Rhododendron mucronulatum*	Genetic similarity of all regenerants between themselves and with the mother seedlings was 99%	Novikova et al. (2020)
Shoot multiplication by axillary buds for 5 years	RAPD, ISSR, SCoT		*Prunus salicina*	Genetic stability	Thakur et al. (2021)
Shoot multiplication	RAPD		*Artocarpus heterophyllus*	No polymorphism detected	Kader et al. (2022)

(continued)

Table 1 (continued)

Type of regeneration/in vitro technique	Genetic variation detection technique	Epigenetic variation detection technique	Species	Genetic/epigenetic variation results	Reference
Soot multiplication (RITA/ partial immersion/semisolid medium)	ISSR		*Citrus* × *latifolia*	26.5% in semisolid medium; 17% partial immersion and 5.42% RITA system polymorphism observed after 6th subculture	Bulbarela-Marini et al. (2022)
Shoot multiplication from nodal explants	RAPD, ISSR		*Mansonia altissimia*	No somaclonal variation	Oseni et al. (2022)
Shoot multiplication from nodal explants	RAPD		*Populus alba*	38.33% polymorphism	Hussien et al. (2022)
Shoot multiplication from nodal explants; stress factors: high salinity, ABA, high kinetin concentration	ISSR		*Orthosiphon stamineus*	70.12% polymorphism	Ali et al. (2022)
Shoot multiplication of virus-free plants obtained from meristem culture	AFLP	MSAP	*Allium sativum*	Genetic and epigenetic polymorphism under field growing conditions, intensified during in vitro culture and increased along time: epigenetic polymorphism reached the major variation after 6 months and then stabilized. Retrotransposons identified	Gimenez et al. (2016)
Shoot propagation, in vitro thermotherapy (37 °C)		MSAP	*Vitis vinifera*	40% of the observed epigenetic changes disappeared within a year after termination of in vitro/thermotherapy conditions. 60% of DNA methylation diversity remained after 1 year	Baranek et al. (2015)
Direct shoot organogenesis					

(continued)

Table 1 (continued)

Type of regeneration/in vitro technique	Genetic variation detection technique	Epigenetic variation detection technique	Species	Genetic/epigenetic variation results	Reference
Direct shoot organogenesis from apical buds and from in vitro-derived leaves, micropropagated for 5 or 7 years	RAPD, AFLP, ISSR		*Pistacia vera*	Low polymorphism among donor plant and its 18 clones	Akdemir et al. (2016)
De novo direct shoot organogenesis	ISSR		*Micranthocereus* spp.	Genetic stability	Civatti et al. (2017)
Direct and indirect shoot organogenesis from apical bud, axillary bud, leaf and internode explants	RAPD		*Guizotia abyssinica*	Somaclonal variation in plantlets arising from direct regeneration as well as from indirect regeneration	Baghel and Bansal (2017)
De novo direct shoot organogenesis	ISSR		*Melocactus glaucescens*	Cultures without PGR produced 25.8% shoots different from mother plant (99.7% similarity); with PGR resulted in a mean of 21.9% of variants (99.4% similarity to mother plant)	Torres-Silva et al. (2018)
Direct shoot organogenesis from cotyledonary node explants	RAPD		*Cucumis sativus*	1.8% polymorphism	Venkatachalam et al. (2018)
De novo direct organogenesis from nodal explants	SCoT		*Nyctanthes arbor-tristis*	Genetic stability	Rath et al. (2020)
De novo direct shoot organogenesis from garlic inflorescences	SSR		*Allium sativum* L.	Genetic stability	Wen et al. (2020)

(continued)

Table 1 (continued)

Type of regeneration/in vitro technique	Genetic variation detection technique	Epigenetic variation detection technique	Species	Genetic/epigenetic variation results	Reference
Shoot multiplication from nodal segments/callus, during 6 subcultures	ISSR, SSR		*Hancornia speciosa*	No somaclonal variation in the subcultured plantlets and calli	da Costa et al. (2022)
In vitro shoots/*ex vitro* regenerated plants from node or leaf explants/plants propagated from cuttings		MSAP	*Vaccinium vitis-idaea*	More methylation events in acclimatized in vitro-derived plants than in those derived from cuttings plants	Sikdar et al. (2022)
Indirect shoot regeneration					
Callus-induced shoots	ISSR		*Vanilla planifolia*	71.66% genetic polymorphism	Ramírez-Mosqueda and Iglesias-Andreu (2015)
Callus-induced shoots, from cotyledon explants	RAPD, SSR		*Sesamum indicum*	Genetic stability	Anandan et al. (2018)
Callus-induced shoots	SNP		*Brassica rapa* ssp. *pekinensis*	39 of the regenerant-specific mutation, five of the SNP candidates occurred in exons, and were maintained in the progeny lines	Park et al. (2020)
Callus-induced shoots, from leaf explant	RAPD		*Humulus lupulus*	9.6% of regenerants showed polymorphisms and were independent of the medium composition and time in culture	Liberatore et al. (2020)
Callus induced shoots (CIS)/meristem induced shoots (MIS)	AFLP		*Olea europaea* L.	Stability was cultivar-dependent. Lower stability from CIS than from MIS (2–32%)	Mirzaei et al. (2021)

(continued)

Table 1 (continued)

Type of regeneration/in vitro technique	Genetic variation detection technique	Epigenetic variation detection technique	Species	Genetic/epigenetic variation results	Reference
Callus-induced shoots from leaf explants	RAPD, ISSR		*Ranunculus wallichianus*	Genetic stability	Srinivasan et al. (2021)
Indirect shoot regeneration	SSR		*Cannabis sativa*	Genetic stability	Ioannidis et al. (2022)
Indirect somatic embryogenesis					
Indirect somatic embryogenesis and organogenesis from leaf-base explants	RAPD		*Iris sibirica*	Genetic stability	Stanisic et al. (2015)
Indirect somatic embryogenesis from nucellar tissues or from ovaries	RAPD		*Citrus jambhiri*	Genetic stability in plantlets from nucellar tissues; plantlets from ovaries explants showed variation in 20–30% plantlets	Savita et al. (2015)
Secondary somatic embryogenesis from zygotic embryos	SSR		*Magnolia dealbata*	Genetic instability in $\leq$ 10% of regenerated plants	Chávez-Cortazar et al. (2020)
Off-type and true-to type field-grown plants originated from immature inflorescence explants through somatic embryogenesis	IRAP		*Phoenix dactylifera*	Retrotransposons activated during the tissue culture process, resulting in somaclonal variants	Mirani et al. (2020)

(continued)

Table 1 (continued)

Type of regeneration/in vitro technique	Genetic variation detection technique	Epigenetic variation detection technique	Species	Genetic/epigenetic variation results	Reference
Embryogenic callus subcultured for 12–13 years, and regenerated plantlets	SNP		*Panax ginseng*	Genes related to processes of cell totipotency and chromosomal stability showed lower mutation rates at both coding and flanking regions	Liu et al. (2022)
Somatic embryos regenerated after 4, 11 and 27 months of embryogenic callus culture	AFLP, SSAP	MSAP	*Coffea arabica*	Abnormal phenotypes only observed from the 11 and 27-month-old cultures, affecting 30 and 94% of regenerated plants, respectively. Aneuploidy (monosomy) was systematically found in variants. No polymorphic fragments were observed with AFLP and SSAP markers	Landey et al. (2015)
Androgenesis					
Androgenesis		metAFLP	× *Triticosecale* spp.	Methylation changes related to the cultured medium used: 5% demethylation, 3% de novo methylation. Changes in DNA sequence detected	Pachota et al. (2022)
Androgenesis and somatic embryogenesis		metAFLP	× *Triticosecale* spp.	Culture technique and genetic background of donor plants affected tissue culture-induced variation	Machczyńska et al. (2015)
Androgenesis and somatic embryogenesis		metAFLP	X *Triticosecale* Wittm. (Triticale)	DNA demethylation increased with Cu(II) concentration. De novo DNA methylation higher in androgenesis than in somatic embryogenesis. Increase of de novo methylation related to reduction in genetic variation. Changes in DNA sequence detected	Pachota (2022)

(continued)

Table 1 (continued)

Type of regeneration/in vitro technique	Genetic variation detection technique	Epigenetic variation detection technique	Species	Genetic/epigenetic variation results	Reference
Androgenesis		metAFLP	*Hordeum vulgare*	10.46% variants, all factors had comparable impact on variation ($CuSO_4$ and $AgNO_3$ concentration and time on induction medium)	Orlowska and Bednarek (2020)
Genetic transformation					
Transgenic plants (Agrobacterium-mediated)	SNP		*Brassica rapa* ssp. *pekinensis*	12 SNPs with gene annotations of interest were selected and validated by RT-PCR methods	Kim et al. (2021)
Transformation and tissue culture protocols		metAFLP	*Arabidopsis thaliana*	Higher methylation values (7.5%) in regenerated and transgenic plants when compared to seed derived plants (3.2%); the main component of the somaclonal variation was genetic rather than epigenetic	Coronel et al. (2018)

For molecular techniques' abbreviations, see text

Amplified Fragment Length Polymorphism (AFLP)

The technique combines the digestion of DNA samples with restriction enzymes and PCR amplification, therefore, it may be considered a mix between RFLP and PCR techniques. In a first step, DNA is digested with restriction endonucleases (usually *Mse*I and *Eco*RI). After digestion, adapters of known sequences are ligated to the fragments obtained. Primers of approximately 20-mer and sequence complementary to adapters are used for PCR amplification. In a second PCR amplification, the products of the first PCR are amplified using primers with the same sequence but with one to three more nucleotides added to make this second amplification more specific. Final amplification products may be analysed by electrophoretic separation, although, in this case, and due to the high number of fragments obtained, a polyacrylamide gel is required. The use of a fluorochrome to label one of the primers and the analysis of the amplification products by sequencing techniques is an alternative procedure. Variations detected through AFLP technique arise from mutations in the restriction enzyme recognition sites or genetic modifications, such as indels, occurring between two recognition sites.

Inter-Simple Sequence Repeat (ISSR)

In this technique primers are designed to be complementary to microsatellite regions (simple sequence regions), and the amplified fragments correspond to the DNA sequence found between two oppositely oriented microsatellites. In this case, primers are short (16–18-mer), although longer than RAPD primers. In order to anchor the primer into the non-repeat adjacent regions, a few selective nucleotides are added to the repeat sequence. Amplification products are easily analysed by electrophoretic separation. Since the presence of repetitive sequences is common in all genomes, the probability to find complementary sequences to the ISSR primers is very high, and, therefore, is an arbitrary process.

Start Codon Targeted (SCoT)

Developed by Collard and Mckill (2009), this technique was initially designed for rice genotypes, although nowadays it is being used in many plant species, even without genome sequence knowledge. The number of available SCoT primers has greatly increased by the research of other authors. Primers are designed to target the region flanking the start codon ATG, a highly conserved region in plants. A single 18-mer primer is used in the PCR amplification. Its simplicity, together with high efficiency and cost-effectiveness, has considerably increased the use of this technique in recent years.

2.2.2 Specific Techniques

Techniques included in this group have the limitation that previous knowledge of the genome is required. Nevertheless, the information obtained from these markers is more precise, as for example, due to their co-dominant character, they allow the differentiation of heterozygous individuals.

Simple Sequence Repeat (SSR)

This technique is also known as microsatellite markers. Contrary to ISSR, previously described, the primers used are complementary to the flanking regions of the repeated motif, instead of the motif, and therefore, the amplified fragment corresponds to the microsatellite itself. The differences in the size of the amplified fragment are due to the variation in the number of repeats of the motif. When the sequence of flanking regions of the microsatellite is available, this technique is very simple and informative.

2.2.3 Mobile Elements Based Techniques

There is a group of molecular markers that, due to their common nature, require a specific classification. They are those markers based on mobile elements sequences. These markers are designed to discover polymorphism associated to the mobility of transposable elements. In particular, different markers have been established based on the long terminal repeats (LTRs) regions of retrotransposons, which are very conservative. Different techniques have been developed depending on the LTR sequence used for primer design. These methods exploit the abundance and widespread random distribution of retrotransposons, making these markers very informative. The main techniques within this group are Inter-Retrotransposon Amplified Polymorphism (IRAP) and Sequence Specific Amplified Polymorphism (SSAP) techniques, although practically the first one is the only one used in genetic stability studies in plant tissue culture.

Inter-Retrotransposon Amplified Polymorphism (IRAP)

This is a method for amplification of genomic DNA between closely located sequences of retrotransposons (Kalendar et al. 1999). Primers are designed to match the LTR sequence, and, since it is very conservative, the technique may be used in different species and without prior knowledge of the genome sequence. Primers are short, 12-13-mer, and a single primer is used for the amplification. One of the weaknesses of the technique is the dominant character of the markers. Polymorphism is caused by a mutation in the binding site of the primer or by retrotransposition resulting from insertion of retrotransposon into a novel site of genomic DNA without loss of the original site (Boronnikovaa and Kalendar 2010).

Sequence Specific Amplified Polymorphism (SSAP)

This method is an adaptation of AFLP markers to detect polymorphisms associated with transposable elements' mobility (Waugh et al. 1997). In a first step, DNA is digested with restriction enzymes, as in AFLP method, and the subsequent PCR amplifications from digested fragments is based on the use of primers complementary to retrotransposon sequences. SSAP requires the design of specific primers for each species/genotype studied. This makes this technique less available for many laboratories and its use is more restrictive in somaclonal variation analyses.

2.3 Sequence Based Techniques

Direct sequence analysis is the most accurate method to detect even the smallest changes in the genome. However, until recently, it has not been possible to widely use these techniques, both because of their cost and their high degree of complexity. The development of new generation sequencing (NGS) has facilitated the study of sequences through the use of new methodologies. The comparison of the studied sequence with the sequence corresponding to the donor or control plant allows the detection of changes in the nucleotide sequence.

Single Nucleotide Polymorphism (SNP) consists on single base pair positional differences in DNA sequences among individuals/samples. There are many methods of SNP genotyping, the most traditional ones are based in extension of primers in PCR, restriction digestion, ligation and allele-specific hybridization. However, NGS offers alternatives to these methods through less complex and labour-intensive sequencing technologies. SNPs detection is accomplished by creating sequence alignments and analyses using sequences published in databases (Amiteye 2021). Although the new resources are promoting the use of this technique, there are still very few studies that address the analysis of genetic stability through the application of sequencing for the detection of SNPs. One of the scarce studies using this methodology is focused on variations of regenerated Chinese cabbage plants (*Brassica rapa* ssp. *pekinensis*) (Park et al. 2020).

Recent improvements in NGS methods and lower costs have resulted in a rapid increase of reference genome and genome sequencing datasets. New technologies available for somaclonal variation detection as RNA sequencing (RNA-seq), not only may detect changes in the sequence, but also analyse which functions are particularly affected.

RNA-seq uses recently developed NGS technology to obtain the transcriptome of a sample. In brief, a RNA sample is converted to a library of cDNA fragments which are then sequenced. Subsequently, the resulting reads are aligned to a reference genome to produce a genome-scale transcription map that consists of the transcriptional structure and/or level of expression for each gene (Wang et al. 2009). This technique is a useful tool to detect SNPs, and, besides, allows the analysis of differences in gene expression associated to changes due to tissue culture. In its application to study somaclonal variation, RNA-seq has been used to detect differences among samples; differences in particular genes are validated through sequencing or analysed through q-PCR. Likewise, the differences detected in gene sequences are used to develop other markers such as Cleaved Amplified Polymorphic Sequence (CAPS). Furthermore, RNA-seq allows detecting differences in gene expression (DEGs, Differentially Expressed Genes); this information may be valuable to analyse the phenotype of somaclonal variants (Skarzyńska et al. 2020; Hou et al. 2022). The latter characteristic makes especially interesting the use of this technique in the evaluation of somaclonal variation occurrence in genetic transformation processes. Rajewski et al. (2021) applied mRNAseq to search for potential differential expression of genes between wild-type and transformed plants of *Datura stramonium*. The technique not only detects genetic differences between transgenic and non-transgenic plants, but also the expression differences that may be associated to these mutations.

2.4 Genetic Marker Technique Selection

The selection of the most appropriate technique for somaclonal variation detection depends, in many cases, on the background knowledge of the genome of the species being studied and the availability of sequence data or DNA probes, plant material, and laboratory facilities. Besides, other aspects must be considered, as the extent and nature of variation, when these data are available (Henry 1998). Although new technologies based on sequencing are very informative, their use is still scarce. They not only offer the capability to detect small genetic changes, but also to associate these changes to specific variations in gene expression of no true-to-type plants. However, the complexity of the techniques and their elevated cost, compare to others, make them not very widely used for this purpose. The analyses of the most employed techniques to detect somaclonal variation in the last years reflect that these aspects may be the key to choose a detection method. The highest number of scientific works that used molecular markers to detect somaclonal variation, published from 2015 to 2022, employed RAPD and ISSR techniques, showing a huge difference with the use of other techniques (Fig. 2a). These two techniques have several common features: they are arbitrary markers, so prior knowledge of the genome is not necessary, their technical simplicity, without the need for large facilities, and their low cost make them affordable for any laboratory. The next most widely used techniques are also arbitrary markers, AFLP and SCoT, along with the specific SSR marker. Although they are also simple techniques, their execution may be somewhat more elaborate than the two previous ones and, above all, the requirements for their analysis imply a higher cost (AFLP and SSR today require fluorescent labelling and detection by sequencing). The use of techniques based on primers matching transposable elements is not very widespread, although their implementation and cost are affordable.

In order to obtain a wider screening of the genome, in most studies two different molecular markers are used. This allows obtaining combined information of different sequences and genomic distribution. In this case, techniques based on arbitrary markers are again the most frequent option, being the combination of RAPD and ISSR the most used one, followed by SCoT combined with ISSR or RAPD (Fig. 2b). The combination of mobile element-based techniques, as IRAP, with other techniques is not usually a common option, similarly to the infrequent use on their own.

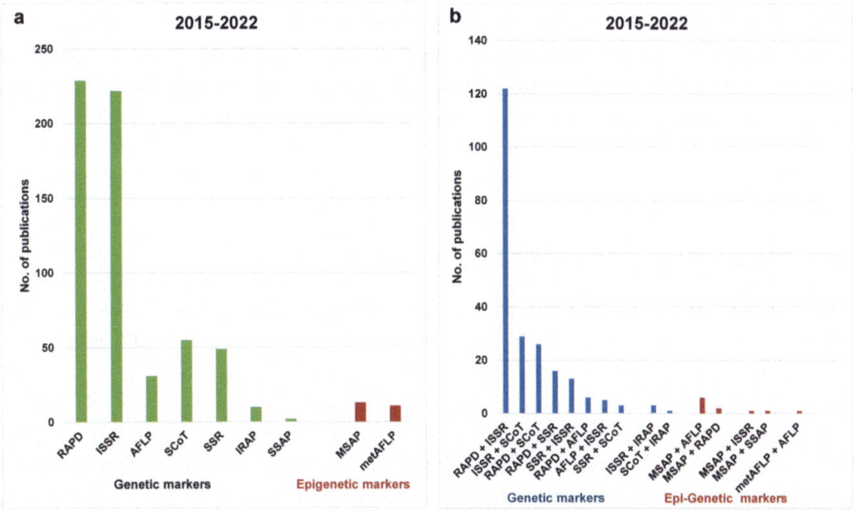

Fig. 2 Number of scientific works published in the period 2015–2022 using PCR-based molecular markers for the detection of somaclonal variation: **a** genetic and epigenetic markers; **b** combination of genetic markers and epigenetic markers. *Source* Web of Science database

3 Molecular Markers for the Detection of Epigenetic Variation

Stress conditions in plant tissue culture may not only affect the genetic but also the epigenetic stability of the plant material. It is well known that epigenetic changes are involved in plant response to stress situations (Finnegan 2001). Besides this role, epigenetic mechanisms are implicated in development and gene expression regulation. The totipotent capacity of plant cells is the base of plant tissue culture. Furthermore, it enables their required plasticity to dedifferentiate and acquire new fates and is, to a certain extent, regulated by epigenetic mechanisms. Throughout this reprogramming process, some epigenetic changes may be permanent and can be then considered part of the somaclonal variation event (Bednarek and Orłowska 2020). DNA methylation is probably the best studied epigenetic phenomenon in eukaryotes, particularly the addition of a methyl group to cytosine residues. The importance of cytosine methylation relies on the fact that it has been associated with numerous biological processes, such as genomic imprinting, transcriptional regulation of genes, and transposable elements and gene silencing (Finnegan et al. 2000). Besides, DNA methylation is considered sensitive to the environment and is involved in the plasticity and adaptative responses to changing environments (González-Benito et al. 2020).

There are diverse available procedures to screen DNA methylation, that have been thoroughly described (Yong et al. 2016). The most widely used techniques for analysing DNA methylation are those based on methylation-sensitive restriction enzymes or on bisulfite modification. However, molecular markers are mainly based on the use of methylation-sensitive restriction enzymes. For most of the genetic molecular markers, a modification to detect DNA methylation changes has been developed when their use is combined with a previous digestion of the DNA using methylation-sensitive restriction enzymes. Modifications of the AFLP technique are the most widely used for epigenetic stability studies in plant tissue culture, in particular, MSAP and metAFLP. Examples of the use of the techniques described below can be found in Table 1.

Methylation Sensitive Amplified Polymorphism (MSAP)

This technique is fundamentally a modification of the AFLP method based on the digestion of genomic DNA with methylation-sensitive restriction endonucleases followed by the amplification of digested fragments. In the MSAP protocol, the extracted genomic DNA is divided into two aliquots, each digested with *Eco*RI, which recognizes the GAATTC target site and is thought to be negligibly influenced by DNA cytosine methylation ("indifferent cutter"). The aliquots are then digested with the methylation-sensitive *Msp*I or *Hpa*II isoschizomers ("methyl-sensitive cutter"), which recognize the same restriction site (CCGG) but show differential sensitivity to cytosine methylation. The DNA samples digested with *Eco*RI and *Msp*I or with *Eco*RI and *Hpa*II are ligated to two dsDNA adapters compatible with *Eco*RI and *Msp*I/*Hpa*II-generated ends. Subsequently, ligated fragments are pre-amplified using non-selective or pre-selective primers complementary to the adapters followed by amplification with a pair of selective primers (these are one to three-base extended variants of non-selective or pre-selective primers at $3'$ ends). Such amplification produces a reduced population of fragments that are separated in order to compare the respective band patterns (González-Benito et al. 2020).

Methylation Sensitive AFLP (metAFLP)

In this AFLP modification, DNA is previously digested with *Acc*65I/*Mse*I or *Kpn*I/*Mse*I. *Acc*65I and *Kpn*I are isoschizomers, which differ in their sensitivity to template methylation. Both restriction enzymes recognise the site GGTACC, but with differences according to the methylation status of the cytosine residues (Bednarek et al. 2007). The protocol and results' analyses are similar to the MSAP method.

There are not many studies focused on the evaluation of epigenetic stability of tissue cultures, but, when these analyses are carried out, these two techniques (MSAP and metAFLP) are the ones that are usually applied. The number of scientific works published between 2015 and 2022 using these techniques is represented in Fig. 2a. The analysis of epigenetic changes is not so common as the study of genetic changes related to in vitro culture. Although is not so extended, the combination of genetic and epigenetic studies to evaluate the fidelity of cultures is of increasing interest. Different combinations of genetic and epigenetic markers have been tested in the last years (Fig. 2b).

4 Study of Somaclonal Variation in Different Tissue Culture Techniques

The explant source and type of tissue culture protocol used can affect the genetic and epigenetic stability of the regenerated plants. One of main objectives in the somaclonal variation studies has been to determine the effect of the in vitro culture technique on the frequency and degree of instability (Table 1). Shoot multiplication from axillary or apical bud elongation is considered a stable way of multiplication. Nevertheless, using mostly RAPD and/or ISSR markers, polymorphism has been found in some studies (see Table 1, e.g. Hussien et al. 2022) while no genetic changes were observed in others (e.g. Goyal et al. 2015; Thakur et al. 2021; Kader et al. 2022; Oseni et al. 2022). In *Vanilla planifolia* (Sharma et al. 2019) and *Citrus* × *latifolia* (Bulbarela-Marini et al. 2022), an increase in the number of variant shoots was found after the 5th-6th subculture cycles. However, not in all studies polymorphism increased with the number of subcultures, for example, in sugarcane the greatest genetic differences among regenerants were found after the 1st, 9th and 10th subculture (Martínez-Estrada et al. 2017).

The tissue culture procedures could have an influence on stability; for example, higher percentage of polymorphism was found when shoot multiplication of *Citrus* × *latifolia* was carried out on semi-solid medium than with temporary immersion (Bulbarela-Marini et al. 2022). Other factors, such as high salinity or high cytokinin concentration could increase the number of polymorphic markers (Chin et al. 2019; Ali et al. 2022).

Using MSAP markers, 60% of the epigenetic changes observed in *Vitis vinifera* in vitro shoots were maintained after one year of cultivation *ex vitro*, indicating mitotically inherited changes (Baranek et al. 2015). In garlic shoots micropropagated from meristem culture, epigenetic polymorphism increased after 6 months in culture and subsequently stabilized (Gimenez et al. 2016); some of the differentially methylated fragments were identified as sequences belonging to retrotransposons. MSAP markers were also used to compare field-grown cuttings and in vitro-grown shoots of cassava (Kitimu et al. 2015), showing a clear divergence between both types of plant material analysed.

Genetic stability has been observed in de novo shoot organogenesis of several species using molecular markers such as RAPD, SCot, SSR or ISSR (Civatti et al. 2017; Sharma et al. 2019; Rath et al. 2020; Wen et al. 2020; Qahtan et al. 2021; da Costa et al. 2022). *Pistacia vera* shoots derived from leaf explants and in vitro micropropagated showed higher instability after 7 years of subculture than after 5 years, using RAPD, AFLP and ISSR markers, revealing that the length of culture time influences variation (Akdemir et al. 2016). In *Vaccinium vitis-idaea*, acclimatized in vitro regenerated plants showed more methylation events than plants obtained from cuttings when studied by MSAP markers (Sikdar et al. 2022).

In indirect embryogenesis of *Citrus sinensis*, extensive increase in variation was detected by SNP during the early dedifferentiation step, although some variations were eliminated during long term culture for 30 years (Wang et al. 2022). Abnormal phenotypes were observed in *Coffea arabica* plants derived from 11 and 27-month-old embryogenic callus cultures; although those plants were aneuploid, no polymorphic fragments were observed with AFLP and SSAP markers (Landey et al. 2015). The type of explant could also have an influence on the genetic instability detected by molecular markers; plantlets obtained from nucellar tissues of *Citrus jambahiri* were identical to the mother plant, while 20–30% of plantlets obtained from ovaries showed variation (Savita et al. 2015).

Analysing the origin of somatic embryos in × *Triticosecale* (triticale), Pachota (2022) found that de novo DNA methylation was higher in androgenesis than in somatic embryogenesis (from immature zygotic embryos), besides, with the increase of de novo methylation, somaclonal variation decreased. Also working with triticale, Machczyńska et al. (2015) observed that the genetic background had an influence on the induced variation.

There are fewer studies related to somaclonal variation in transgenic plants. DNA of *Brassica rapa* Agrobacterium-mediated transgenic plants and non-transgenic plants was studied by whole-genome sequencing and SNPs were detected and validated by qPCR (Kim et al. 2021). Likewise, analyses of callus culture stability are scarcely published. One of the few works approaching this kind of cultures studied genetic instability in *Tetrastigma hemsleyanum* callus, and found an increase related to the number of subcultures, with a peak in the 4th-6th subcultures; subsequently, variation frequency was almost constant after the 8th subculture (Peng et al. 2015).

The results obtained in the reviewed works could have been due to multiple factors, some concerning to the plant material (genotype, type of explant, chimerism, …), others due to the different in vitro procedures (regeneration system, culture procedures, PGR, …). The genetic and/or epigenetic marker employed could also have an influence in the proportion of the genome explored. Another factor to be considered is the sampling method and number of samples studied. In most works exploring somaclonal variation with genetic markers, around 10 in vitro plants are compared to the mother plant (e.g. Akdemir et al. 2016; Bhattacharyya et al. 2017; Martínez-Estrada et al. 2017; Venkatachalam et al. 2018; Thakur et al. 2021; da Costa et al. 2022). In some studies, somaclonal variation was detected even when the number of sampled in vitro plants was somehow low, e.g. three replicates per treatment (Bulbarela-Marini et al. 2022). When epigenetic markers are employed, the nature of the sample could also have an influence on the results. For example, cytosine methylation is involved in gene silencing, and different DNA methylation status could reflect different developmental stages or response to environmental stresses. Therefore, if in vitro and *ex vitro* plants are compared with epigenetic markers, the differences showed could be due to different developmental stages.

5 Conclusions

Detection of somaclonal variation in tissue culture processes is necessary to ensure the fidelity of regenerated plants. Advances in molecular biology knowledge and methodology have allowed the development of molecular markers' techniques that are increasingly simple to apply or more informative. The first studies to analyse true-to-type nature of regenerated plants focused on the genetic stability of material from in vitro culture and have given way, each time in greater depth, to epigenetic analysis. New generation sequencing techniques are opening a possibility to detect somaclonal variation and, at the same time, to understand the effect that this variation may pose on the gene expression and phenotypes of the somaclones. There is not a clear relationship between the type of culture and the degree of instability detected by molecular markers, but it is clear that in any plant tissue culture process the study of the stability is a necessary step. There is a range of techniques available and the selection of the most appropriate one depends on the species and facilities of the laboratory. Probably due to equipment and budget restrictions in many laboratories working on somaclonal variation detection, the most frequent techniques used are arbitrary molecular markers, such as RAPD and ISSR, since these techniques are the simplest and most affordable ones of all. An important key point in the assessment of stability is the choice of the control plant material and number of samples (when material is available). The physiological and developmental state of samples must be carefully considered especially in epigenetic studies.

References

Akdemir H, Suzerer V, Tilkat E, Onay A, Ciftci YO (2016) Detection of variation in long-term micropropagated mature pistachio via DNA-based molecular markers. Appl Biochem Biotech 180:1301–1312. https://doi.org/10.1007/s12010-016-2168-7

Ali H, Sallahuddin N, Shamsudin NHA, Zain NAM, Ibrahim MH, Yaacob JS (2022) Abiotic stress induces morphological, physiological, and genetic changes in *Orthosiphon stamineus* Benth. In vitro cultures. Horticulturae 8:153. https://doi.org/10.3390/horticulturae8020153

Amiteye S (2021) Basic concepts and methodologies of DNA marker systems in plant molecular breeding. Heliyon 7:e08093. https://doi.org/10.1016/j.heliyon.2021.e08093

Anandan R, Prakash M, Deenadhayalan T, Nivetha R, Kumar NS (2018) Efficient in vitro plant regeneration from cotyledon-derived callus cultures of sesame (*Sesamum indicum* L.) and genetic analysis of true-to-type regenerants using RAPD and SSR markers. S Afr J Bot 119:244–251. https://doi.org/10.1016/j.sajb.2018.09.020

Baghel S, Bansal YK (2017) In vitro regeneration of *Guizotia abyssinica* Cass. and evaluation of genetic fidelity through RAPD markers. S Afr J Bot 109:294–307. https://doi.org/10.1016/j.sajb.2017.01.002

Baranek M, Cechova J, Raddova J, Holleinova V, Ondrusikova E, Pidra M (2015) Dynamics and reversibility of the DNA methylation landscape of grapevine plants (*Vitis vinifera*) stressed by in vitro cultivation and thermotherapy. PLoS ONE 10:e0126638. https://doi.org/10.1371/journal.pone.0126638

Bednarek PT, Orłowska R (2020) Plant tissue culture environment as a switch-key of (epi)genetic changes. Plant Cell Tissue Organ Cult 140:245–257. https://doi.org/10.1007/s11240-019-01724-1

Bednarek PT, Orłowska R, Koebner RMD, Zimny J (2007) Quantification of the tissue-culture induced variation in barley (*Hordeum vulgare* L.). BMC Plant Biol 7:10. https://doi.org/10.1186/1471-2229-7-10

Bhat SR, Srinivasan S (2002) Molecular and genetic analyses of transgenic plants: considerations and approaches. Plant Sci 163:673–681

Bhattacharyya P, Kumar V, Van Staden J (2017) Assessment of genetic stability amongst micropropagated *Ansellia africana*, a vulnerable medicinal orchid species of Africa using SCoT markers. S Afr J Bot 108:294–302. https://doi.org/10.1016/j.sajb.2016.11.007

Boronnikovaa SV, Kalendar RN (2010) Using IRAP markers for analysis of genetic variability in populations of resource and rare species of plants. Russ J Genet 46:36–42. https://doi.org/10.1134/S1022795410010060

Brown PTH, Kyozuka J, Sukekiyo Y, Kimura Y, Shimamoto K, Lörz H (1990) Molecular changes in protoplast-derived rice plants. Molec Gen Genet 223:324–328. https://doi.org/10.1007/BF00265070

Bulbarela-Marini JE, Gomez-Merino FC, Galindo-Tovar ME, Pastelin-Solano MC, Murguia-Gonzalez J, Nunez-Pastrana R, Castaneda-Castro O (2022) Ratio of somaclonal variation and the phytohormonal content of *Citrus* x *latifolia* in three in vitro culture systems. J Plant Growth Regul. https://doi.org/10.1007/s00344-022-10796-x

Chávez-Cortazar A, Mata-Rosas M, Oyama K, Samain MS, Quesada M (2020) Induction of somatic embryogenesis and evaluation of genetic stability in regenerated plants of *Magnolia dealbata*. Biol Plant 64:224–233. https://doi.org/10.32615/bp.2020.006

Chin CK, Lee ZH, Mubbarakh SA, Antony JJJ, Chew BL, Subramaniam S (2019) Effects of plant growth regulators and activated charcoal on somaclonal variations of protocorm-like bodies (PLBs) of *Dendrobium* Sabin Blue orchid. Biocatal Agr Biotechnol 22:101426. https://doi.org/10.1016/j.bcab.2019.101426

Civatti LM, Marchi MNG, Schnadelbach AS, Bellintani MC (2017) In vitro multiplication and genetic stability of two species of *Micranthocereus* Backeb. (Cactaceae) endemic to Bahia, Brazil. Plant Cell Tissue Organ Cult 131:537–545. https://doi.org/10.1007/s11240-017-1304-6

Collard BC, Mckill DJ (2009) Start codon targeted (SCoT) polymorphism: a simple novel DNA marker technique for generating gene-targeted markers in plants. Plant Mol Biol Rep 27:86–93. https://doi.org/10.1007/s11105-008-0060-5

Coronel CJ, Gonzalez AI, Ruiz ML, Polanco C (2018) Analysis of somaclonal variation in transgenic and regenerated plants of *Arabidopsis thaliana* using methylation related metAFLP and TMD markers. Plant Cell Rep 37:137–152. https://doi.org/10.1007/s00299-017-2217-x

da Costa GF, Cabral PDS, Silva FG, Neto AR, Mendonca MAC (2022) Clonal fidelity and genetic diversity of micropropagated *Hancornia speciosa* Gomes (*Apocynaceae*) as evaluated by molecular markers. Forests 13:1645. https://doi.org/10.3390/f13101645

Finnegan EJ (2001) Epialleles—a source of random variation in times of stress. Curr Opin Genet Dev 5:101–106

Finnegan E, Peacock W, Dennis E (2000) DNA methylation, a key regulator of plant development and other processes. Curr Opin Genet Dev 10:217–223

Garrido-Cardenas JA, Mesa-Valle C, Manzano-Agugliaro F (2018) Trends in plant research using molecular markers. Planta 247:543–557. https://doi.org/10.1007/s00425-017-2829-y

Gimenez MD, Yanez-Santos AM, Paz RC, Quiroga MP, Marfil CF, Conci VC, Garcia-Lampasona SC (2016) Assessment of genetic and epigenetic changes in virus-free garlic (*Allium sativum* L.) plants obtained by meristem culture followed by in vitro propagation. Plant Cell Rep 35:129–141. https://doi.org/10.1007/s00299-015-1874-x

González-Benito ME, Ibáñez MÁ, Pirredda M, Mira S, Martín C (2020) Application of the MSAP technique to evaluate epigenetic changes in plant conservation. Int J Mol Sci 21:7459. https://doi.org/10.3390/ijms21207459

Goyal AK, Pradhan S, Basistha BC, Sen A (2015) Micropropagation and assessment of genetic fidelity of *Dendrocalamus strictus* (Roxb.) Nees using RAPD and ISSR markers. 3 Biotech 5:473–482. https://doi.org/10.1007/s13205-014-0244-7

Henry RJ (1998) Molecular and biochemical characterization of somaclonal variation. In: Jain SM, Brar DS, Ahloowalia BS (eds) Somaclonal variation and induces mutations in crop improvement. Kluwer Academic Publishers, Dordrecht, p 615

Hou BH, Tsai YH, Chiang MH, Tsao S-M, Huang S-H, Chao C-P, Chen H-M (2022) Cultivar-specific markers, mutations, and chimerisim of Cavendish banana somaclonal variants resistant to *Fusarium oxysporum* f. sp. *cubense* tropical race 4. BMC Genomics 23:470. https://doi.org/10.1186/s12864-022-08692-5

Hsie BS, Brito JZ, Nova MXV, Borges-Paluch LR, Silva MV, Donato VMST (2015) Determining the genetic stability of micropropagated sugarcane using intersimple sequence repeat markers. Genet Mol Res 14:17651–17659. https://doi.org/10.4238/2015.December.21.38

Hussien ET, Ahmed MF, Ahmed EZ (2022) Biometric measurements and genetic instability assessment of in vitro micro-shoots culture of *Populus alba*. Plant Physiol Rep 27:398–406. https://doi.org/10.1007/s40502-022-00682-9

Ioannidis K, Tomprou I, Mitsis V, Koropouli P (2022) Genetic evaluation of in vitro micropropagated and regenerated plants of *Cannabis sativa* L. using SSR molecular markers. Plants 11:2569. https://doi.org/10.3390/plants11192569

Jaligot E, Beule T, Rival A (2002) Methylation-sensitive RFLPs: characterisation of two oil palm markers showing somaclonal variation-associated polymorphism. Theor Appl Genet 104:1263–1269. https://doi.org/10.1007/s00122-002-0906-4

Kader A, Sinha SN, Ghosh P (2022) Clonal fidelity investigation of micropropagated hardened plants of jackfruit tree (*Artocarpus heterophyllus* L.) with RAPD markers. J Genet Eng Biotechnol 20:145. https://doi.org/10.1186/s43141-022-00426-0

Kalendar R, Grob T, Regina M, Suoniemi A, Schulman A (1999) IRAP and REMAP: two new retrotransposon-based DNA fingerprinting techniques. Theor Appl Genet 98:704–711

Kim SJ, Park JS, Shin YH, Park YD (2021) Identification and validation of genetic variations in transgenic Chinese cabbage plants (*Brassica rapa* ssp. *pekinensis*) by Next-Generation Sequencing. Genes 12:621. https://doi.org/10.3390/genes12050621

Kitimu SR, Taylor J, March TJ, Tairo F, Wilkinson MJ, Lopez CMR (2015) Meristem micropropagation of cassava (*Manihot esculenta*) evokes genome-wide changes in DNA methylation. Front Plant Sci 6:590. https://doi.org/10.3389/fpls.2015.00590

Krishna H, Alizadeh M, Singh D, Singh U, Chauhan N, Eftekhari M, Sadh R-K (2016) Somaclonal variations and their applications in horticultural crops improvement. 3 Biotech 6:54. https://doi.org/10.1007/s13205-016-0389-7

Landey RB, Cenci A, Guyot R, Bertrand B, Georget F, Dechamp E, Herrera JC, Aribi J, Lashermes P, Etienne H (2015) Assessment of genetic and epigenetic changes during cell culture ageing and relations with somaclonal variation in *Coffea arabica*. Plant Cell Tissue Organ Cult 122:517–531. https://doi.org/10.1007/s11240-015-0772-9

Larkin P, Scowcroft WR (1981) Somaclonal variation—a novel source of variability from cell cultures for plant improvement. Theor Appl Genet 60:197–214

Leva A, Rinaldi LMR (2017) Somaclonal variation. In: Thomas B, Murray B, Murphy D (eds) Encyclopedia of applied plant sciences, 2nd edn. Academic Press, pp 468–473. https://doi.org/10.1016/B978-0-12-394807-6.00150-7

Liberatore CM, Rodolfi M, Beghe D, Fabbri A, Ganino T, Chiancone B (2020) In vitro leaf-derived organogenesis and somaclonal variant detection in *Humulus lupulus* L. In Vitro Cell Dev Biol-Plant 56:865–874. https://doi.org/10.1007/s11627-020-10088-7

Liu ST, Wang XF, Ding N, Liu YT, Li N, Ma YQ, Zhao J, Wang ZH, Li XM, Fu XQ, Li LF (2022) Nucleotide sequence variation in long-term tissue cultures of Chinese ginseng (*Panax ginseng* C. A. Mey.). Plants 11:79. https://doi.org/10.3390/plants11010079

Manchanda P, Kaur A, Gosal SS (2018) Somaclonal variation for sugarcane improvement. In: Gosal SS, Wani SH (eds) Biotechnologies of crop improvement, vol 1. Springer International Publishing, pp 299–326. https://doi.org/10.1007/978-3-319-78283-6_9

Machczyńska J, Zimny J, Bednarek PT (2015) Tissue culture-induced genetic and epigenetic variation in triticale (x *Triticosecale* spp. Wittmack ex A. Camus 1927) regenerants. Plant Mol Biol 89:279–292. https://doi.org/10.1007/s11103-015-0368-0

Martínez-Estrada E, Caamal-Velazquez JH, Salinas-Ruiz J, Bello-Bello JJ (2017) Assessment of somaclonal variation during sugarcane micropropagation in temporary immersion bioreactors by intersimple sequence repeat (ISSR) markers. In Vitro Cell Dev Biol-Plant 53:553–560. https://doi.org/10.1007/s11627-017-9852-3

Mirani AA, Teo CH, Markhand GS, Abul-Soad AA, Harikrishna JA (2020) Detection of somaclonal variations in tissue cultured date palm (*Phoenix dactylifera* L.) using transposable element-based markers. Plant Cell Tissue Organ Cult 141:119–130. https://doi.org/10.1007/s11240-020-01772-y

Mirzaei L, Yadollahi A, Kermani MJ, Naderpour M, Zeinanloo AA, Farsi M, Davoodi D (2021) Evaluation of genetic stability in olive callus-induced and meristem-induced shoots using flow cytometry and amplified fragment length polymorphism techniques. Plant Methods 17:31. https://doi.org/10.1186/s13007-021-00724-7

Nasri A, Baklouti E, Ben Romdhane A, Maalej M, Schumacher HM, Drira N, Fki L (2019) Large-scale propagation of myrobolan (*Prunus cerasifera*) in RITA ® bioreactors and ISSR-based assessment of genetic conformity. Sci Hortic 245:144–153. https://doi.org/10.1016/j.scienta.2018.10.016

Neelakandan AK, Wang K (2012) Recent progress in the understanding of tissue culture-induced genome level changes in plants and potential applications. Plant Cell Rep 31:597–620. https://doi.org/10.1007/s00299-011-1202-z

Novikova TI, Asbaganov SV, Ambros EV, Zaytseva YG (2020) TDZ-induced axillary shoot proliferation of *Rhododendron mucronulatum* Turcz and assessment of clonal fidelity using DNA-based markers and flow cytometry. In Vitro Cell Dev Biol-Plant 56:307–317. https://doi.org/10.1007/s11627-019-10049-9

Orlowska R, Bednarek PT (2020) Precise evaluation of tissue culture-induced variation during optimisation of in vitro regeneration regime in barley. Plant Mol Biol 103:33–50. https://doi.org/10.1007/s11103-020-00973-5

Oseni OM, Nailwal TK, Pande V (2022) Assessment of somaclonal variation occurrence in micropropagated *Mansonia altissimia* (A. Chev.) A. Chev. using molecular marker. S Afr J Bot 149:511–515. https://doi.org/10.1016/j.sajb.2022.06.041

Pachota KA (2022) Comparison of tissue culture-induced variation in triticale regenerants obtained by androgenesis and somatic embryogenesis. Cereal Res Commun. https://doi.org/10.1007/s42976-022-00300-2

Pachota KA, Orlowska R, Bednarek PT (2022) Medium composition affects the tissue culture-induced variation in triticale regenerants. Plant Cell Tissue Organ Cult 151:35–46. https://doi.org/10.1007/s11240-022-02327-z

Park JS, Park JH, Kim SJ, Park YD (2020) Genome analysis of tissue culture-derived variations in regenerated *Brassica rapa* ssp. *pekinensis* plants using next-generation sequencing. Hortic Environ Biotechnol 61:549–558. https://doi.org/10.1007/s13580-020-00237-7

Peng X, Zhang TT, Zhang J (2015) Effect of subculture times on genetic fidelity, endogenous hormone level and pharmaceutical potential of *Tetrastigma hemsleyanum* callus. Plant Cell Tissue Organ Cult 122:67–77. https://doi.org/10.1007/s11240-015-0750-2

Qahtan AA, Faisal M, Alatar AA, Abdel-Salam EM (2021) High-frequency plant regeneration, genetic uniformity, and flow cytometric analysis of regenerants in *Ruta chalepensis* L. Plants 10:2820. https://doi.org/10.3390/plants10122820

Rajewski A, Carter-House D, Stajich J, Lit A (2021) Datura genome reveals duplications of psychoactive alkaloid biosynthetic genes and high mutation rate following tissue culture. BMC Genomics 22:201. https://doi.org/10.1186/s12864-021-07489-2

Ramírez-Mosqueda MA, Iglesias-Andreu LG (2015) Indirect organogenesis and assessment of somaclonal variation in plantlets of Vanilla planifolia Jacks. Plant Cell Tissue Organ Cult 123:657–664. https://doi.org/10.1007/s11240-015-0868-2

Ranghoo-Sanmukhiya VM (2021) Somaclonal variation and methods used for its detection. In: Siddique I (ed) Propagation and genetic manipulation of plants. Springer Nature, Singapore Pte Ltd., pp 1–18

Rath SC, Seth S, Mishra SK, Yadav PK, Gupta AK, Panigrahi J (2020) Genetic homogeneity assessment of in vitro-regenerated plantlets of *Nyctanthes arbor-tristis* L. and comparative evaluation of bioactive metabolites and antioxidant activity. In Vitro Cell Dev Biol-Plant 56:72–87. https://doi.org/10.1007/s11627-019-10004-8

Savita PPK, Virk GS, Nagpal A (2015) An efficient somatic embryogenesis protocol for *Citrus jambhiri* and assessment of clonal fidelity of plantlets using RAPD markers. J Plant Growth Regul 34:309–319. https://doi.org/10.1007/s00344-014-9465-6

Sharma U, Rai MK, Shekhawat NS, Kataria V (2019) Genetic homogeneity revealed in micropropagated *Bauhinia racemosa* Lam. using gene targeted markers CBDP and SCoT. Physiol Mol Biol Plants 25:581–588. https://doi.org/10.1007/s12298-018-00639-z

Sikdar A, Sharma U, Barua R, Igamberdiev AU, Debnath SC (2022) Epigenomic insight of lingonberry and health-promoting traits during micropropagation. Sci Rep 12:12487. https://doi.org/10.1038/s41598-022-16530-7

Skarzyńska A, Pawełkowicz M, Pląder W (2020) Genome-wide discovery of DNA variants in cucumber somaclonal lines. Gene 736:144412. https://doi.org/10.1016/j.gene.2020.144412

Solano MCP, Ruiz JS, Arnao MTG, Castro OC, Tovar MEG, Bello JJB (2019) Evaluation of in vitro shoot multiplication and ISSR marker based assessment of somaclonal variants at different subcultures of vanilla (*Vanilla planifolia* Jacks). Physiol Mol Biol Plants 25:561–567. https://doi.org/10.1007/s12298-019-00645-9

Srinivasan P, Raja HD, Tamilvanan R (2021) Efficient in vitro plant regeneration from leaf-derived callus and genetic fidelity assessment of an endemic medicinal plant *Ranunculus wallichianus* Wight & Arnn by using RAPD and ISSR markers. Plant Cell Tissue Organ Cult 147:413–420. https://doi.org/10.1007/s11240-021-02134-y

Stanisic M, Raspor M, Ninkovic S, Milosevic S, Calic D, Bohanec B, Trifunovic M, Petric M, Subotic A, Jevremovic S (2015) Clonal fidelity of *Iris sibirica* plants regenerated by somatic embryogenesis and organogenesis in leaf-base culture—RAPD and flow cytometer analyses. S Afr J Bot 96:42–52. https://doi.org/10.1016/j.sajb.2014.10.014

Thakur M, Rakshandha SV, Chauhan A (2021) Genetic fidelity assessment of long term in vitro shoot cultures and regenerated plants in Japanese plum cvs Santa Rosa and Frontier through RAPD, ISSR and SCoT markers. S Afr J Bot 140:428–433. https://doi.org/10.1016/j.sajb.2020.11.005

Torres-Silva G, Resende SV, Lima-Brito A, Bezerra HB, de Santana JRF, Schnadelbach AS (2018) In vitro shoot production, morphological alterations and genetic instability of *Melocactus glaucescens* (Cactaceae), an endangered species endemic to eastern Brazil. S Afr J Bot 115:100–107. https://doi.org/10.1016/j.sajb.2018.01.001

Venkatachalam P, Jinu U, Sangeetha P, Geetha N, Sahi SV (2018) High frequency plant regeneration from cotyledonary node explants of *Cucumis sativus* L. cultivar 'Green Long' via adventitious shoot organogenesis and assessment of genetic fidelity by RAPD-PCR technology. 3 Biotech 8:60. https://doi.org/10.1007/s13205-018-1083-8

Wang Z, Gerstein M, Snyder M (2009) RNA-Seq: a revolutionary tool for transcriptomics. Nat Rev Genet 10:57–63. https://doi.org/10.1038/nrg2484

Wang X, Ke LL, Wang ST, Fu JL, Xu JD, Hao YJ, Kang CY, Guo WW, Deng XX, Xu Q (2022) Variation burst during dedifferentiation and increased CHH-type DNA methylation after 30 years of in vitro culture of sweet orange. Hortic Res 9:uhab036. https://doi.org/10.1093/hr/uhab036

Waugh R, McLean K, Flavell AJ, Pearce SR, Kumar A, Thomas BBT, Powell W (1997) Genetic distribution of Bare–1-like retrotransposable elements in the barley genome revealed by sequence-specific amplification polymorphisms (S-SAP). Mol Gen Genet 253:687–694

Wen YB, Liu XX, Liu HJ, Wu CN, Meng HW, Cheng ZH (2020) High-frequency direct shoot organogenesis from garlic (*Allium sativum* L.) inflorescence and clonal fidelity assessment in regenerants. Plant Cell Tissue Organ Cult 141:275–287. https://doi.org/10.1007/s11240-020-01785-7

Yong W-S, Hsu F-M, Chen P-Y (2016) Profiling genome-wide DNA methylation. Epigenet Chromatin 9:26. https://doi.org/10.1186/s13072-016-0075-3

Analysis of Somaclonal Variation by Metabolite Profiling

Fredy Carrera

Abstract Somaclonal variation is defined as genetic and epigenetic transformations that appear in tissues and organs of plants grown in vitro. These changes can be determined by applying a technology such as metabolomics, which is a tool for the study and elucidation of the relationships of metabolites produced by plants. Metabolomics is important to understand chemical variations as plants grow and develop. A biological system can be analyzed by metabolomics through the combination of chromatographic techniques coupled with mass spectrometry to separate and identify plant-derived compounds. Plant metabolites present a wide variety of chemical structures, being possible to distinguish primary and secondary metabolites. Primary compounds are basic for plant growth and development, while secondary metabolites have more specific purposes. Both primary and secondary compounds are part of the plant response mechanisms.

Keywords Somaclonal variation · In vitro culture · Gas chromatography · Metabolomics

1 Introduction

In vitro cell and tissue culture is an important tool for plant production and conservation. However, in vitro culture conditions and manipulation of plant material can cause various genetic and epigenetic changes in the regenerated plants. This phenomenon was called "somaclonal variation" by Larkin and Scowcroft (1981).

Somaclonal variation (SV) refers to alterations observed in in vitro cultures and their offsprings, regardless of whether their origin is genetic or epigenetic. In a broad sense, SV encompasses both phenotypic and genotypic variability (Israeli et al. 1996). In general, it affects between 1 and 32% of cultures (Roux et al. 2021), but in banana, it affects up to 88% of the plants produced in vitro (Sahijram et al. 2003). Musa plants

F. Carrera (✉)
Departamento de Ciencias de la Vida y la Agricultura, Carrera de Agropecuaria, Universidad de las Fuerzas Armadas ESPE, P.O. Box 171-5-231B, Santo Domingo 230118, Ecuador
e-mail: fpcarrera@espe.edu.ec

C. Sánchez-Romero (ed.), *Somaclonal Variation: Basic and Practical Aspects*,
https://doi.org/10.1007/978-3-031-51626-9_5

produced in vitro give rise to somaclonal variants with molecular changes that often exhibit morphological alterations.

Metabolomics can be defined as a new field of study, for the general analysis and identification of qualitative and/or quantitative interactions of all metabolites at a given time within a biological system under specific conditions (Nam et al. 2017). The metabolome are chemical compounds responsible for a large number of physiological functions in a biological system. Metabolites contribute to the structural and functional integrity of the plant cell, serving as mobile signals within and between cells, and participating in defensive responses (passive and active) of the plant (Xia and Wishart 2016).

Metabolomics has become a fundamental field of systems biology and an important technique for understanding plant responses to external stimuli at the metabolite level. Metabolomics is fundamental to understand chemical signals as plants grow and develop. However, it is difficult to estimate the full importance of metabolomics because the metabolome is not necessarily associated to the genome of the organism under study, unlike the proteome and transcriptome which are substantial parts of the organism's genome (Alseekh et al. 2018). Compounds analyzed in metabolomics are small molecules that undergo modifications during cell metabolism. The identification and quantification of metabolites can provide the chemical fingerprint of a phenotype and the biochemical response of a plant under particular situations (Chaudhary et al. 2019).

Plants utilize a wide variety of primary and secondary metabolites with diverse chemical structures. Primary metabolites are essential for plant growth and development, whereas secondary metabolites have more specific targets. Both types of compounds are constitutive parts of the plants' response to stress produced by in vitro culture (Hong et al. 2016). Secondary compounds are very different from each other, for example, alkaloids, terpenoids and polyphenols show great variability among plant species, while primary compounds, such as amino acids, sugars and lipids, have very well preserved structures (Scossa et al. 2016).

Metabolomic analysis in plants has generated innumerable amounts of data, especially on key compounds and metabolic pathways (Kumar et al. 2017). Metabolomics is currently an important technology in several disciplines or areas of study, such as biomarker identification, toxicology, molecular epidemiology and plant analysis (Sharma et al. 2018). Recently, metabolomics has been used in applied research, such as functional metabolomics, which investigates the functions of specific metabolites (Yan and Xu 2018). Metabolomics can contribute to increase crop yield and quality (Sweetlove et al. 2017); and the combination of metabolomics with other technologies related to quantitative genetics, proteomics, transcriptomics, or genetic modification, has boosted plant breeding. (Kumar et al. 2017). Moreover, metabolomics has strengthened the study of the physiological, phenotypic and morphological responses of plants to interactions with other organisms and also to environmental perturbations (Peters et al. 2018).

2 Metabolic Analysis

Metabolomic studies are usually classified into non-targeted and targeted. The objective of non-targeted metabolomics is to find a broad group of metabolites with the aim of reaching patterns or fingerprints regardless of whether or not the detected metabolites have been identified. Meanwhile, targeted metabolomics is based on the quantification and identification of metabolites of a specific chemical group.

Metabolite research requires the use of several processes that integrate aspects of biochemistry, bioinformatics, chemistry and statistics (Cambiaghi et al. 2017). Depending on the purpose of the study, targeted and non-targeted metabolomics research involves several steps; including experimental design, replication and randomization, sample preparation, separation and detection of metabolites, followed by data processing and statistical analysis (Rodrigues et al. 2022).

2.1 Experimental Design

Experimental design is an important phase in metabolomics studies. The purpose of experimental design is to obtain true data that provide reliable conclusions while reducing errors, so that researchers can finally analyze the effect of the independent variables on the dependent variables (Jacyna et al. 2019).

Variations in metabolomics analysis can be found during sample collection, storage, sample preparation, analysis and data pre-treatment (Dudzik et al. 2018). Therefore, in metabolomic analysis it is essential to take into account factors that can generate variation because they can alter the assessment of the biological hypothesis. Among them, the stage of growth and development of the plant, the collection, extraction and transport of the sample, the analytical platform used (e.g. liquid chromatography-mass spectrometry (LC-MS)) and finally the statistical analysis can be included (Ivanisevic and Want 2019).

2.2 Replication and Randomization

The choice of the correct number of biological and technical replicates is significant for running a robust statistical test and achieving a convenient biological explanation of the results (Rodrigues et al. 2019). Biological replicates provide information on the natural variation of the individuals studied and are specific to the study population and research conditions, whereas technical replicates are capable of assessing instrumental or protocol variations (Story 2011).

Replicates in metabolomics studies are typically six, but statistical power should be estimated prior to performing metabolic research in order to establish an adequate simple size (Rodrigues et al. 2019; Rodríguez-Calcerrada et al. 2018). Field investigations generally have greater variability, while studies carried out under controlled conditions result in less variability, thus requiring a smaller sample size than field investigations. However, the use of a robust statistic is often discarded and the sample size is limited by the number of usable individuals or the surface of the greenhouse. The evaluation of statistical power can be carried out by executing preparatory tests on a small number of samples or data (Blaise et al. 2016; Xia and Wishart 2016).

Randomization is key for reducing experimental error and biological variability, and for maintaining sample uniformity (Rodrigues et al. 2019). When the work is carried out under controlled environmental conditions, such as a growth chamber, it is advisable to rotate the plants during the research so that all plants receive a homogenous amount of light and ventilation, factors that could affect their metabolism and data progression. When randomization is not possible, an alternative frequently used is to arrange the plants in uniform blocks and assess the dependent and independent variables within each block (Rodrigues et al. 2019).

2.3 Sampling and Inactivation of Metabolites

Collection, sampling and brief freezing stages are critical preparatory steps to achieve metabolomic data. In experiments with a high number of samples, it is essential to prepare a large number of operators to reduce the collection time. It is interesting to estimate that primary and secondary metabolites can be modified in response to natural cycles (Bénard et al. 2015). After harvest, repression of plant metabolism is vital, in particular for the estimation of metabolites with high regeneration rates, such as glycolytic intermediates and sugar phosphates (Kim and Verpoorte 2010). Reduction of water content and temperature are the most common procedures to stop metabolic reactions. Temperature decrease is achieved by freezing samples under liquid nitrogen shortly after harvesting, then samples can be stored frozen at − 80 °C until analysis (Jorge et al. 2016b). Another strategy to suppress metabolism involves moisture reduction by freeze-drying (Glauser et al. 2013).

2.4 Metabolite Extraction

One important step in metabolomic studies is metabolite extraction. The purpose of metabolite extraction is to maximize the number and exuberance of metabolites in the sample. Hydrophilic and lipophilic metabolites are found in plants. Polar solvents, such as methanol or a combination of methanol and water, are used to extract hydrophilic compounds; while non-polar solvents, such as chloroform, are used to extract the lipophilic compounds (Dunn and Ellis 2005). One of the most common solvent combinations is methanol, water and chloroform. This solvent mixture allows the separation of hydrophilic and lipophilic metabolites in a single extraction (Maridueña-Zavala et al. 2017).

2.5 Derivatization

Derivatization is a procedure by which a chemical compound is modified in one or more of its functional groups by reaction with a derivatizing reagent to produce a chemically related product (Obata and Fernie 2012). In plant metabolomics, derivatization is often used prior to the separation of metabolites by gas chromatography (GC). The purpose of derivatization is to improve thermal equilibrium, volatility, weaken polarity, decrease analyte adsorption on the GC system and increase detector response, thus obtaining better peak resolution and superior detection efficiency compared to non-derivatized metabolites (Jorge et al. 2016b). One of the drawbacks of this treatment is the possible creation of artifacts associated with the derivatization agent used (Shinbo et al. 2006).

There are two steps in the derivatization process: first, transformation of the carbonyl group (aldehyde or ketone) to an oxime or methoxyamine to reduce the tautomerism of the monosaccharides (Gullberg et al. 2004); and second, to increase volatility, silylation of active hydrogens, such as H, COOH, NH and SH, and functional groups, originating trimethylsilane ($C_3H_{10}Si$) thus producing minimally polar and more thermostable trimethylsilyl groups (TMS). (Dettmer et al. 2007). The most commonly applied compounds for silylation of plant samples are *N*-methyl-trimethylsilyltrifluoroacetamide (MSTFA) and N,O-bis-trimethylsilyltrifluoroacetamide (BSTFA).

2.6 Separation and Detection

Following metabolite extraction, techniques such as GC, LC or capillary electrophoresis (CE) are applied to separate the metabolites present in the sample matrix. Technologies such as LC and GC are the most frequently applied for plant metabolomics due to their high separation efficiency of metabolites with different degrees of polarity and volatility. LC has been suggested to separate metabolites based on their mass-to-charge ratio (m/z). Among LC technologies, high performance thin layer chromatography (HPTLC) has been greatly employed due to its simplicity and low cost. Other similar techniques are HPLC and UHPLC, both used for studies of semi-polar, polar and hydrophilic analytes (Shawky and Selim 2017).

After the separation of metabolites, detection, quantification and identification are normally the subsequent steps. NMR (Liu et al. 2017) and MS are the most widely applied detection technology in metabolomics (Deborde et al. 2017). In particular, MS devices can be coupled to LC, GC or EC separation equipment in a single system, thus enabling the separation and detection of metabolites in a single run.

GC-MS has been the most exploited method for plant metabolomics (Jorge et al. 2016b) and has demonstrated very efficient metabolite separation, as well as high resolution, sensitivity and reproducibility (Wolfender et al. 2015). CG-MS is used for the analysis of volatile and thermostable compounds, which are very abundant in plant samples and can provide ample information on the physiological state of the plant. GC-MS has also been used to estimate metabolite fluxes in plants (Shih and Morgan 2020). A variation of this technique, gas chromatography-real time mass spectrometry (GC-TOF-MS) is applied to the study of compounds present in forest species (Rodríguez-Calcerrada et al. 2018).

LC-MS can be used to analyze thermolabile, polar and high molecular weight compounds, which are not detectable by GC-MS. Derivatization is not needed in this case (Jorge et al. 2016a; Fernie and Tohge 2017).

Samples for the use of the LC technique must have a considerable volume and solvents are required; however, CE requires a small sample volume and no solvents. This technology is considered a non-chromatographic platform that contributes to the discovery of a large number of metabolites, which have short response times (Ramautar et al. 2019). The derivatization used in the GC-MS technique excludes several metabolites, but EC is a technique that is applied to identify the metabolites eliminated during the derivatization (Jorge et al. 2016a). Separation of plant compounds in an EC run is achieved by their m/z ratio and detection is carried out by MS.

NMR is a fingerprinting technology that does not destroy metabolites and allows to detect different types of metabolites regardless of their size, charge, volatility or stability (Markley et al. 2017; Emwas et al. 2019). However, due to its reduced sensitivity (Kumar et al. 2017), NMR-assisted plant metabolomics is only adequated to detect and quantify metabolites at high concentrations (Markley et al. 2017) while compounds present at low concentrations are an inconvenient for this technology (Pan and Raftery 2007). See in Table 1 characteristics and applications of the techniques used in metabolomics studies.

2.7 Data Processing

In metabolomics studies, the data of most interest are retention or migration time, m/z ratios and intensity values of each peak. However, background noise due to column dragging and instrumental electronic noise may be present. Therefore, analysis of metabolomics data involves a series of steps including quality control, metabolite identification, data conversion, intensity normalization, retention time alignment, feature detection, and noise reduction (Cambiaghi et al. 2017).

File format of the data obtained from the instrument normally needs to be changed to a more general format in order to perform a proper analysis of data, such as mzXML and NetCDF, which are the most commonly used file configurations for SM data storage (Pedrioli et al. 2004). The use of ad-hoc software such as Thermo Scientific's SIEVE (Cambiaghi et al. 2016), XCMS (Treutler and Neumann 2016) or MZmine

Table 1 Characteristics and applications of techniques used in metabolomics

Technique	Technical characteristics	Application	Bibliography
HPTLC	Simplicity and low cost	Used for the development of fingerprint profiles of plant extracts	Cevallos-Cevallos et al. (2012)
HPLC and UPLC	Techniques similar to HTPLC	Used for the analysis of semipolar, polar and hydrophilic analytes	Cevallos-Cevallos et al. (2012)
LC-MS	Separation of metabolites is performed by means of a stationary phase incorporated into a column	Investigation of thermolabile, polar and high molecular weight metabolites	Dettmer et al. (2007)
GC-MS	Resolution, sensitivity and reproducibility	Analysis of volatile and thermostable compounds	Wolfender et al. (2015)
GC-QTOF-MS	Accuracy and sensitivity	Metabolic study of forest species	Rodríguez-Calcerrada et al. (2018)
GC-Q-MS and GC-oTOF-MS	Sensitive and reliable	For large samples	Rodrigues et al. (2019)
GC-MS (MFA)	Assesses the management and regulation of metabolic networks	Cell cultures and embryo growth	Allen and Young (2013)
LTQ-MS QIT-MS FT-ICR-MS	Structures of unknown plant metabolites elucidated	Molecular transformations inside plant tissues investigated	Matros and Mock (2013), Rodrigues et al. (2019)
MALDI-TOF MALDI-qTOF MALDI-IT-Orbitrap MALDI-FT-ICR	Optimal sensitivity and spatial resolution	Identify and quantify a large diversity of metabolites	Bjarnholt et al. (2014), Lee et al. (2012)
CE-MS	More dynamic than the usual LC, it shows a supremacy of resolution	Distinguishes an increased variety of metabolites	Ramautar et al. (2019)
NMR	Detects primary and secondary metabolites	Abundant metabolomic profile	Liu et al. (2017)

2 (Pluskal et al. 2010), serve for noise removal, metabolite detection, alignment and transformation. Finally, the data must be normalized in order to reduce variability caused by unequal sample concentrations as well as analytical errors affecting the intensity of the revealed peaks (Wu and Li 2016).

2.8 Statistical Analysis

A wide variety of statistical procedures and techniques have been used in metabolomics studies. The classical strategy includes multivariate and univariate analysis with the aim of assessing the effect of treatment on the overall metabolite profile and individual metabolites.

The purpose of multivariate studies is to reduce the dimensionality associated with the number of variables and to evaluate variability in the metabolite profile between research groups, thus obtaining the metabolites that contribute most to the separation of the groups (Sas et al. 2015). The most commonly applied multivariate techniques in plant metabolomics are partial least squares with discriminant analysis (PLS-DA), partial least squares (PLS) and principal component analysis (PCA). PCA and PLS moderate the dimensionality of the data into individual components that explain part of the total variance of the samples (Moncayo-Martín and Cáceres-Gianni 2017). PLS maximizes the correlation between metabolite levels and treatments (Gromski et al. 2015). PLS-DA is an extension of PLS that is used when the treatment variable is categorical instead of continuous.

Other procedures such as univariate techniques have generally been applied in metabolomics studies to assess significant metabolite variation after treatment. These techniques have enabled the reduction of a large number of analytes by keeping only those significantly accumulated at a given p-value (Bartel et al. 2013). Analysis of variance (ANOVA) and t-test are the most used univariate techniques in metabolomics.

3 Metabolomics in the Study of Somaclonal Variation

Recently, Zhang et al. (2023) characterized the volatile compounds of strawberry 'Benihoppe' and its somaclonal mutant 'Xiaobai', at all four stages of fruit development. In vitro culture was established from stolon tip apex tissues. The techniques used were GC-MS and solid-phase microextraction. The results revealed that the highest relative abundances were of ketones and aldehydes in the red, green and white fruit phases. It was also shown that the single esters of 'Xiaobai' were much higher than those of 'Benihoppe'. In addition, the content and odorant activity values, of ethyl butyrate, linalool, nerolidol, ethyl isovalerate, ethyl hexanoate and ethyl pentanoate, were significantly higher in 'Xiaobai' than 'Benihoppe'. However, the eugenol content in 'Benihoppe' was higher than that in 'Xiaobai'. In the red fruit phase, the relative contents of aldehydes, ketones, alcohols, furan, isoprenoids, benzene, volatile phenols, acids, acetate esters, and ethyl esters in 'Xiaobai' strawberry were significantly higher than those in 'Benihoppe' strawberry. However, ketones and aldehydes made up only 7.52% of the total volatile compounds, whereas acetate, ethyl and other esters made up 70.45% of the total volatile compounds in 'Xiaobai' red fruits. Therefore, the contents of esters and the percentages of esters in the total volatile compounds in 'Xiaobai' were much higher than in 'Benihoppe' at the red fruit stage. Similarly, the study reported that diversification in esters were the fundamental cause of aroma between 'Benihoppe' and 'Xiaobai' (Zhang et al. 2023).

In a recent investigation on the metabolic profile of *Gardenia jasminoides* Ellis, it was determined that all differentiated shoot cultures correlated with genetic variability compared to undifferentiated in vitro cultures of callus and cell suspension. The technique used for separation and detection was GC/MS and HPLC. It was also shown that the shoot cultures stored phenols in concentrations higher or similar to those found in the plant leaves, however, the cell suspension and callus cultures presented a significantly lower biosynthetic potential. The in vitro culture types elicited metabolites not detectable in plant leaves, p-coumaric acids and protocatechuic acids were localized in shoots, while catechin was found in callus and cell suspension. The data of this investigation provides evidence that there is a strong correlation of hesperidin, rutin, salicylic acid and chlorogenic acid with the antioxidant activities of methanol extracts of *Gardenia jasminoides* leaves, shoots, callus and cell cultures (Krasteva et al. 2022).

In a metabolomics study of strawberry 'Sachinoka' and its somaclonal mutant 'Mixue', it was revealed that differences in fruit volatile metabolites, cyanidin-3-O-glucoside and pelargonidin-3-O-glucoside contents in the red fruits of 'Mixue' were significantly decreased compared to 'Sachinoka' (Bian et al. 2022).

Another recent study, Deng et al. (2021) investigated the compounds related to pseudostem elongation in a wild type (WT) of a dwarf banana 'Aifen No. 1' (Musa spp. Pisang Awak subgroup ABB) and its semi-dwarf mutant (MT) obtained from in vitro tissue culture. The results revealed that metabolites corresponding to phenolic acids, organic acids, flavonoids, lignans and coumarins, were more abundant in WT; however, the metabolites less accumulated in WT were, isochlorogenic acid, lumazine, coumarin, asiatic acid, *N*-phenylacetylglycine, cimidahurinine. Among the most regulated metabolites in MT, terpineol monoglucoside, inositol, tricine 5-O-hexoside, sudachiine B and C, the lithium salt of D-xylonic acid, scutellarin, among others, predominated. Finally, they related the differentially accumulated metabolites in the MT banana pseudostem to the Kyoto Encyclopedia of Genes and Genomes (KEEG) database and found that the most significantly enriched metabolic pathways were tyrosine metabolism, phenylalanine and phenylpropanoid biosynthesis. The elongation and growth of MT banana pseudo-stem not only depended on the differential accumulation of several metabolites but also depended on the presence of two hormones, gibberellic acid (GA) and brassinosteroids (BR). These hormones have the function of regulating plant cell elongation (Deng et al. 2021).

In another study, Carrera et al. (2021) analyzed the changes in the metabolite profile during micropropagation of normal and somaclonal variatns plants of Musa (AAA) cv. 'Williams'. The in vitro cultures were stablished from meristematic tissue, the technique used was GC-MS, and the samples for analysis were micro-shoots and meristematic tissue. The results showed that seedlings in the in vitro phases of proliferation and rooting, accumulated insufficient metabolites in dwarf plants. Among the metabolites were found several apoplast precursors, including arabinose and galactose or deoxygalactose, while mannose and fucose accumulated insufficiently in phase two of acclimatization. In all three in vitro phases studied, all differentially accumulated carboxylic acids were more abundant in dwarf plants, while most of the differentially accumulated monosaccharides, all amino acids and nucleoside, were more abundant in normal plants. The abundance of carboxylic acids in dwarf plants hints at a mineral imbalance. The deficient accumulation of monosaccharides, amino acids and nucleoside in dwarf plants would be related to reduced plant growth (Carrera et al. 2021).

In *Catharanthus roseus*, Lee et al. (2020) studied the phytochemical composition, antioxidant capacity and enzyme inhibitory activity in callus, somaclonal variants and normal tissues of green shoot, using the UHPLC/ESI-MS/MS technique. This analysis determined that phenolic compounds and alkaloids were present in the axillary shoots that showed SV.

In dwarf banana plants, Cevallos-Cavallos (2018), investigated metabolic profiles using GC-MS. They observed significant differences between sample groups in 82 metabolites. Rhamnose was found only in SV dwarf plants, but trehalose and allothreonin were found in all samples except in somaclonal variants. Cellobiose was only detected in normal plants, while 45 other metabolites, including L-lysine, phenylalanine, lactose, allopyranose and methylglucopyranoside, were detected in all samples, ruling out those infected with cucumber mosaic virus (CMV). See Table 2 for recent metabolomics studies in plants.

Table 2 Recent metabolomics studies for analysis of SV in plants

Species	In vitro culture technique	Extraction	Separation and detection technique	Metabolites of SV	Bibliography
Malaxis acuminata	Transverse thin cell layer explants	Ethanol	HPTLC	p-coumaric acid, quercetin and eugenol	Bhattacharyya et al. (2023)
Strawberry Benihoppe and Xiaobai	Apex tissues	Methanol	HS-SPME-GC-MS	Linalool, nerolidol, ethyl hexanoate, among others	Zhang et al. (2023)
Gardenia jasminoides Ellis	Sprouts, callus and cell suspension	Methanol, ethyl acetate	GC/MS and HPLC	Chlorogenic acid, salicylic acid, rutin and hesperidin	Krasteva et al. (2022)
Musa spp. Pisang Awak subgroup ABB	Tissue culture	Not specified	No mention of	Coumarins, lignans and flavonoids	Deng et al. (2021)
Musa AAA cv. Williams	Tissue culture	Methanol, chloroform and water	GC/MS	Arabinose, galactose mannose and fucose	Carrera et al. (2021)
Catharanthus roseus	Axillary buds	Methanol	UHPLC/ESI-MS/MS	Phenolic compounds and alkaloids	Lee et al. (2020)
Strawberry 'Sachinoka' and your mutant 'Mixue'	Not specified	Not specified	Not specified	Cyanidin-3-O-glucoside and pelargonidin-3-O-glucoside	Bian et al. (2022)
Dwarf plantain	Tissue culture	Methanol, chloroform and water	GC/MS	Rhamnose	Cevallos-Cevallos et al. (2018)

References

Allen DK, Young JD (2013) Carbon and nitrogen provisions alter the metabolic flux in developing soybean embryos. Plant Physiol 161:1458–1475. https://doi.org/10.1104/pp.112.203299

Alseekh S, Bermudez L, de Haro LA et al (2018) Crop metabolomics: from diagnostics to assisted breeding. Metabolomics 14:0. https://doi.org/10.1007/s11306-018-1446-5

Bartel J, Krumsiek J, Theis FJ (2013) Statistical methods for the analysis of high-throughput metabolomics data. Comput Struct Biotechnol J 4:e201301009. https://doi.org/10.5936/csbj.201301009

Bénard C, Bernillon S, Biais B et al (2015) Metabolomic profiling in tomato reveals diel compositional changes in fruit affected by source-sink relationships. J Exp Bot 66:3391–3404. https://doi.org/10.1093/jxb/erv151

Bhattacharyya P, Kumar S, Lalthafamkimi L et al (2023) Molecular and phytomedicinal stability of long term micropropagated Malaxis acuminata: an endangered terrestrial orchid of biopharmaceutical importance. South African J Bot 155:372–382. https://doi.org/10.1016/j.sajb.2023.01.051

Bian R, Yu S, Song X, Yao J, Zhang J, Zhang Z (2022) An integrated metabolomic and gene expression analysis of 'Sachinoka' strawberry and its somaclonal mutant reveals fruit color and volatiles differences. Plants 12(1):82

Bjarnholt N, Li B, D'Alvise J, Janfelt C (2014) Mass spectrometry imaging of plant metabolites-principles and possibilities. Nat Prod Rep 31:818–837. https://doi.org/10.1039/c3np70100j

Blaise BJ, Correia G, Tin A et al (2016) Power Analysis and sample size determination in metabolic phenotyping. Anal Chem 88:5179–5188. https://doi.org/10.1021/acs.analchem.6b00188

Cambiaghi A, Ferrario M, Masseroli M (2016) Analysis of metabolomic data: tools, current strategies and future challenges for omics data integration. Brief Bioinform 498–510. https://doi.org/10.1093/bib/bbw031

Cambiaghi A, Ferrario M, Masseroli M (2017) Analysis of metabolomic data: tools, current strategies and future challenges for omics data integration. Brief Bioinform 18:498–510. https://doi.org/10.1093/bib/bbw031

Carrera FP, Noceda C, Maridueña-Zavala MG et al (2021) Changes in the metabolite profile during micropropagation of normal and somaclonal variants of banana musa AAA CV. Williams Horticulturae 7:1–14. https://doi.org/10.3390/horticulturae7030039

Cevallos-Cevallos JM, Futch DB, Shilts T et al (2012) GC-MS metabolomic differentiation of selected citrus varieties with different sensitivity to citrus huanglongbing. Plant Physiol Biochem 53:69–76. https://doi.org/10.1016/j.plaphy.2012.01.010

Cevallos-Cevallos JM, Jines C, Maridueña-Zavala MG et al (2018) GC-MS metabolite profiling for specific detection of dwarf somaclonal variation in banana plants. Appl Plant Sci 6. https://doi.org/10.1002/aps3.1194

Chaudhary J, Khatri P, Singla P et al (2019) Advances in omics approaches for abiotic stress tolerance in tomato. Biology (Basel) 8:1–19. https://doi.org/10.3390/biology8040090

Deborde C, Moing A, Roch L et al (2017) Plant metabolism as studied by NMR spectroscopy. Prog Nucl Magn Reson Spectrosc 102–103:61–97. https://doi.org/10.1016/j.pnmrs.2017.05.001

Deng G, Bi F, Liu J et al (2021) Transcriptome and metabolome profiling provide insights into molecular mechanism of pseudostem elongation in banana. BMC Plant Biol 21:1–14. https://doi.org/10.1186/s12870-021-02899-6

Dettmer K, Aronov PA, Hammock BD (2007) Mass spectrometry-based metabolomics. Mass Spectrom Rev 26. https://doi.org/10.1002/mas.20108

Dudzik D, Barbas-Bernardos C, García A, Barbas C (2018) Quality assurance procedures for mass spectrometry untargeted metabolomics. a review. J Pharm Biomed Anal 147:149–173. https://doi.org/10.1016/j.jpba.2017.07.044

Dunn WB, Ellis DI (2005) Metabolomics: current analytical platforms and methodologies. TrAC Trends Anal Chem 24:285–294. https://doi.org/10.1016/j.trac.2004.11.021

Emwas AH, Roy R, McKay RT et al (2019) Nmr spectroscopy for metabolomics research. Metabolites 9. https://doi.org/10.3390/metabo9070123

Fernie AR, Tohge T (2017) The genetics of plant metabolism. Annu Rev Genet 51. https://doi.org/10.1146/annurev-genet-120116-024640

Glauser G, Boccard J, Wolfender JL, Rudaz S (2013) Metabolomics: application in plant sciences. Metabolomics Pract Success Strateg to Gener Anal Metab Data 313–343. https://doi.org/10.1002/9783527655861.ch13

Gromski PS, Muhamadali H, Ellis DI et al (2015) A tutorial review: metabolomics and partial least squares-discriminant analysis—a marriage of convenience or a shotgun wedding. Anal Chim Acta 879:10–23. https://doi.org/10.1016/j.aca.2015.02.012

Gullberg J, Jonsson P, Nordström A et al (2004) Design of experiments: an efficient strategy to identify factors influencing extraction and derivatization of *Arabidopsis thaliana* samples in metabolomic studies with gas chromatography/mass spectrometry. Anal Biochem 331:283–295. https://doi.org/10.1016/j.ab.2004.04.037

Hong J, Yang L, Zhang D, Shi J (2016) Plant metabolomics: an indispensable system biology tool for plant science. Int J Mol Sci 17

Israeli Y, Ben-Bassat D, Reuveni O (1996) Contribution from the Agricultural Research Organization. The Volcani Center

Ivanisevic J, Want EJ (2019) From samples to insights into metabolism: uncovering biologically relevant information in LC-HRMS metabolomics data. Metabolites 9:1–30. https://doi.org/10.3390/metabo9120308

Jacyna J, Kordalewska M, Markuszewski MJ (2019) Design of experiments in metabolomics-related studies: an overview. J Pharm Biomed Anal 164:598–606. https://doi.org/10.1016/j.jpba.2018.11.027

Jorge TF, Mata AT, António C (2016a) Mass spectrometry as a quantitative tool in plant metabolomics. Philos Trans R Soc A Math Phys Eng Sci 374. https://doi.org/10.1098/rsta.2015.0370

Jorge TF, Rodrigues JA, Caldana C, et al (2016b) Mass spectrometry-based plant metabolomics: metabolite responses to abiotic stress. Mass Spectrom Rev 35. https://doi.org/10.1002/mas.21449

Kim HK, Verpoorte R (2010) Sample preparation for plant metabolomics. Phytochem Anal 21. https://doi.org/10.1002/pca.1188

Krasteva G, Berkov S, Pavlov A, Georgiev V (2022) Metabolite profiling of *Gardenia jasminoides* ellis in vitro cultures with different levels of differentiation. Molecules 27:1–14. https://doi.org/10.3390/molecules27248906

Kumar R, Bohra A, Pandey AK et al (2017) Metabolomics for plant improvement: status and prospects. Front Plant Sci 8:1–27. https://doi.org/10.3389/fpls.2017.01302

Larkin PJ, Scowcroft WR (1981) Somaclonal variation—a novel source of variability from cell cultures for plant improvement. Theor Appl Genet 60:197–214. https://doi.org/10.1007/BF02342540

Lee YJ, Perdian DC, Song Z et al (2012) Use of mass spectrometry for imaging metabolites in plants. Plant J 70:81–95. https://doi.org/10.1111/j.1365-313X.2012.04899.x

Lee ON, Ak G, Zengin G et al (2020) Phytochemical composition, antioxidant capacity, and enzyme inhibitory activity in Callus, Somaclonal variant, and normal green shoot tissues of *Catharanthus roseus* (L.) G Don. Molecules 25:1–23. https://doi.org/10.3390/molecules25214945

Liu Y, Fan G, Zhang J et al (2017) Metabolic discrimination of sea buckthorn from different Hippophaë species by 1H NMR based metabolomics. Sci Rep 7:1–11. https://doi.org/10.1038/s41598-017-01722-3

Maridueña-Zavala MG, Freire-Peñaherrera A, Cevallos-Cevallos JM, Peralta EL (2017) GC-MS metabolite profiling of Phytophthora infestans resistant to metalaxyl. Eur J Plant Pathol 149:563–574. https://doi.org/10.1007/s10658-017-1204-y

Markley JL, Brüschweiler R, Edison AS et al (2017) The future of NMR-based metabolomics. Curr Opin Biotechnol 43:34–40. https://doi.org/10.1016/j.copbio.2016.08.001

Matros A, Mock HP (2013) Mass spectrometry based imaging techniques for spatially resolved analysis of molecules. Front Plant Sci 4:1–7. https://doi.org/10.3389/fpls.2013.00089
Moncayo-Martín S, Cáceres-Gianni J (2017) Desarrollo y aplicación de métodos quimiométricos para el estudio de muestras mediante Espectroscopia de Ablación Láser (LIBS). Univ Complut Madrid
Nam KH, Kim YJ, Moon YS et al (2017) Salinity affects metabolomic profiles of different trophic levels in a food chain. Sci Total Environ 599–600:198–206. https://doi.org/10.1016/j.scitotenv.2017.05.003
Obata T, Fernie AR (2012) The use of metabolomics to dissect plant responses to abiotic stresses. Cell Mol Life Sci 69:3225–3243. https://doi.org/10.1007/s00018-012-1091-5
Pan Z, Raftery D (2007) Comparing and combining NMR spectroscopy and mass spectrometry in metabolomics. Anal Bioanal Chem 387:525–527. https://doi.org/10.1007/s00216-006-0687-8
Pedrioli PGA, Eng JK, Hubley R et al (2004) A common open representation of mass spectrometry data and its application to proteomics research. Nat Biotechnol 22:1459–1466. https://doi.org/10.1038/nbt1031
Peters K, Worrich A, Weinhold A et al (2018) Current challenges in plant eco-metabolomics. Int J Mol Sci 19:1–38. https://doi.org/10.3390/ijms19051385
Pluskal T, Castillo S, Villar-Briones A, Orešič M (2010) MZmine 2: modular framework for processing, visualizing, and analyzing mass spectrometry-based molecular profile data. BMC Bioinformatics 11. https://doi.org/10.1186/1471-2105-11-395
Ramautar R, Somsen GW, de Jong GJ (2019) CE-MS for metabolomics: developments and applications in the period 2016–2018. Electrophoresis 40:165–179. https://doi.org/10.1002/elps.201800323
Rodrigues AM, Miguel C, Chaves I, António C (2019) Mass spectrometry-based forest tree metabolomics. Mass Spectrom Rev 1–32. https://doi.org/10.1002/mas.21603
Rodrigues PHV, Oliveira EL, Demetrio CA et al (2022) Effects of different light spectra on the slow-grown in vitro storage and quality of banana plantlets cv. Prata Catarina (AAB). Plant Cell Tissue Organ Cult 150:479–485. https://doi.org/10.1007/s11240-022-02280-x
Rodríguez-Calcerrada J, Rodrigues AM, Perdiguero P et al (2018) A molecular approach to drought-induced reduction in leaf CO_2 exchange in drought-resistant *Quercus ilex*. Physiol Plant 162:394–408. https://doi.org/10.1111/ppl.12649
Roux N, Chase R, Van den Houwe I et al (2021) Somaclonal variation in clonal crops: containing the bad, exploring the good. In: Mutation breeding, genetic diversity and crop adaptation to climate change. CABI International, pp 355–365
Sahijram L, Soneji JR, Bollamma KT (2003) Analyzing somaclonal variation in micropropagated bananas (Musa spp.). Vitr Cell Dev Biol Plant 39:551–556
Sas KM, Karnovsky A, Michailidis G, Pennathur S (2015) Metabolomics and diabetes: analytical and computational approaches. Diabetes 64:718–732. https://doi.org/10.2337/db14-0509
Schiffmann C, Hansen R, Baumann S et al (2014) Comparison of targeted peptide quantification assays for reductive dehalogenases by selective reaction monitoring (SRM) and precursor reaction monitoring (PRM). Anal Bioanal Chem 406:283–291. https://doi.org/10.1007/s00216-013-7451-7
Scossa F, Brotman Y, de Abreu e Lima F et al (2016) Genomics-based strategies for the use of natural variation in the improvement of crop metabolism. Plant Sci 242:47–64. https://doi.org/10.1016/j.plantsci.2015.05.021
Sharma K, Sarma S, Bohra A et al (2018) Plant metabolomics: an emerging technology for crop improvement. New Visions Plant Sci. https://doi.org/10.5772/intechopen.76759
Shawky E, Selim DA (2017) Evaluation of the effect of extraction solvent and organ selection on the chemical profile of *Astragalus spinosus* using HPTLC-multivariate image analysis. J Chromatogr B Anal Technol Biomed Life Sci 1061–1062:134–138. https://doi.org/10.1016/j.jchromb.2017.07.025
Shih ML, Morgan JA (2020) Metabolic flux analysis of secondary metabolism in plants. Metab Eng Commun 10

Shinbo Y, Sakaguchi S, Nakamura Y et al (2006) Species-metabolite database (KNApSAcK): elucidating diversity of flavonoids. J Comput Aided Chem 7:94–101. https://doi.org/10.2751/jcac.7.94 (Special Issue: Fact Databases and Freewares)
Story W (2011) Chapter 1 股関節 概念 Chapter 1 股関節. An Autom Irrig Syst Arduino Microcontroller 1908:2–6
Sweetlove LJ, Nielsen J, Fernie AR (2017) Engineering central metabolism—a grand challenge for plant biologists. Plant J 90:749–763. https://doi.org/10.1111/tpj.13464
Treutler H, Neumann S (2016) Prediction, detection, and validation of isotope clusters in mass spectrometry data. Metabolites 6. https://doi.org/10.3390/metabo6040037
Wolfender JL, Marti G, Thomas A, Bertrand S (2015) Current approaches and challenges for the metabolite profiling of complex natural extracts. J Chromatogr A 1382:136–164. https://doi.org/10.1016/j.chroma.2014.10.091
Wu Y, Li L (2016) Sample normalization methods in quantitative metabolomics. J Chromatogr A 1430:80–95. https://doi.org/10.1016/j.chroma.2015.12.007
Xia J, Wishart DS (2016) Using metaboanalyst 3.0 for comprehensive metabolomics data analysis. Curr Protoc Bioinforma 2016:14.10.1–14.10.91. https://doi.org/10.1002/cpbi.11
Yan M, Xu G (2018) Current and future perspectives of functional metabolomics in disease studies—a review. Anal Chim Acta 1037:41–54. https://doi.org/10.1016/j.aca.2018.04.006
Zhang Z, Yu S, Zhang Z et al (2023) Comparative characterization of fruit volatiles and volatile-related genes expression of 'Benihoppe' strawberry and its somaclonal mutant. Plants 12. https://doi.org/10.3390/plants12051109

Somaclonal Variation in Fruit Crop Improvement

Manoj K. Rai, Roshni Rathour, Srishti Yadav, Amit Singh, and Sandeep Kaushik

Abstract Genetic variability induced by somaclonal variation is a viable tool to generate new traits in plants and create a new variety. During the last three-four decades, somaclonal variation has been utilized efficiently in selecting variants with desirable characteristics in many commercially important plants, including fruit crops. Somaclonal variation offers immense scope for crop improvement in fruit plants, as most fruit plants are trees that have many breeding concerns. Several valuable somaclonal variants, mainly associated with crop improvement traits like disease resistance, abiotic stress tolerance, fruit quality, growth characteristics, etc., have been selected in many fruit plants through in vitro selection techniques. Few of them are released as commercial cultivars. This chapter emphasizes the potential and perspectives of somaclonal variation in fruit crop improvement. Further, the chapter also includes a list of useful variants selected through somaclonal variation in fruit plants.

Keywords Crop improvement · Disease resistance · Fruit plants · In vitro selection · Plant tissue culture · Somaclonal variation

1 Introduction

With the gradual rise in the global population, producing sufficient food to meet the demand is one of the major challenges that the world is facing. Further, in the era of global climate change, the balance of natural resources is being threatened by the limited availability of water resources, heat and drought stresses, and poor soil fertility, which restrict food crop production (Varshney et al. 2017; Rai et al. 2017). Under these constraints, agricultural production must increase to ensure food and nutritional security, particularly in poor and developing countries. Despite such limitations, there has been a notable increase in agricultural productivity over the

M. K. Rai (✉) · R. Rathour · S. Yadav · A. Singh · S. Kaushik
Department of Environmental Science, Indira Gandhi National Tribal University, Amarkantak, Madhya Pradesh 484887, India
e-mail: manoj.rai@igntu.ac.in

C. Sánchez-Romero (ed.), *Somaclonal Variation: Basic and Practical Aspects*,
https://doi.org/10.1007/978-3-031-51626-9_6

past 50 years through better agronomic practices and the development of improved cultivars through biotechnology (Sharma et al. 2002).

Many fruit plants are an integral part of the human diet as they are excellent sources of antioxidants and bioactive phytoconstituents, thus contributing to nutritional and health security (Penna and Jain 2023). Since the olden time, several indigenous fruit crops have been exploited for their nutritional value. Most of these indigenous fruits can potentially solve the deficiency of vitamins and micronutrients in rural populations and address the food insecurity problem in many poor and developing countries (Rai et al. 2021). Despite their significant role in food security, most indigenous fruit crops remain considered underutilized. Therefore, selecting fruit crops with improved fruit quality and the ability to survive against biotic and abiotic stresses is necessary to boost the fruit industries. Some important desirable crop improvement traits of fruit crops include fruits with high nutrient and low seed content, early flowering, longer shelf life of fruits, and resistance to many biotic and abiotic stresses. Although conventional breeding has achieved some success in fruit plants to address such issues, the improvement of fruit trees by this approach is hindered by factors like long juvenile periods, reproductive obstacles, and complex heterozygosity (Litz and Padilla 2012; Rai and Shekhawat 2014). In addition, it is a multi-generational program that involves hybridization and selection of progeny with the trait of interest (Dalla Costa et al. 2017). Alternatively, new biotechnological tools (NBT), i.e., plant tissue culture, transgenesis, cisgenesis and gene editing, have played a critical role in fruit crop improvement. Generation of variations in plants regenerated from tissue culture coupled with in vitro selection are now well-recognized practices for crop improvement. Occurrence of genetic variation in tissue culture-raised plant, termed somaclonal variation (Larkin and Scowcroft 1981), has been widely applied in several agricultural and horticultural crops to induce genetic variation for desirable traits and generation of new cultivars with many important agronomic and horticultural traits (Rai 2021).

Usually, somaclonal variation is undesirable in commercial micropropagation of a specific genotype, particularly in perennial fruit crops, in which the regenerated plants are expected to be true to type. However, many studies clearly revealed that it could be exploited to generate disease-resistant and abiotic stress-tolerant clones via in vitro selection (Brar and Jain 1998; Jain 2001; Rai et al. 2011). Earlier, many review papers have been published on the molecular mechanisms of somaclonal variation, factors responsible for somaclonal variation, generation of somaclones via in vitro selection, and different methods applied for the detection of somaclonal variations (Evans 1989; Karp 1995; Kaeppler et al. 2000; Bairu et al. 2011; Miguel and Marum 2011; Smulders and de Klerk 2011; Neelakandan and Wang 2012; Krishna et al. 2016; Rai 2021). The main causes of somaclonal variations are categorized either as pre-existing or induced. During tissue culture of any plant species, different factors like mode of regeneration, culture conditions, imbalance concentrations of plant growth regulators, additives, nutrients or other stress-causing chemicals, duration of culture period and the number of subcultures, etc. determine the frequency of variation (Bairu et al. 2011; Bradaï et al. 2016). The variation in regenerants may either be permanent, genetic and heritable or temporary, epigenetic and non-heritable (Kaeppler et al. 2000; Bairu et al. 2011). Therefore, early detection of somaclonal variation is essential to identify variants with useful traits. Nowadays, a number of PCR-based molecular markers have been used to detect nucleotide variation between source plants and somaclones. Molecular techniques are easy to use and able to detect variations at the early growth phase of regenerants and, therefore, preferred over the somaclones detection methods based on biochemical or morphological characteristics. Several PCR-based molecular markers, i.e., RAPD (random amplified polymorphic DNA), ISSR (inter simple sequence repeat), SSR (simple sequence repeat), and SCoT (start codon targeted) polymorphism, have now been efficiently used to detect somaclonal variations (Rani and Raina 2000; Bairu et al. 2011; Rai 2023). Thus, different types of DNA-based molecular markers have been applied in the analysis of genetic stability of tissue culture-raised plants of several fruit plants, like apple (Modgil et al. 2005; Pathak and Dhawan 2012), banana (Ray et al. 2006; Uma et al. 2021), grapevine (Nookaraju and Agrawal 2012; Carra et al. 2016), guava (Rai et al. 2012; Kamle et al. 2014), kiwifruit (Martins et al. 2004), pineapple (da Silva et al. 2016), *Prunus* (Helliot et al. 2002), etc.

Fruit plants are a suitable target for crop improvement programs through somaclonal variation, as most fruit plants are trees and have many breeding concerns. During the past two–three decades, somaclonal variation has been detected in numerous fruit plants like apple, banana, *Citrus*, pineapple, grapes, peach, pears, strawberries etc. (Table 1). Somaclonal variation has been observed in many fruit plants, mainly associated with crop improvement traits like abiotic stress tolerance, disease resistance, and fruit quality characteristics (Table 2). Few of them are released as commercial cultivars. This paper provides an overview of the application of somaclonal variation in fruit plants, mainly focusing on selecting somaclones linked with crop improvement traits.

2 Somaclonal Variation in Fruit Crop Improvement: Case Studies

2.1 *Apple (*Malus *sp.)*

Apple is one of the most commercially and nutritionally important fruit crops cultivated in temperate climatic regions. However, numerous diseases caused by different types of pathogens are a serious problem faced by apple growers, as these diseases reduce apple production and its trade (Bhatti and Jha 2010). The major diseases affecting apple production include apple scab, collar rot, root rot, fire blight, and powdery mildew.

In vitro selection using culture filtrate as a selective agent has been proven effective in obtaining disease-resistant somaclonal variants in many fruit plants, including apple. Many workers have reported the generation of somaclonal variants in apple

Table 1 Selected reports on somaclonal variation in fruit plants and their causes and detection methods

Plant species	Mode of regeneration/explants	Source/causes of somaclonal variation	Molecular marker-based detection method	References
Apple (*Malus* sp.)	Axillary buds of apple (*Malus pumila*) rootstock MM106	Off-types plants, 23.2% polymorphism	RAPD	Modgil et al. (2005)
	Nodes from M7 and M9 rootstocks	Molecular (qualitative) and genome size (quantitative) changes, 53% or 46% polymorphism	ISSR	Noormohammadi et al. (2015)
Banana (*Musa* sp.)	Adventitious shoots from excised sucker shoot tips	Tetraploid clone	RAPD	Giménez et al. (2001)
	Procured tissue cultured plants	*Fusarium* wilt-resistant and susceptible somaclones, 38% polymorphism	IRAP and RAPD	Muhammad and Othman (2005)
	Obtained in vitro plants from African Biotechnologies Ltd., South Africa	Type and concentration of plant growth regulators and sub-culturing	RAPD	Bairu et al. (2006)
	Variant phenotype created in vitro from banana shoot cultures	Higher rearrangement and mutation rates in visible culture-induced off-type and normal banana plants	RDA to isolate unique fragments of DNA	Oh et al. (2007)
	Shoot tip cultures	Age of culture and PGRs levels (2,4-D)	SSR and MSAP	Sales and Butardo (2014)
Citrus sp.	Somatic embryogenesis in calamondin (*Citrus madurensis*)	Effect of three synthetic phenylurea derivatives, 3.7 and 7.4% polymorphism	ISSR	Siragusa et al. (2007)
	Protoplast from embryogenic calli	Inter-specific protoplast fusion	SSR and RFLP	Ding et al. (2009)
	Shoot culture of persian lime (*Citrus* × *latifolia*)	Effect of three in vitro culture systems and subcultures	ISSR	Bulbarela-Marini et al. (2023)

(continued)

Table 1 (continued)

Plant species	Mode of regeneration/ explants	Source/causes of somaclonal variation	Molecular marker-based detection method	References
Grapevine (*Vitis vinifera*)	Anther derived direct and indirect somatic embryogenesis in two cultivars	Age of culture	AFLP	Popescu et al. (2002)
	Somatic embryogenesis in two distinct cultivars	DNA methylation	AFLP	Schellenbaum et al. (2008)
	Somatic embryogenesis in three cultivars	Effect of genotypes	SSR	Prado et al. (2010)
Kiwifruit (*Actinidia deliciosa*)	Bud meristems and fruit peduncles	Variations in genotypes, 55.04 and 85.7% polymorphism	RAPD and SSR	Palombi and Damiano (2002)
Mango (*Mangifera indica*)	Embryogenic cultures of two cultivars	Cultures selected for resistance to *Colletotrichum gloeosporioides* culture filtrate	RAPD	Jayasankar et al. (1998)
Papaya (*Carica papaya*)	Somatic embryogenesis from cotyledonary explants	Plants derived from callus mediated indirect somatic embryogenesis	RAPD	Homhuan et al. (2008)
Peach (*Prunus persica*)	Embryo derived callus cultures	Effect of genotypes	RAPD	Hashmi et al. (1997)
Pear (*Pyrus* sp.)	Axillary buds of genotype (ISF 61) of *Pyrus pyraster*	Selective treatment utilized MS medium with FeSO4 with either $KHCO_3$ or $NaHCO_3$, 33.8% polymorphism	RAPD	Palombi et al. (2007)
Pineapple (*Ananas comosus*)	Buds and leaf explants	Comparison of two somaclones P3R5 and Dwarf with donor cv. Red Spanish Pinar	AFLP	Pérez et al. (2009)
	Axillary bud explants	Regeneration methods, long-term culture and subcultures, 75–95% polymorphism	RAPD	Roostika et al. (2015)

AFLP—amplified fragment length polymorphism, *IRAP*—inter retrotransposon amplified polymorphism, *ISSR*—inter simple sequence repeat, *MSAP*—methylation-sensitive amplified polymorphism, *RAPD*—random amplified polymorphic DNA, *RFLP*—restriction fragment length polymorphism, *RDA*—representational difference analysis, *SSR*—simple sequence repeat

Table 2 Selection of somaclonal variants for crop improvement traits through somaclonal variation in fruit crops

Plant species	Characteristics/description of somaclonal variants	Mode of generation of somaclonal variation	References
Apple (*Malus* sp.)	Apple rootstocks M 26 and MM 106 (*Malus pumila*) resistant to *Phytophthora cactorum*	Regenerated shoots from leaves treated with FCF of *Phytophthora cactorum*	Rosati et al. (1990)
	Assessment of somaclonal variants of apple cv. Greensleeves for resistance to the fire blight pathogen, *Erwinia amylovora*	Pre-treated leaf disc explants with pathogen, *Erwinia amylovora*	Donovan et al. (1994)
	Regeneration of four somaclonal variants of apple cv. Greensleeves resistant to fire blight (*Erwinia amylovora*)	Use of various inoculation techniques i.e., in vitro leaves, microcuttings, greenhouse plants and in field conditions	Chevreau et al. (1998)
	Assessment of somaclonal variation for the morphological and reproductive traits among tissue culture-derived 'Gala' and 'Royal Gala' apples (*Malus domestica* Borkh.)	Gala' and 'Royal Gala' apple trees obtained via axillary and adventitious bud formation	McMeans et al. (1998)
	Apple root stock Malling 7 resistance to white root rot caused by *Dematophora necatrix*	In vitro selection using FCF as a selective agent	Modgil et al. (2012)
	Apple rootstock MM106 somaclones resistant to collar rot (*Phytophthora cactorum*)	In vitro selection using FCF for screening of collar rot tolerance among regenerants	Verma et al. (2021)
Banana and plantain (*Musa* sp.)	Assessment of field performance of somaclonal variants of plantain cultivar 'Agbagba' by measuring growth and yield variables	Somaclonal variants derived through standard shoot-tip culture	Vuylsteke et al. (1996)
	Somaclonal variant CIEN BTA-03 resistant to yellow Sigatoka	Generation of somaclonal variant from a susceptible banana clone i.e., Williams clone using high concentrations of BAP and increasing the production of adventitious buds	Vidal and García (2000)

(continued)

Table 2 (continued)

Plant species	Characteristics/description of somaclonal variants	Mode of generation of somaclonal variation	References
	A tetraploid somaclonal variant of banana CIEN BTA-03 resistant to yellow Sigatoka disease	Somaclonal variant obtained by induction of adventitious shoots from excised sucker shoot tips	Giménez et al. (2001)
	Release of a *Fusarium* wilt resistant cultivar 'Tai Chiao No. 1' (GCTCV215–1) for commercial planting in Taiwan	Screening and selection of the plantlets from tissue culture for resistance to *Fusarium* wilt	Hwang and Ko (2004)
	Selection and release of a new cultivar 'Tai-Chiao No. 5' resistant to *Fusarium* wilt in Chinese Taipei	Mass planting of tissue culture-raised plantlets in heavily infested banana orchards and identification of somaclonal variants	Lee et al. (2011)
	Characterization of four somaclonal variants of banana cv. Rasthali resistant to *Fusarium* wilt by cDNA-RAPD	Callus-mediated plantlet regeneration and two-month-old hardened plants subjected to Foc resistance bioassays	Ghag et al. (2014)
	Field trial of four somaclonal variants of banana resistant to *Fusarium* wilt in the Philippines	Identification of somaclonal variants after mass screening plantlets in infested fields	Molina et al. (2016)
	Selection of two somaclonal variants T2-1 and T2-2 of the cultivar 'Prata-Anã' resistant to *Fusarium oxysporum f. sp. cubense* race 1 in Brazil	Identification of micropropagated seedlings of the somaclones resistant to *Fusarium* wilt	Ferreira et al. (2020)
	Identification of SNVs via RNA-sequencing	Tissue culture-derived Foc TR4–resistant somaclonal variants	Hou et al. (2022)
Blackberry (*Rubus laciniatus*)	Field evaluation of thornless evergreen (TE) blackberry over two fruiting seasons for various vegetative traits and fruiting characteristics	Somaclonal variants derived from shoot tip cultures of thornless evergreen blackberry	McPheeters and Skirvin (1989)
Citrus sp.	Selection of NaCl-tolerant plant of *Citrus sinensis*	Plant regenerated from NaCl-selected callus through somatic embryogenesis	Ben-Hayyim and Goffer (1989)

(continued)

Table 2 (continued)

Plant species	Characteristics/description of somaclonal variants	Mode of generation of somaclonal variation	References
	Selection of disease-resistant plant of *Citrus limon* caused by *Phoma tracheiphila*	Protoplasts isolated from an embryogenic callus line and incubated with toxin of *Phoma tracheiphila* for the selection of disease-resistant plant	Deng et al. (1995)
	Field trial and evaluation of more than 2000 sweet orange somaclones of two sweet orange cultivars i.e., 'Hamlin' and 'Valencia' for fruit quality and release of nine new cultivars of sweet orange for the commercial production	Somaclones regenerated from protoplasts, nucellar seedlings or callus	Grosser et al. (1997), Germanà et al. (2020)
	Regeneration of plants of *Citrus jambhiri* resistance to *Phytophthora parasitica*	In vitro selection using culture filtrate of *Phytophthora parasitica*	Savita et al. (2011)
	The genomic changes associated with dedifferentiation and somaclonal variation during long-term in vitro culture of sweet orange i.e., transcriptome reprogramming and DNA methylation	Analysis of somaclonal variation after 30 years of in vitro callus culture	Wang et al. (2022)
Grapevine (*Vitis vinifera*)	Selection of grapevine var. 'Chardonnay' (Clone 02Ch) for anthracnose resistance caused by *Elsinoe ampelina*	In vitro selection by exposure of proembryogenic masses to the culture filtrate of *E. ampelina*	Jayasankar et al. (2000)
Guava (*Psidium guajava*)	Selection of clones tolerant to guava wilt disease caused by *Penicillium vermoesenii*	Cell-free filtrates derived from *Penicillium vermoesenii* used as a selective agent for in vitro selection	Vos et al. (2000)
Kiwifruit (*Actinidia deliciosa*)	Selection of somaclones tolerant to NaCl and their molecular and in vivo ecophysiological characterization	In vitro selection of callus on NaCl-containing medium for six subcultures	Caboni et al. (2003)

(continued)

Table 2 (continued)

Plant species	Characteristics/description of somaclonal variants	Mode of generation of somaclonal variation	References
Mango (*Mangifera indica*)	Generation of anthracnose resistance clones in two polyembryonic mango cultivars, 'Hindi' and 'Carabao'	In vitro selection of embryogenic nucellar cultures resistance to the culture filtrate and phytotoxin of a virulent strain of *Colletotrichum gloeosporioides*	Jayasankar and Litz (1998)
Myrobalan plum (*Prunus cerasifera*)	Rootstock somaclonal variants (S.4 clone) tolerant to waterlogging	Somaclones generated in vitro from leaf explants of rootstock Mr.S.2/5	Pistelli et al. (2012)
Peach (*Prunus persica*)	Somaclonal variants of peach resistant to leaf spot disease caused by *Xanthomonas campestris* pv. *pruni*	In vitro selection using toxic culture filtrate of *X. campestris* pv. *pruni* as a selective agent and regenerating resistant plants from selected cell line	Hammerschlag (1988)
	Screening for somaclonal variation from a leaf spot susceptible and moderately resistant cultivar	Regenerants derived from callus cultures of cultivars Sunhigh (susceptible to leaf spot) and Redhaven (moderately resistant to leaf spot) for resistance to *X. campestris* pv. *pruni*	Hammerschlag (1990)
	Somaclonal variation in peach for resistance to *Xanthomonas camperstris* pv. *pruni* and *Pseudomonas syringae* pv. *syringae*	In vitro selection of somaclonal variants using toxic culture filtrate of *X. campestris* pv. *pruni* and *Pseudomonas syringae* pv. *syringae* as selective agents	Hammerschlag and Ognjanov (1990)
	Somaclonal variants resistance to the root-knot nematode, *Meloidogyne incognita*	Somaclones derived from callus cultures of cultivars Sunhigh and Redhaven exposed to *Meloidogyne incognita*	Hashmi et al. (1995)
	Response of selected Somaclone 122-1 to *Xanthomonas campestris* pv. *pruni* and to *Pseudomonas syringae* pv. *syringae*	Somaclonal variant 122-1 derived from callus generated on an immature Redhaven' embryo	Hammerschlag (2000)
Pear (*Pyrus* sp.)	Somaclonal variants resistant to fire blast caused by *Erwinia amylovora*	Somaclonal variant derived from callus of pear cultivar 'Durondeau'	Viseur (1990)

(continued)

Table 2 (continued)

Plant species	Characteristics/description of somaclonal variants	Mode of generation of somaclonal variation	References
	NaCl-tolerant somaclones of pear cvs. 'Conference' and 'Abbé Fetel'	Shoot regeneration induced in leaf explants and selection of somaclones on NaCl-containing medium	Marino and Molendini (2005)
	Wild pear (*Pyrus pyraster*) clones tolerant to Fe-chlorosis	In vitro selection of Fe-chlorosis tolerant plant by the replacement of Fe-EDTA in medium from $FeSO_4$ with either $KHCO_3$ or $NaHCO_3$	Palombi et al. (2007)
	Somaclonal variants obtained from pear rootstock (*P. communis*) 'Old Home 9 Farmingdale (OHF 333)' resistant to the fire blast	Somaclones regenerated from adventitious buds obtained from leaf segments and their assessment for fire blast resistance	Nacheva et al. (2014)
Pineapple (*Ananas comosus*)	Regenerated somaclonal variants with spined and spineless characteristics	Somaclonal variation observed in micropropagated dormant axillary buds excised from crowns	Soneji et al. (2002)
	Two new pineapple somaclones (P3R5 and Dwarf) differ in fruit color, growth habit, fruit size and length of plant generation cycle	Both somaclones derived from in vitro culture of cv. Red Spanish Pinar	Pérez et al. (2011)
	Production of dwarf variants with increased HDAC activity	ABA and salinity stress induced somaclonal variation in var. MD2	Halim et al. (2018)

(continued)

Table 2 (continued)

Plant species	Characteristics/description of somaclonal variants	Mode of generation of somaclonal variation	References
	Large variability in the somaclones and parent cv. 'Mauritius' in terms of fruit and yield traits	Somaclones derived from in vitro cultures multiplied up to the eighth subculture	Dhurve et al. (2021)
	Analysed DNA methylation levels and the expression of cytosine DNA methyltransferase genes in four somaclonal variants, namely, yellow-striped, narrow, green-striped, and spiny leaves	All four variants derived from callus culture	Lin et al. (2022)
	Identification of a cold-tolerant variant	Cold-tolerant variant developed through in vitro cold selection of somaclonal variations	Zhang et al. (2022)
Pomegranate (*Punica granatum*)	High anthocyanin content in fruits and better fruit quality in identified clones than parent cultivar 'Bhagwa'	Somaclonal variants identified from an orchard having tissue-cultured pomegranate plants	Sridhar andHarish Babu (2019)
Strawberry (*Fragaria* × *ananassa*)	Selection of somaclonal variants resistant to *Fusarium* wilt disease	Transplanting the plants regenerated from leaf-derived callus tissues to a field heavily infested with the pathogen and identification of somaclonal variants	Toyoda et al. (1991)
	Selected plants with resistance to *Rhizoctonia fragariae* and *Botrytis cinerea*	Purified pectic enzymes produced by *R. fragariae* used as a selective agent for in vitro selection	Orlando et al. (1997)
	Selection of somaclones tolerant to salt stress	In vitro selection of strawberry on NaCl-containing media during germination and early seedling growth	Dziadczyk et al. (2003)
	Selection of somaclonal variants resistant to *Verticillium dahliae* and their *ex vitro* establishment and greenhouse performance	Selection of somaclonal variants by cocultivation of shoots with a homogenate of *Verticillium dahliae*	Sowik et al. (2003)

(continued)

Table 2 (continued)

Plant species	Characteristics/description of somaclonal variants	Mode of generation of somaclonal variation	References
	In vitro screening of strawberry plants resistant to anthracnose disease caused by *Colletotrichum acutatum*	Incubation of in vitro derived leaf explants of three anthracnose susceptible and resistance cultivars with pathogen *C. acutatum* for selection of resistant variants	Hammerschlag et al. (2006)
	Selection of somaclonal variants for improved horticultural traits after field evaluation	Induction of somaclonal variation by high concentration of BAP in culture medium	Biswas et al. (2009)
	Transcriptome and hormones analysis of elongated fruit somaclonal mutant, i.e., 'finger strawberry' and compared with cultivated 'Akihime' strawberry	Somaclonal mutant 'finger strawberry' derived from 'Akihime' strawberry using tissue culture	Li et al. (2023)

ABA—abscisic acid, *BAP*—6-benzylaminopurine, *FCF*—fungal culture filtrate, *HDAC*—histone deacetylase, *PEG*—polyethelene glycol, *RAPD*—random amplified polymorphic DNA, *SNVs*—single-nucleotide variants linked with disease resistance. Collar rot, caused by *Phytophthora cactorum*, is a common disease of apple. Rosati et al. (1990) attempted shoot regeneration from leaf explants of apple rootstocks M26 and MM111 and selected somaclonal variants from shoot cultures treated with *P. cactorum* culture filtrate. They also found that selected somaclones showed higher resistance to the culture filtrate of *P. cactorum* when compared with non-selected somaclones. In another study, Verma et al. (2021) generated somaclones of apple rootstock MM106 tolerant to collar rot by exposing shoot cultures to culture filtrate of *P. cactorum*. Some rosaceous species, including many commercial apple cultivars, are susceptible to another fire blast disease caused by *Erwinia amylovora*. Donovan et al. (1994) reported somaclonal variation in the apple cultivar 'Greensleeves' for fire blight resistance. They selected four somaclones and found that all were more resistant than the control. Later, Chevreau et al. (1998) examined the field behavior and stability of the fire blight resistance in these four apple somaclones and confirmed the existence of somaclonal variation among regenerants. Modgil et al. (2012) screened out the regenerated shoots of apple rootstock Malling 7 (M7) against the culture filtrate of *Dematophora necatrix*, a white root rot disease causal agent. Selected disease-resistant variants were planted in pathogen-infested soil in pots and showed resistance to white root rot disease. Besides disease resistance, McMeans et al. (1998) also assessed somaclonal variation in two apple cultivars, 'Gala' and 'Royal Gala', for the morphological and reproductive traits.

*2.2 Banana and Plantain (*Musa *sp.)*

Banana and plantain (*Musa* sp.) are important fruit crops in terms of total production and trade worldwide. These are mainly produced in tropical and subtropical regions and provide one of the leading staple foods to the large population of many countries (Vuylsteke et al. 1996; Bairu et al. 2006). However, the production of banana is threatened by many diseases, mainly *Fusarium* wilt and black Sigatoka leaf spot, caused by *Fusarium oxysporum* and *Mycosphaerella fijiensis*, respectively (Vuylsteke et al. 1996). Banana is a model fruit crop in which plant tissue culture technique is extensively practiced on a commercial scale (Sahijram 2015). Using in vitro culture technique, several somaclonal variants have been generated in banana and plantain. However, most of them were associated with *Fusarium* wilt resistance (Table 2).

After a mass screening of tissue culture-raised plantlets in infested fields, a *Fusarium* wilt-resistant cultivar 'Tai Chiao No. 1' (GCTCV215–1) was released for commercial planting in Taiwan (Hwang and Ko 2004). Similarly, after a field trial of identified somaclonal variants in heavily infested banana orchards, a new cultivar 'Tai-Chiao No. 5' resistant to *Fusarium* wilt was selected for commercialization in Chinese Taipei (Lee et al. 2011). Compared to their parent cultivar, 'Giant Cavendish', both cultivars were superior in many other horticultural traits and resistant to *Fusarium* wilt. Molina et al. (2016) conducted a field trial of four somaclonal variants of banana resistant to *Fusarium* wilt in the Philippines. Results of this study showed that all four somaclones were more resistant to Foc TR4 than the local sensitive cultivar 'Grand Naine'. In another study, Ferreira et al. (2020) selected two *Fusarium* (Foc race 1) resistant somaclonal variants, T2-1 and T2-2, from the susceptible cultivar 'Prata-Anã' of Brazil. Ghag et al. (2014) selected four somaclonal variants from the sensitive cultivar 'Rasthali', which showed enhanced resistance to Foc infection. Further, they identified one differentially expressed gene, i.e., lipoxygenase, in the generated somaclonal variants through cDNA-RAPD analysis. Besides *Fusarium* wilt resistance, another somaclone resistant to yellow Sigatoka disease was also been developed in banana. Vidal and García (2000) obtained a somaclone, CIEN BTA-03, from the susceptible banana clone Williams clone. They observed that the selected somaclonal variant showed higher resistance against the yellow Sigatoka disease in the field for five successive years. Later, Giménez et al. (2001) characterized the somaclonal variant, CIEN BTA-03, by flow cytometry and RAPD analyses and confirmed that this variant was a tetraploid clone. Recently, Hou et al. (2022) performed RNA-seq analysis of three somaclonal variants resistant to Foc TR4 and their susceptible progenitor. Based on RNA-seq data, they identified unique single-nucleotide variants (SNVs) in these somaclonal variants and developed cultivar-specific CAPS/dCAPS markers. These cultivar-specific markers may be useful in detecting somaclonal variation among regenerants and identifying local and international cultivars of banana.

Based on morphological variations in micropropagated plantain, Vuylsteke et al. (1991) identified five stable variant phenotypes, two with off-type inflorescence morphology, while three exhibited abnormal foliage. Further, Vuylsteke et al. (1996) compared the field performance of these somaclones, and field trials revealed that all variants differed from true-to-type plantain cultivars in horticultural traits. Out of four, three somaclonal variants were inferior to the original clone, and only one, 'French reversion' variant, resembled the existing cultivar.

2.3 Citrus *spp.*

Genus *Citrus* (family Rutaceae), which includes a number of fruit species with a rich source of vitamin C and many macro- and micronutrients, is one of the most valuable and leading fruit crops of the world (Germanà et al. 2020). *Citrus* species are vulnerable to various pathogens and abiotic stresses during cultivation. Several somaclonal variants have been selected in many *Citrus* species with the aim of crop improvement, particularly for disease resistance, salt tolerance, and fruit quality traits.

Mal secco disease, caused by *Phoma tracheiphila*, is one of the major concerns of the lemon (*C. limon*). Using *P. tracheiphila* toxin as a selective agent, Deng et al. (1995) selected two toxin-tolerant cell lines from the protoplast of the embryogenic callus of a susceptible cultivar 'Femminello siracusano' and regenerated plants from each line. *Phytophthora* sp. is one of the severe disease-causing fungi in *Citrus*, including rough lemon (*C. jambhiri*). Savita et al. (2011) successfully regenerated *Phytophthora*-resistant rough lemon by selecting cotyledon-derived embryogenic callus against the culture filtrate obtained from the causal agent *P. parasitica*. In vitro selection technique coupled with somaclonal variation has also been used to generate salt-tolerant somaclones in sweet orange (Ben-Hayyim and Goffer 1989).

After the screening and massive field trial of more than 2000 sweet orange somaclones at the *Citrus* improvement program at the University of Florida, nine cultivars were released for commercial production (Grosser et al. 1997; Germanà et al. 2020). All these somaclones were selected from the two most promising sweet oranges cultivars, i.e., 'Hamlin' and 'Valencia'. These cultivars were superior in many fruit quality traits like earlier maturity, seedlessness, and improved fruit/juice quality.

2.4 *Grapevine (***Vitis Vinifera***)*

Grapevine is a prized fruit crop commercialized worldwide for many value-added products like wine, juice, jelly, jam, etc. (Dhekney et al. 2020). In grapevine, only one report is available on the generation of somaclonal variants. Anthracnose disease, caused by *Elsinoe ampelina*, is a destructive fungal disease affecting grapevine production worldwide. Jayasankar et al. (2000) screened embryogenic cultures of *V. vinifera* cultivar 'Chardonnay' against culture filtrate produced by fungal pathogen *Elsinoe ampelina* and reported that regenerated plants showed enhanced resistance to anthracnose.

2.5 *Peach (***Prunus Persica***)*

Peach (family Rosaceae) is an important stone fruit crop, extensively grown in the world with temperate climates. Peaches are well-known for their nutritional value, providing vitamins, minerals and antioxidants (Scorza and Petri 2009). Earlier, considerable efforts have been made on the generation of somaclonal variants in peach for disease resistance, particularly for resistance to leaf spot disease bacterial canker. Hammerschlag (1988) selected somaclonal variants of peach resistant to leaf spot disease caused by *Xanthomonas campestris* pv. *pruni*. They used toxic culture filtrate of *X. campestris* pv. *pruni* as a selective agent for the in vitro selection of callus derived from immature embryos of a susceptible 'Sunhigh' peach. Selected clones were propagated in vitro, and regenerated plantlets showed higher resistance against *X. campestris* pv. *pruni* than the parental cultivar 'Sunhigh'. Later, Hammerschlag (1990) used callus derived from zygotic embryos of two different cultivars, one susceptible to leaf spot disease, 'Sunhigh', and another moderately resistant to leaf spot, 'Redhaven', for the generation of somaclones resistant to *X. campestris* pv. *Pruni*. In another study, Hammerschlag and Ognjanov (1990) reported somaclonal variation in peach for the resistance to leaf spot and bacterial canker using toxic culture filtrate of *X. campestris* pv. *pruni* and *Pseudomonas syringae* pv. *syringae*, respectively as selective agents. The in vitro screening of cultures against the culture filtrates of causal pathogens of leaf spot and bacterial canker generated a promising somaclonal variant 122-1 (Hammerschlag 2000). The somaclonal variant 122-1 was more resistant to leaf spot and bacterial canker than moderately resistant cultivar 'Redhaven', and this resistance was retained after two years of cycles of propagation and establishment in the greenhouse. Besides bacterial diseases, Hashmi et al. (1995) reported the selection of somaclones in peaches from two cultivars, 'Sunhigh' (susceptible) and 'Redhaven' (moderately resistant), that were resistant to the root-knot nematode *Meloidogyne incognita.* All somaclones developed in this study showed different growth characteristics and root-knot nematode resistance.

*2.6 Pear (***Pyrus** *sp.)*

Pear (*Pyrus* sp.), cultivated mostly in temperate regions, is a fruit crop with high commercial value and consumer demand. Most cultivated pear (*P. communis*) rootstocks and cultivars are susceptible to a fungal disease fire blight caused by *Erwinia amylovora* (Nacheva et al. 2014). Viseur (1990) screened calluses derived from roots of the pear cultivar 'Durondeau' against the virulent strain of *E. amylovora.* Out of the selected four somaclonal variants from pear cultivar 'Durondeau', two were identified as tetraploids with reduced growth and lower susceptibility to fire blight. They reported increased ploidy levels in pears as a result of somaclonal variation. Nacheva et al. (2014) obtained somaclones of pear rootstock 'Old Home × Farmingdale' (OHF 333) by artificial inoculation with *E. amylovora,* which showed resistance to the fire blight. Besides disease resistance, shoot cultures derived from leaf calluses of two pear cultivars, 'Conference' and 'Abbé Fetel', and quince (*Cydonia oblonga*) cultivar 'BA 29', a widely-used dwarfing pear rootstock, generated NaCl-tolerant somaclones after selecting in vitro on a NaCl-containing medium (Marino and Molendini 2005). In wild pear (*P. pyraster*), Palombi et al. (2007) developed somaclones tolerant to Fe-chlorosis. They selected Fe-chlorosis tolerant plants by replacing in the culture medium Fe-EDTA by equimolar concentration of $FeSO_4$ with either $KHCO_3$ or $NaHCO_3$ for in vitro selection.

*2.7 Pineapple (***Ananas Comosus***)*

Pineapple is one of the important commercial tropical fruits valued due to its delicious taste, nutrition, and many processed fruit products like juices, jams or candies (Dhurve et al. 2021). In pineapple, somaclonal variation has been well characterized with the main emphasis on fruit quality and spine characteristics. In a few reports, variations resulting from micropropagation of pineapple had no benefit or produced non-commercial attributes. Soneji et al. (2002) reported somaclonal variations in proliferated shoots of pineapple cultivar 'Queen' developed from dormant axillary buds and maintained in liquid medium. They found a few variants with anthocyanin-rich leaves and spineless characteristics. However, several unusual phenotypic variants like albinos, white streaked shoots and shoots with elongated internodes were also observed. Pérez et al. (2011) characterized two pineapple somaclones (P3R5 and Dwarf) derived from in vitro culture of cultivar 'Red Spanish Pinar'. Both somaclones differed in fruit color, size, and length of plant generation cycle. In another study, Halim et al. (2018) induced somaclonal variation in pineapple var. MD2 through abscisic acid (ABA) and salinity stress. ABA and salinity stress generated dwarf variants in pineapple, which showed increased histone deacetylase (HDAC) activity. They also reported phenotypic reversion of the dwarf variants by removing the stress treatment, probably due to epigenetic modulation. Recently, Dhurve et al. (2021) obtained somaclones of pineapple cultivar 'Mauritius' derived from cultures maintained up to the eighth subculture. After field evaluation, they revealed massive variation in the fruit and yield traits and selected 12 superior lines for further evaluation trials and variety release. More recently, Zhang et al. (2022) developed a cold-tolerant variant of pineapple using low temperature (0 °C for 72 h) as a selective agent for in vitro selection. Compared to the non-selected control, the selected variant showed a higher survival rate under cold stress, increased proline content, and higher superoxide dismutase (SOD) activity.

2.8 *Strawberry (*Fragaria x ananassa*)*

Strawberry is another popular fruit crop in temperate and sub-tropical regions of the world. The production of strawberries is significantly hindered by many pathogens and abiotic stresses, especially salt stress (Dziadczyk et al. 2003; Biswas et al. 2009). In strawberry, several somaclonal variants have been selected regarding disease resistance and abiotic stress tolerance traits (Table 2).

Toyoda et al. (1991) demonstrated the existence of somaclonal variation in strawberry cultivar 'Hoko-Wase' for resistance to *Fusarium* wilt disease, caused by *F. oxysporum* f. sp. *fragariae*. They identified *Fusarium* wilt-resistant somaclonal variants by transplanting the plants developed from leaf-derived callus to the field severely infested with the pathogen. Similarly, exposure of cultures to the pathogen itself also generated strawberry somaclones resistant to *Verticillium dahliae* (Sowik et al. 2003) and *Colletotrichum acutatum* (Hammerschlag et al. 2006). In another investigation, Orlando et al. (1997) used purified pectic enzymes produced by *Rhizoctonia fragariae* as a selective agent for in vitro selection of strawberry plants resistant to *R. fragariae* and *Botrytis cinerea.*

Strawberry is considered a fruit crop with low tolerance to soil salinity. Dziadczyk et al. (2003) selected a clone, Pau/27, from strawberry cultivar 'Paula' cultured on a medium containing 200 mM NaCl. Subsequently, they obtained 18 clones of the second generation after the selection from self-pollinated Pau/27. Some genotypes selected from Pau/27 as salt resistant performed better than their parent.

More recently, Li et al. (2023) performed transcriptome and hormones analysis of elongated fruit somaclonal mutant, i.e., 'finger' strawberry, and compared it with cultivated 'Akihime' strawberry. Several auxin-related genes and transcription factors were differentially expressed in finger strawberries that might be responsible for elongated fruit mutant phenotype.

2.9 Other Species

Somaclonal variation has also been reported related to crop improvement traits in some other fruit crops like blackberry (McPheeters and Skirvin 1989), guava (Vos et al. 2000), kiwifruit (Caboni et al. 2003), mango (Jayasankar and Litz 1998), myrobalan plum (Pistelli et al. 2012), pomegranate (Sridhar and Harish Babu 2019), etc. (Table 2).

3 Conclusion

In vitro culture technologies offer an alternative means of crop improvement by induction of genetic variability and selection of novel variants through somaclonal variation. Since tissue culture procedures trigger this variation, thus, varieties developed through somaclonal variation do not have any socioethical hurdles and are not regulated as genetically modified organisms (GMO). Somaclonal variation offers an opportunity to induce DNA sequence changes and introduce new target traits of interest without altering the genetic combinations, particularly for crops with a narrow genetic base. However, crop improvement through somaclonal variation has some restrictions, such as epigenetic variations, selection of undesirable traits, and loss of regeneration potential. Therefore, the validation of genetically stable and desired variants through accurate detection methods is required to utilize the technique effectively. Many somaclonal variants have been selected in many fruit plants in past years using in vitro selection techniques. Many desirable characteristics, particularly disease resistance, abiotic stress tolerance, and fruit quality traits, have been obtained through somaclonal variation. The selected somaclones have been released as new cultivars in many agricultural crops for commercial exploitation; however, only a few in fruit plants like banana and *Citrus*. Therefore, more somaclonal variation-based research is needed in commercially important and underutilized fruit crops to select genetically stable variants and develop superior cultivars, especially linked with many agronomic and horticultural traits.

References

Bairu MW, Fennell CW, van Staden J (2006) The effect of plant growth regulators on somaclonal variation in Cavendish banana (*Musa* AAA cv. 'Zelig'). Sci Hortic 108:347–351

Bairu MW, Aremu AO, Van Staden J (2011) Somaclonal variation in plants: causes and detection methods. Plant Growth Regul 63:147–173

Ben-Hayyim G, Goffer Y (1989) Plantlet regeneration from NaCl-selected salt-tolerant callus culture of Shamouti oranges (*Citrus sinesis* L. Osbeck). Plant Cell Rep 7:680–683

Bhatti S, Jha G (2010) Current trends and future prospects of biotechnological interventions through tissue culture in apple. Plant Cell Rep 29:1215–1225

Biswas MK, Dutt M, Roy UK, Islam R, Hossain M (2009) Development and evaluation of in vitro somaclonal variation in strawberry for improved horticultural traits. Sci Hortic 122:409–416

Bradaï F, Pliego-Alfaro F, Sanchez-Romero C (2016) Somaclonal variation in olive (*Olea europaea* L.) plants regenerated via somatic embryogenesis: influence of genotype and culture age on phenotypic stability. Sci Hortic 213:208–215

Brar DS, Jain SM (1998) Somaclonal variation: mechanism and application in crop improvement. In: Jain SM, Brar DS, Ahloowalia BS (eds) Somaclonal variation and induced mutations in crop improvement. Kluwer Academic Publishers, Dordrecht, pp 15–37

Bulbarela-Marini JE, Gómez-Merino FC, Galindo-Tovar ME, Pastelín-Solano MC, Murguía-González J, Núñez-Pastrana R, Castañeda-Castro O (2023) Ratio of somaclonal variation and the phytohormonal content of *Citrus* × *latifolia* in three in vitro culture systems. J Plant Growth Regul 42:3356–3364

Caboni E, Anselmi S, Donato E, Manes F (2003) In vitro selection of *Actinidia deliciosa* clones tolerance to NaCl and their molecular and in vivo ecophysiological characterisation. Acta Hortic 618:77–83

Carra A, Sajeva M, Abbate L, Siragusa M, Pathirana R, Carimi F (2016) Factors affecting somatic embryogenesis in eight Italian grapevine cultivars and the genetic stability of embryo-derived regenerants as assessed by molecular markers. Sci Hortic 204:123–127

Chevreau E, Brisset MN, Paulin JP, James DJ (1998) Fire blight resistance and genetic trueness-to-type of four somaclonal variants from the apple cultivar greensleeves. Euphytica 104:199–205

da Silva RL, Ferreira CF, da Silva Ledo CA, de Souza EH, da Silva PH, de Carvalho Costa MAP, Souza FVD (2016) Viability and genetic stability of pineapple germplasm after 10 years of in vitro conservation. Plant Cell Tiss Org Cult 127:123–133

Dalla Costa L, Malnoy M, Gribaudo I (2017) Breeding next generation tree fruits: technical and legal challenges. Hort Res 4:17067

Deng ZN, Gentile A, Domina F, Nicolosi E, Tribulato E (1995) Selecting lemon protoplasts for insensitivity to *Phoma tracheiphila* toxin and regenerating tolerant plants. J Am Soc Hortic Sci 120:902–905

Dhekney SA, Basford AT, Chhatre VE, Rosenberg MB, Claflin C, Sessions SK, Li ZJ, Gray DJ (2020) *Vitis* spp. Grape. In: Litz RE, Pliego-Alfaro F, Homaza JI (eds.) Biotechnology of fruit and nut crops, 2nd edn. CAB International, Wallingford, pp 655–674

Dhurve L, Ajith Kumar K, Bhaskar J, Sobhana A, Francies RM, Mathew D (2021) Wide variability among the 'Mauritius' somaclones demonstrates somaclonal variation as a promising improvement strategy in pineapple (*Ananas comosus* L.). Plant Cell Tiss Org 145:701–705

Ding J, Deng XX, Zhang H, Cheng Y, Xu Q, Liu Q (2009) Somaclonal variation with early juice-sac granulation obtained by inter-specific protoplast fusion in citrus. J Hortic Sci Biotechnol 84:567–573

Donovan AM, Morgan R, Valobra-Piagnani C, Ridout MS, James DJ, Garrett CM (1994) Assessment of somaclonal variation in apple. I. Resistance to the fire blight pathogen *Erwinia amylovora*. J Hort Sci 69:105–113

Dziadczyk P, Bolibok H, Tyrka M, Hortynski JA (2003) In vitro selection of strawberry (*Fragaria* × *ananassa* Duch.) clones tolerant to salt stress. Euphytica 132:49–55

Evans DA (1989) Somaclonal variation—genetic basis and breeding applications. Trends Genet 5:46–50

Ferreira MD, Moura ÉR, Lino LS, Amorim EP, Santos-Serejo JA, Haddad F (2020) Selection of somaclonal variants of the cultivar 'Prata-Anã' for resistance to *Fusarium oxysporum* f. sp. *cubense* race 1. Rev Bras Frutic 42:e-620

Germana MA, Aleza P, Grosser JW, Dutt M, Wang N, Cuenca J, Kaur P (2020) Citrus biotechnology. In: Talon M, Caruso M, Gmitter FG (eds) The genus citrus. Woodhead Publishing, Cambridge, pp 171–192

Ghag SB, Shekhawat UK, Ganapathi TR (2014) Characterization of *Fusarium* wilt resistant somaclonal variants of banana cv. *Rasthali* cDNA-RAPD. Mol Biol Rep 41:7929–7935

Giménez C, de García E, de Enrech NX, Blanca I (2001) Somaclonal variation in banana: cytogenetic and molecular characterization of the somaclonal variant CIEN BTA-03. In Vitro Cell Dev Biol Plant 37:217–222

Grosser JW, Gmitter FG, Chandler JL (1997) Development of improved sweet orange cultivars using tissue culture methods. Proc Fla State Hort Soc 110:13–16

Halim NAA, Tan BC, Midin MR, Madon M, Khalid N, Yaacob JS (2018) Abscisic acid and salinity stress induced somaclonal variation and increased histone deacetylase (HDAC) activity in *Ananas comosus* var. MD2. Plant Cell Tiss Org Cult 133:123–135

Hammerschlag FA (1988) Selection of peach cells for insensitivity, to culture filtrates of *Xanthomonas campestris* pv. *pruni* and regeneration of resistant plants. Theor Applied Genet 76:865–869

Hammerschlag FA (1990) Resistance responses of plants regenerated from peach callus cultures to *Xanthomonas campestris* pv.*pruni*. J Amer Soc Hortic Sci 115:1034–1037

Hammerschlag FA (2000) Resistant responses of peach somaclone 122–1 to *Xanthomonas campestris* pv. *pruni* and to *Pseudomonas syringae* pv. *syringae*. HortScience 35:141–143

Hammerschlag FA, Ognjanov V (1990) Somaclonal variation in peach: screening for resistance to *Xanthomonas camperstris* pv. *pruni* and *Pseudomonas syringae* pv. *syringae*. Acta Hortic 280:403–408

Hammerschlag F, Garces S, Koch-Dean M, Stephanie R, Lewers K, Maas J, Smith BJ (2006) In vitro response of strawberry cultivars and regenerants to *Colletotrichum acutatum*. Plant Cell Tiss Org Cult 84:255–261

Hashmi G, Hammerschlag FA, Huettel RN, Krusberg LR (1995) Growth, development and response of peach somaclones to the root-knot nematode, *Meloidogyne incognita*. J Am Soc Hortic Sci 120:932–937

Hashmi G, Huettel R, Meyer R, Krusberg L, Hammerschlag F (1997) RAPD analysis of somaclonal variants derived from embryo callus cultures of peach. Plant Cell Rep 16:624–627

Helliot B, Madur D, Dirlewanger E, De Boucaud MT (2002) Evaluation of genetic stability in cryopreserved *Prunus*. In Vitro Cell Dev Biol Plant 38:493–500

Homhuan S, Kijwijan B, Wangsomnuk P, Bodhipadma K, Leung DWM (2008) Variation of plants derived from indirect somatic embryogenesis in cotyledon explants of papaya. Sci Asia 34:347–352

Hou BH, Tsai YH, Chiang MH, Tsao SM, Huang SH, Chao CP, Chen HM (2022) Cultivar-specific markers, mutations, and chimerisim of Cavendish banana somaclonal variants resistant to *Fusarium oxysporum* f. sp. *cubense* tropical race 4. BMC Genomics 23:470

Hwang SC, Ko WH (2004) Cavendish banana cultivars resistant to *Fusarium* wilt acquired through somaclonal variation in Taiwan. Plant Dis 88:580–588

Jain SM (2001) Tissue culture-derived variation in crop improvement. Euphytica 118:153–166

Jayasankar S, Litz RE (1998) Characterization of resistance in mango embryogenic cultures selected for resistance to *Colletotrichum gloeosporioides* culture filtrate and phytotoxin. Theor Appl Genet 96:823–831

Jayasankar S, Li Z, Gray DJ (2000) In vitro selection of *Vitis vinifera* 'Chardonnay' with *Elsinoe ampelina* culture filtrate is accompanied by fungal resistance and enhanced secretion of chitinase. Planta 211:200–208

Jayasankar S, Litz RE, Schnell RJ, Hernandez A-C (1998) Embryogenic mango cultures selected for resistance to *Colletotrichum gloeosporioides* culture filtrate show variation in random amplified polymorphic DNA (RAPD) markers. In Vitro Cell Dev Biol Plant 34:112–116
Kaeppler SM, Kaeppler HF, Rhee Y (2000) Epigenetic aspects of somaclonal variation in plants. Plant Mol Biol 43:179–188
Kamle M, Kumar P, Bajpai A, Kalim S, Chandra R (2014) Assessment of genetic fidelity of in vitro regenerated guava (*Psidium guajava* L.) plants using DNA based markers. N Z J Crop Hortic Sci 42:1–9
Karp A (1995) Somaclonal variation as a tool for crop improvement. Euphytica 85:295–302
Krishna H, Alizadeh M, Singh D, Singh U, Chauhan N, Eftekhari M, Sadh RK (2016) Somaclonal variations and their applications in horticultural crops improvement. 3 Biotech 6:54
Larkin PJ, Scowcroft WR (1981) Somaclonal variation a new source of variability from cell cultures for crop improvement. Theor Appl Genet 60:197–214
Lee SY, Su YU, Chou CS, Liu CC, Chen CC, Chao CP (2011) Selection of a new somaclone cultivar 'Tai-Chiao No. 5' (AAA, Cavendish) with resistance to *Fusarium* wilt of banana in Chinese Taipei. Acta Hortic 897:391–397
Li Z, Song X, Li H, Zhang Z, Zhang J (2023) Transcriptome and hormones analysis provide insights into elongated fruit somaclonal mutant in 'Akihime' strawberry. Sci Hortic 309:111608
Lin W, Xiao XO, Sun W, Liu S, Wu Q, Yao Y, Zhang H, Zhang X (2022) Genome-wide identification and expression analysis of cytosine DNA methyltransferase genes related to somaclonal variation in pineapple (*Ananas comosus* L.). Agronomy 12:1039
Litz RE, Padilla G (2012) Genetic transformation of fruit trees. In: Priyadarshan PM, Schnell RJ (eds) Genomics of tree crops. Springer, Berlin, pp 117–153
Marino G, Molendini L (2005) *In vitro* leaf-shoot regeneration and somaclone selection for sodium chloride tolerance in quince and pear. J Hort Sci Biotechnol 80:561–570
Martins M, Sarmento D, Oliveira MM (2004) Genetic stability of micropropagated almond plantlets, as assessed by RAPD and ISSR markers. Plant Cell Rep 23:492–496
McMeans O, Skirvin RM, Otterbacher A, Mitiku G (1998) Assessment of tissue culture-derived 'Gala' and 'Royal Gala' apples (*Malus* × *domestica* Borkh.) for somaclonal variation. Euphytica 103(2):251–257
McPheeters KD, Skirvin RM (1989) Somaclonal variation among ex vitro 'Thornless Evergreen' trailing blackberries. Euphytica 42:155–162
Miguel C, Marum L (2011) An epigenetic view of plant cells cultured in vitro somaclonal variation and beyond. J Exp Bot 62:3713–3725
Modgil M, Mahajan K, Chakrabarti SK, Sharma DR, Sobti RC (2005) Molecular analysis of genetic stability in micropropagated apple rootstock MM106. Sci Hortic 104:151–160
Modgil M, Guleria N, Ghani M, Sharma JN (2012) Identifying somaclonal variants of the apple rootstock Malling 7 resistant to white root rot. Sci Hortic 137:148–155
Molina AB, Sinohin VO, Fabregar EG, Ramillete EB, Loayan MM, Chao CP (2016) Field resistance of Cavendish somaclonal variants and local banana cultivars to tropical race 4 of *Fusarium* wilt in the Philippines. Acta Hortic 1114:227–230
Muhammad AJ, Othman RY (2005) Characterization of *Fusarium* wilt-resistant and *Fusarium* wilt-susceptible somaclones of banana cultivar rastali (*Musa* AAB) by random amplified polymorphic DNA and retrotransposon markers. Plant Mol Biol Report 23:241–249
Nacheva LR, Gercheva PS, Andonova MY, Panayotova DV, Dzhuvinov VT (2014) Somaclonal variation: a useful tool to improve disease resistance of pear rootstock 'Old Home × Farmingdale' (OHF 333) (*Pyrus communis* L.). Acta Hortic 1056:253–258
Neelakandan A, Wang K (2012) Recent progress in the understanding of tissue culture-induced genome level changes in plants and potential applications. Plant Cell Rep 31:597–620
Nookaraju A, Agrawal DC (2012) Genetic homogeneity of in vitro raised plants of grapevine cv. Crimson seedless revealed by ISSR and microsatellite markers. S Afr J Bot 78:302–306
Noormohammadi Z, Fazeli S, Sheidai M, Farahani F (2015) Molecular and genome size analyses of somaclonal variation in apple rootstocks Malling 7 and Malling 9. Acta Biol Szeged 59:139–149

Oh TJ, Cullis MA, Kunert K, Engelborghs I, Swennen R, Cullis CA (2007) Genomic changes associated with somaclonal variation in banana (*Musa* spp.). Physiol Plant 129:766–774

Orlando R, Magro P, Rugini E (1997) Pectic enzymes as a selective pressure tool for *in vitro* recovery of strawberry plants with fungal disease resistance. Plant Cell Rep 16:272–276

Palombi M, Damiano C (2002) Comparison between RAPD and SSR molecular markers in detecting genetic variation in kiwifruit (*Actinidia deliciosa* A. Chev). Plant Cell Rep 20:1061–1066

Palombi MA, Lombardo B, Caboni E (2007) In vitro regeneration of wild pear (*Pyrus pyraster* Burgsd) clones tolerant to Fe-chlorosis and somaclonal variation analysis by RAPD markers. Plant Cell Rep 26:489–496

Pathak H, Dhawan V (2012) ISSR assay for ascertaining genetic fidelity of micropropagated plants of apple rootstock Merton 793. In Vitro Cell Dev Biol Plant 48:137–143

Penna S, Jain SM (2023) Fruit crop improvement with genome editing, in vitro and transgenic approaches. Horticulturae 9:58

Pérez G, Mbogholi A, Sagarra F, Aragón C, González J, Isidrón M, Lorenzo JC (2011) Morphological and physiological characterization of two new pineapple somaclones derived from in vitro culture. In Vitro Cell Dev Biol Plant 47:428–433

Pérez G, Yanez E, Isidrón M, Lorenzo JC (2009) Phenotypic and AFLP characterization of two new pineapple somaclones derived from in vitro culture. Plant Cell Tiss Org Cult 96:113–116

Pistelli L, Iacona C, Miano D, Cirilli M, Colao MC, Mensuali-Sodi A, Muleo R (2012) Novel *Prunus* rootstock somaclonal variants with divergent ability to tolerate waterlogging. Tree Physiol 32:355–368

Popescu CF, Falk A, Glimelius K (2002) Application of AFLPs to characterize somaclonal variation in anther-derived grapevines. Vitis 41:177–182

Prado M, Rodriguez E, Rey L, González M, Santos C, Rey M (2010) Detection of somaclonal variants in somatic embryogenesis-regenerated plants of *Vitis vinifera* by flow cytometry and microsatellite markers. Plant Cell Tiss Org Cult 103:49–59

Rai MK (2021) Somaclonal variation in improvement of agricultural crops: recent progress. In: Srivastava DK, Thakur AK, Kumar P (eds) Agricultural biotechnology: latest research and trends. Springer, Singapore, pp 129–146

Rai MK (2023) Start codon targeted (SCoT) polymorphism marker in plant genome analysis: current status and prospects. Planta 257:34

Rai MK, Shekhawat NS (2014) Recent advances in genetic engineering for improvement of fruit crops. Plant Cell Tiss Org Cult 116:1–15

Rai MK, Kalia RK, Singh R, Gangola MP, Dhawan AK (2011) Developing stress tolerant plants through *in vitro* selection—an overview of the recent progress. Environ Exp Bot 71:89–98

Rai MK, Phulwaria M, Harish GAK, Shekhawat NS, Jaiswal U (2012) Genetic homogeneity of guava plants derived from somatic embryogenesis using SSR and ISSR markers. Plant Cell Tiss Org Cult 111:259–264

Rai MK, Shekhawat JK, Kataria V, Shekhawat NS (2017) *De novo* assembly of leaf transcriptome, functional annotation and genomic resources development in *Prosopis cineraria*, a multipurpose tree of Indian Thar Desert. Plant Gene 12:88–97

Rai MK, Rathour R, Kaushik S (2021) Recent advances in transcriptomics: an assessment of recent progress in fruit plants. In: Kumar A, Kumar R, Shukla P, Patel HK (eds) Omics technologies for sustainable agriculture and global food security (Vol II). Springer, Singapore, pp 95–122

Rani V, Raina SN (2000) Genetic fidelity of organized meristem derived micropropagated plants: a critical reappraisal. In Vitro Cell Dev Biol Plant 36:319–330

Ray T, Dutta I, Saha P, Das S, Roy SC (2006) Genetic stability of three economically important micropropagated banana (*Musa* spp.) cultivars of lower Indo-Gangetic plains, as assessed by RAPD and ISSR markers. Plant Cell Tiss Org Cult 85:11–21

Roostika I, Khumaida N, Ardie SW (2015) RAPD analysis to detect somaclonal variation of pineapple in vitro cultures during micropropagation. Biotropia 22(2):109–119

Rosati P, Mezzetti B, Ancherani M, Foscolo S, Predieri S, Fasolo F (1990) In vitro selection of apple rootstock somaclones with *Phytophthora cactorum* culture filtrate. Acta Hortic 280:409–416

Sahijram L (2015) Somaclonal variation in micropropagated plants. In: Bahadur B, Rajam MV, Sahijram L, Krishnamurthy KV (eds) Plant biology and biotechnology. Springer, New Delhi, pp 407–416

Sales EK, Butardo NG (2014) Molecular analysis of somaclonal variation in tissue culture derived bananas using MSAP and SSR markers. Int J Biol Vet Agric Food Eng 8:63–610

Savita, Gurdeep SV, Avinash N (2011) In vitro selection of calli of *Citrus jambhiri* Lush. for tolerance to culture filtrate of *Phytophthora parasitica* and their regeneration. Physiol Mol Biol Plant 17:41–47

Schellenbaum P, Mohler V, Wenzel G, Walter B (2008) Variation in DNA methylation patterns of grapevine somaclones (*Vitis vinifera* L.). BMC Plant Biol 8:78

Scorza R, Petri C (2009) Peach. In: Kole C, Hall TC (eds) A compendium of transgenic crop plants: transgenic temperate fruits and nuts. Blackwell Publishing, Oxford, pp 79–92

Sharma HC, Crouch JH, Sharma KK, Seetharama N, Hash CT (2002) Applications of biotechnology for crop improvement: prospects and constraints. Plant Sci 163:381–395

Siragusa M, Carra A, Salvia L, Puglia A, De Pasquale F, Carimi F (2007) Genetic instability in calamondin (*Citrus madurensis* Lour.) plants derived from somatic embryogenesis induced by diphenylurea derivatives. Plant Cell Rep 26:1289–1296

Smulders M, de Klerk G (2011) Epigenetics in plant tissue culture. Plant Growth Regul 63:137–146

Soneji JR, Rao PS, Mhatre M (2002) Somaclonal variation in micropropagated dormant axillary buds of pineapple (*Ananas comosus* L., Merr.). J Hortic Sci Biotechnol 77:28–32

Sowik I, Wawrzyńczak D, Michalczuk L (2003) Ex vitro establishment and greenhouse performance of somaclonal variants of strawberry selected for resistance to *Verticillium dahliae*. Acta Hortic 616:497–500

Sridhar R, Harish Babu BN (2019) A novel pomegranate (*Punica granatum* L.) variant with high anthocyanin content. Acta Hortic 1255:149–152

Toyoda H, Horikoshi K, Yamano Y, Ouchi S (1991) Selection of *Fusarium* wilt disease resistance from regenerant derived from callus of strawberry. Plant Cell Rep 10:167–170

Uma S, Kumaravel M, Backiyarani S Saraswathi SM, Dura P, Karthic R (2021) Somatic embryogenesis as a tool for reproduction of genetically stable plants in banana and confirmatory field trials. Plant Cell Tiss Org Cult 147:181–188

Varshney RK, Roorkiwal M, Sorrells ME (2017) Genomic selection for crop improvement: an introduction. In: Varshney RK (ed) Genomic selection for crop improvement: new molecular breeding strategies for crop improvement. Springer, New York, pp 1–6

Verma S, Modgil M, Patidar S (2021) *In vitro* screening of apple rootstock MM106 somaclones with *Phytophthora cactorum* culture filtrate. J Plant Pathol 103:231–240

Vidal MDC, De García E (2000) Analysis of a *Musa* spp. somaclonal variant resistant to yellow Sigatoka. Plant Mol Biol Rep 18:23–31

Viseur J (1990) Evaluation of fire blight resistance of somaclonal variants obtained from the pear cultivar 'Durondeau.' Acta Hortic 273:275–284

Vos JE, Schoeman MH, Berjak P, Watt MP, Toerien AJ (2000) In-vitro selection and commercial release of guava wilt resistant rootstocks. Acta Hortic 513:69–75

Vuylsteke DR, Swenne R, De Langhe EA (1991) Somaclonal variation in plantains (*Musa* spp. AAB group) derived from shoot tip culture. Fruits 46:429–439

Vuylsteke D, Swennen R, De Langhe E (1996) Field performance of somaclonal variants of plantain (*Musa* spp., AAB group). J Am Soc Hortic Sci 121:42–46

Wang X, Ke L, Wang S, Fu J, Xu J, Hao Y, Kang C, Guo W, Deng X, Xu Q (2022) Variation burst during dedifferentiation and increased CHH-type DNA methylation after 30 years of in vitro culture of sweet orange. Horti Res 9:uhab036

Zhang YF, Xu ZQ, Xie T, Zhang W, He YH, Liu CY (2022) In vitro selection and identification of a cold-tolerant variant in pineapple (*Ananas comosus*). Hortic Environ Biotech 63:275–286

Somaclonal Variation in Conifers

Mostafa Khoshhal Sarmast, Zahra Rezaei Ghaleh, and Mehdi Alizadeh

Abstract Conifers are a diverse and important group of seed-bearing gymnosperm plants that include commercially valuable species, such as cedars, firs, cypresses, junipers, pines, and redwoods, to name a few important constituents. By studying diverse facets of conifer tissue culture, our understanding of the key factors responsible for enhancing the quality and quantity of production might potentially be elucidated in this group. Furthermore, this can prevent unnoticed and unwanted variations in regenerated offspring. The plant tissue culture procedure, the composition of the culture media, the type and origin of explants, or the number of subculture passages, may all contribute to in vitro-induced variation in conifers. While most somaclonal variation will likely not be useful in terms of being applicable to conifer breeding purposes, it is also possible that there may be as-of-yet undiscovered changes that may be useful for conifer breeding. This will necessitate a combination of phenotypical analyses, molecular marker expression, cytogenetic surveys, hypo/hypermethylation assessments, acetylation percentage assessments, and a general consideration of the involved processes, which together may help to control the frequency of tissue culture-induced variation.

Keywords Micropropagation · Subculture duration · Plant growth regulators · Molecular markers · Somaclonal variation

1 Introduction

The seed plants include five living groups: the conifers, the cycads, *Ginkgo*, the gnetophytes, and the angiosperms. The four first groups comprise the gymnosperms, which are collectively characterized by an exposure of their ovules during the pollination process. Conifers, arguably the most varied subgroup of the gymnosperms, are a group of commercially valuable cone-bearing seed plants. Typically, conifers

M. K. Sarmast (✉) · Z. R. Ghaleh · M. Alizadeh
Department of Horticultural Science, Faculty of Plant Production, Gorgan University of Agricultural Sciences and Natural Resources, 49138-43464 Basij, SQ, Gorgan, Golestan, Iran
e-mail: mkhsarmast@gau.ac.ir

C. Sánchez-Romero (ed.), *Somaclonal Variation: Basic and Practical Aspects*,
https://doi.org/10.1007/978-3-031-51626-9_7

have needle-shaped evergreen leaves, male and female unisexual flowers, and seeds attached to the scales of a woody, bracted cone (Bhardwaj et al. 2020). All presently extant conifers are perennial woody plants with botanically secondary growth. The vast majority of conifers are trees, though a few species are shrubs. Taxonomically, they comprise eight families, namely Taxaceae, Pinaceae, Cupressaceae, Araucariaceae, Cephalotaxaceae, Phyllocladaceae, Sciadopityaceae and Podocarpaceae, which include 70 genera and 630 species (Bhardwaj et al. 2021). Specifically, the cedars, Douglas-firs, cypresses, firs, junipers, kauri, larches, pines, hemlocks, redwoods, spruces, and yews are the best-known examples of conifers (Farjon and Page 1999).

Conifers are incredibly important with respect to the global industrial wood supply and they are considered as being one of the world's most important resources for timber. Specifically, conifers alone have been associated with supplying over 50% of the world's timber harvest (Cooper 2003). The second-most economically significant use of conifers is in horticulture, the history of which has been well documented in previous publications (Flanagan 2003). Additionally, there is a prosperous industry in many countries producing "dwarf conifers" suitable for small gardens and even high-rise balconies. Moreover, conifers are often considered to be the backbone of the winter landscape, although many deciduous plants can add interest to the winter landscape in the form of colorful bark, unique branching patterns, and bright berries and seed capsules. This is primarily because conifers are the largest group of plants that hold their foliage in the cold winter months, and so conifers can be very useful, long-lasting design elements in a garden or landscape. Furthermore, this means that conifers can be useful as decorative trees that provide year-round cover in the landscape. Conifers can also be used in hedgerows on their own, or combined with deciduous options in mixed hedgerows. As previously mentioned, coniferous trees can also come in dwarf or ground-growing varieties, both of which can be useful in mixed beds, borders, or pots. In addition to their myriad of uses in a horticultural setting, conifers have long been recognized for their therapeutic potential in the treatment of different disorders. Alkaloids, terpenes, and polyphenols are the most abundant naturally occurring phytochemicals in these plants (Ghorbanzadeh et al. 2021). Bhardwaj et al. (2021) have already provided an overview of the phytochemistry and related commercial products obtained from conifers.

The main methods used to propagate conifers include seeds, cuttings, and grafting, as well as micropropagation techniques (Sarmast 2016). Seed propagation is the principal method and it is important for the mass production of plants for forestry. Several methods of clonal propagation have been practiced with conifers, including rooting of cuttings, grafting, and tissue-culture techniques (Jinks 2003). Currently, grafting is not used for the mass-propagation of conifers (Wendling et al. 2014).

Plant tissue culture is one of the most effective tools in modern plant science research. The use of this technique has commercial value and it is effectively used in many breeding programs (Neelakandan and Wang 2012). The commercial micropropagation industry is highly relevant to agroindustry, as tissue culture techniques have several elite practical applications leading to crop improvement, which are not currently possible using more conventional methods.

Tissue culture methods offer an alternative means of plant vegetative propagation. Different methods are available for propagation of plants in vitro, including multiplication of shoots from axillary buds, adventitious organogenesis and somatic embryogenesis (SE) (George and Debergh 2008).

SE has been previously determined to have promising attributes that may help to scale up the production of improved varieties in conifers (Lelu-Walter et al. 2013; Klimaszewska et al. 2015). As such, the application of SE on an industrial scale in conifers has long been a desired goal because it has the potential to increase forest productivity dramatically. SE in conifers can also benefit tree improvement programs, but the challenge remains to achieve this on a cost-effective, industrial scale for many conifer species (Bonga et al. 2018). Nevertheless, for some species, primarily *Picea glauca*, *Pinus radiata*, *Sequoia sempervirens*, and *Pinus taeda*, industrial applications have been discovered, and, presumably, these techniques could be used to refine our understanding of the specific growth characteristics of other commercially valuable conifers (Boulay 1987). Moreover, SE has been used to exploit synseed production in conifers, specifically in *Picea glauca* (Sutton et al. 2004) and *Pinus radiata* (Aquea et al. 2008).

However, the high cost of production, the risk of culture contamination, the risk of explant browning, hyperhydricity (production of vitrified plants), and low survival rates of tissue culture-raised plantlets during the hardening stage are the most common problems associated with this technology (Bhojwani and Dantu 2013; Sarmast et al. 2012b, 2012c). In addition, the survival and yield performance of plantlets in the field after transplanting are not guaranteed and are not uniform (Alizadeh et al. 2010). Furthermore, in commercial micropropagation, it is mandatory to regularly check the clonal fidelity or genetic uniformity of the micropropagated plantlets, as induced variations of any kind may multiply very fast and lead to loss of the chief characteristics of the parent genotypes (Alizadeh and Singh 2009).

2 Somaclonal Variation: Definition and Concept

The term *somaclonal variation* (*soma* = vegetative, *clone* = matching copy) was pioneered by Larkin and Scowcroft (1981), and refers to the genetic and epigenetic changes in *in-vitro* cell cultures, regenerated plants and their progenies. Thus, somaclonal variation yields genotypes which are not true-to-type from their donor mother plants (Sarmast et al. 2012a; Mahdavi-Darvari et al. 2015). *Gametoclonal variation* is other term for explaining variation observed in plants regenerated from gametic cells (Bhojwani and Dantu 2013). In addition to these, *calliclones* and *protoclones* are terms used to describe somaclones that arise from callus and protoplast cultures, respectively. The occurrence of somaclonal variation in cultures, depending on the goal of a particular research project, may be considered either beneficial or detrimental. Clearly, the occurrence of somaclonal variation is often considered to be undesirable when the objective of plant regeneration is to produce high numbers of true-to-type plantlets; however, somaclonal variation may also potentially yield desirable phenotypes or genotypes for the development of new plant varieties. Ideally, these new somaclones may have enhanced tolerance to biotic and/or abiotic stressors or even a higher production rate of industrially valuable phytochemicals. Thus, changes due to somaclonal variation are rarely used in crop improvement programs because of its uncrontrollable nature and the loss of desirable traits in crop plants, but several commercial cultivars have been introduced through tissue culture induction due to the creation of novel variants (Karp 1994; Katiyar and Chopra 1995; Sarmast 2016).

Unfortunately, due to woody plant species frequently have a long lifespan, somaclonal variation is often considered to be a drawback for plant tissue culture, because occasional changes go unnoticed until very late in development or even until the following generation of offspring can be analyzed (Fourré et al. 1997). Therefore, there is a great need to develop novel approaches that can detect these variations as quickly as possible during the in vitro regeneration process.

Somaclonal variation can manifest as mitotically and meiotically heritable changes, but some of these changes can also be potentially reversible (Kaeppler et al. 2000; Us-Camas et al. 2014).

3 Factors Involved in Somaclonal Variation

The causes and origins of somaclonal variation are often related to intrinsic and extrinsic issues in plant tissue culture. Genotype, the presence of chimeras, and the origin of the explants are among the internal factors influencing genetic variation in regenerants. Genetic variation within the somatic cells obtained from the original explants is one of the causes for the observation of somaclonal variation in plants regenerated from in vitro culture (D'Amato 1985). However, more important for the induction of somaclonal variation are the specific conditions of the in vitro culture. Some of the key external factors influencing the genetic stability of tissue culture regenerants involve the in vitro culture procedure, prolonged culture maintenance, the number of subculture passages, and the culture conditions (Haisel et al. 2001), including the types of minerals that may be present, the type of plant growth regulators (PGRs) (mainly auxins and cytokinins), and the concentration of those PGRs (Sarmast 2016).

Information about somaclonal variation in conifers is summarized in Table 1.

3.1 Genotype, Type of Explant and In Vitro Procedure

Somaclonal variation largely depends on the type of the tissue source (Bairu et al. 2011) and the genotype (Gulyaeva et al. 2020). Furthermore, these factors can respond differently when the plants are exposed to stressful conditions. Different types of explants are sourced from the stem of juvenile or mature conifer trees,

Table 1 Selected works on somaclonal variation in conifers species

Plant	Explants source	Source of variation	Detection method	Presence of variation (±)	References
Abies alba Mill	ECL	Nitrogen organic source (glutamine and casein hydrolysate)	Cytological study	+	Roth et al. (1997)
Abies alba Mill	SE	PGR	RAPD	+	Nawrot-Chorabik (2009)
Abies cephalonica	ECL	Cryopreservation	RAPD	-	Aronen et al. (1999)
Abies cephalonica Loud	ECL	Cryopreservation	RAPD	–	Krajňáková et al. (2011)
Araucaria excelsa R. Br. var. *glauca*	Orthotropic stem	PGR	Morphological and RAPD analysis	+	Sarmast et al. (2012a)
Cedrus atlantica and *Cedrus libani*	Microcuttings	Explant, genotype	Reversed-phase HPLC analysis and RAPD	+	Renau-Morata et al. (2005a)
Cedrus atlantica and *Cedrus libani*	Juvenile and mature explants	Explant, PGR	Morphological and RAPD analysis	–	Renau-Morata et al. (2005b)
Chamaecyparis pisifera Sieb. & Zucc	SE	Somatic embryogenesis micropropagation	RAPD	–	Hosoi et al. (2015)
Juniperus phoenicea L	Microcuttings with axillary buds	Number of subcultures, PGR	Flow cytometry techniques	–	Loureiro et al. (2007)
Larix gmelinii var. *principis-rupprechtii*	ECL	Long-time cryopreservation		–	Xiaoyi et al. (2021)
Larix sibirica	ECL	Duration in culture	FISH and morphometric analyses of karyotypes, RAPD	+	Goryachkina et al. (2018)

(continued)

Table 1 (continued)

Plant	Explants source	Source of variation	Detection method	Presence of variation (±)	References
Larix sibirica Ledeb	Megagametophytes	PGRs	Cytological and cytogenetic analysis, genotyping of 11 microsatellite loci	+	Krutovsky et al. (2014)
Picea abies (L.) Karst	Embryogenic clones	Embryogenic culture	Chromosome counting and RAPD	+	Fourré et al. (1997)
Picea abies (L.) Karst and *P. omorika* (Pančić) Purk	SE	Cryopreservation, PGR	Microsatellite markers	+	Hazubska-Przybył and Dering (2017)
Picea glauca Moench	ECL	Duration in culture	RAPD	+	Isabel et al. (1996)
Picea glauca (Moench) Voss	Embryogenic clones	Long-time cryopreservation	RAPD	+	De Verno et al. (1999)
Picea glauca and *Picea mariana*	SE	Explant, duration in culture	Morphology, chromosome count	+	Tremblay et al. (1999)
Pinus nigra Arn	ECL	Embryogenic culture, duration of culture	Morphology, Genomic DNA methylation analysis, HPCE analysisand polyamine analysis	+	Noceda et al. (2009)
Pinus pinaster (Ait.)	SE	5-azaC	HPLC, HPCE analysis and MSAP	+	Klimaszewska et al. (2009)
Pinus pinaster Ait	SE	PGRs	Flow cytometric and morphological analyses	+	Marum et al. (2009b)
Pinus radiata	ECL	Cryopreservation	Microscopic study	+	Lineros et al. (2018)

(continued)

Table 1 (continued)

Plant	Explants source	Source of variation	Detection method	Presence of variation (±)	References
Pinus resinosa	Embryonic explants	PGR	Isoenzyme analysis	–	Noh et al. (1988)
Pinus silvestris	ECL	Cryopreservation	Microscopical observation and RAPD	–	Häggman et al. (1998)
Pseudotsuga menziesii	SE	PGR	Histological-cytological observations and protein identification by mass spectrometry	+	Lelu-Walter et al. (2018)
Sequoia sempervirens	Basal sprouts or branch segments	PGR	Morphological study	–	Boulay (1977)

Embryogenic cell lines *ECL*, somatic embryo *SE*, random amplified polymorphic DNA *RAPD* analysis, fluorescent in situ hybridization *FISH*, methylation-sensitive amplification polymorphism *MSAP* immature megagametophytes, mature and immature zygotic embryos, cotyledons, and even the tips of adult conifer tree shoots and roots. In early reports, the plantlets originated from axillary buds of *Pseudotsuga menziesii* showed very few changes in appearance in the first year of field growth. Three or 4 years after the initial field planting, the initially observed variability was completely lost and plantlets began showing normal growth and vigor (Boulay 1976). *Araucaria excelsa* R. Br. produced from stem explant-derived axillary bud exhibited a high affinity compared to their donor plants, according to RAPD analysis. This was particularly true in culture media supplemented with low concentrations of cytokinins (Sarmast et al. 2012a, 2012c). This observation is in accordance with the fact that the lowest amount of variability is observed when axillary bud cultures, shoot tip cultures, and meristem cultures are used. Commercial micropropagation of *Pinus taeda*, *Pinus pinaster*, and *Pinus radiata* adventitious buds isolated from excised cotyledons did not lead to somaclonal variation (David et al. 1978; Aitken-Christie and Thorpe 1984; Boulay 1987). Meanwhile, callus-mediated somatic embryogenesis originated from protoplasts and adventitious shoots was previously demonstrated to induce somaclonal variation. The highest degree of variation was observed in callus cultures; however, in roots and leaves, not all the observed morphological differences within micropropagated plants were due to somaclonal variation. For example, in vitro propagation of *Pinus taeda* through adventitious budding produced 50% less shoot growth than those that were conventionally propagated. The authors of that study concluded that the absence of a taproot and fewer branched roots in micropropagated *P. taeda* caused changes to the innate ion absorption capacity of the plant, which ultimately lead to a suppression of in vitro plantlet growth (Boulay 1976).

In contrast, callus-mediated somatic embryogenesis can lead to somaclonal variation, especially in long-term cultures as observed in *Picea glauca* (Isabel et al. 1996), *P. mariana* and *P. glauca* (Tremblay et al. 1999), *P. abies* and *P. omorika* (Hazubska-Przybył and Dering 2017), and *Larix sibirica* (Goryachkina et al. 2018). The occurrence of somaclonal variation in initial stages of somatic embryogenesis process is higher mostly because of the vigorous growth of genetically heterogeneous callus tissue (Gulyaeva et al. 2020).

It is generally accepted that the degree of variability created from different explants can be significantly different (Skirvin et al. 1994). Meanwhile, some authors have highlighted the genotype-dependency of somaclonal variation in conifers, whereby different conifer cell lines showed variable somaclonal changes. For example, in a number of previous studies, variation was observed in 2 out of 7 *Picea glauca* cell lines (Isabel et al. 1996), 2 out of 17 *Pinus pinaster* cell lines (Marum et al. 2009a), 10 out of 19 *Picea mariana* cell lines (Tremblay et al. 1999), and 6 out of 10 *Picea glauca* cell lines (Tremblay et al. 1999). Gulyaeva et al. (2020) also shown the instability in the chromosome number ($2n = 24$–30) during the course of somatic embryogenesis among different embryogenic cell lines of *Larix sibirica* Ledeb.. Indeed, the formation of a callus in the culture medium has a strong relationship with the manifestation of somaclonal variation, but the induction of variation is not restricted to the existence of a callus phase (Skirvin et al. 1994). The original tissue source may have somatic mutations that can cause changes in the genome during the de-differentiation and re-differentiation processes (Lee and Phillips 1988). This is especially important in breeding programs, where plant material gained from in vitro culture or cryopreserved samples must be genetically stable.

3.2 PGRs and Subculture Duration

Long-term maintenance under stressful in vitro culture conditions has been previously theorized to be a possible factor of detected somaclonal variation in regenerated plants, especially when high concentrations of PGRs are added to media (Sahavacharin 1982). It has been reported that the successive subcultures of mature *Sequoia sempervirens* axillary buds cultivated on media supplemented with cytokinin influenced the mean rooting time, growth vigor, and plagiotropic habit (de la Goublaye de Nantois 1981). Even one year after transferring the plantlets to soil, the mean weight of stems, the mean number of orthotropic shoots, and the rooting ability of cuttings taken from in vitro copies were observed to have been influenced. In another report on *Sequoia sempervirens*, MS media supplemented with N6-benzylaminopurine (BA) at 0.5–1 mg/l and *a*-naphthalene acetic acid (NAA) at 0.01 mg/l resulted in phenotypic differences when basal sprouts or branch segments were used as explants for clonal micropropagation (Boulay 1977).

The concentration of PGRs in tissue culture has been clearly shown to influence the rate of somaclonal variation observed and it has been indicated that increasing the PGR concentration increases variation (Sarmast 2016). Indeed, a high concentration of cytokinins, especially BA, in the tissue culture medium has been shown to be effective in inducing genetic variation. This has been confirmed in different conifers species, such as *A. excelsa* R. Br., *A. cunninghamii* Ait. and *A. alba* (Burrows et al. 1988, Nawrot-Chorabik 2009; Sarmast et al. 2012c). In *A. excelsa* R. Br., MS media supplemented with a high concentration of cytokinin resulted in abnormal axillary shoot growth that exhibited yellow coloration and suppressed growth rate (Sarmast et al. 2012c). The same result was reported on *A. cunninghamii* Ait., in which BA concentrations of 1 and 10 μM led to the development of distorted buds and inhibition of bud expansion, respectively (Burrows et al. 1988). As previously mentioned, the type of cytokinin hormones has been shown to influence somaclonal variation differently (Nawrot-Chorabik 2009).

Auxins can induce somaclonal variation via hypermethylation. It has been previously stated that changes in the methylation of cytosine through the production of ethylene is due to the inclusion of an uncontrolled amount of 2,4-D in the culture media, thus leading to genetic abnormalities during the course of somatic embryogenesis (Ahmed et al. 2004; Mohanty et al. 2008; Us-Camas et al. 2014; Shemer et al. 2015).

Of course, it cannot always be said with certainty that this increase in hormone concentration always leads to somaclonal variation (Ahmed et al. 2004; Venkatachalam et al. 2007).

The genetic changes revealed by microsatellite markers in cotyledonary-stage somatic embryos of *P. abies* and *P. omorika* could be due to a long cultivation period of embryonic tissues in media supplemented with picloram/BA, or possibly because of the stressful conditions induced by a specific osmoticum or the presence of ABA during somatic embryo maturation (Hazubska-Przybył and Dering 2017). Indeed, it has been reported that high ABA concentration (higher than 60 μM) induces developmentally abnormal somatic embryos in *Pseudotsuga menziesii* (Lelu-Walter et al. 2018). Somatic embryo-derived plantlets of *Picea glauca* Moench. exhibited genetic variability after long-term in vitro culture when analyzed with RAPD markers (Isabel et al. 1996). Similarly, long-term maintenance of embryogenic cell lines of *Larix sibirica* resulted in a high degree of observed mixoploidy (Goryachkina et al. 2018).

Nevertheless, it cannot be said with certainty that this increase in hormone concentration always leads to somaclonal variation (Ahmed et al. 2004; Venkatachalam et al. 2007).

3.3 Cryopreservation

Cryopreservation is a method whereby structurally intact living cells and tissues are frozen at very low temperatures to preserve that material for an extended time. In the context of tissue culture, the biological material is initially exposed to different concentrations of sucrose, maltose, sorbitol, dimethyl sulfoxide (DMSO), and/or proline to prevent cells from perishing inside the liquid nitrogen in which they are stored for preservation, at a temperature of − 196 °C. The effects of cryopreservation on somaclonal variation have proven to be controversial, to some extent. For example, Krajňáková et al. (2011) and Häggman et al. (1998) did not observe any somaclonal variation after cryopreservation of *Abies cephalonica* and *Pinus silvestris* embryogenic cell lines, respectively; however, Aronen et al. (1999) reported great genetic changes in embryogenic cultures of *Abies cephalonica* treated with DMSO (Table 1). It has been also determined that embryogenic tissues exposed to PGD (a mixture of polyethylene glycol, glucose and DMSO) did not result in changes in the DNA fragments pattern during cryopreservation in the way that tissues exposed to DMSO did (Aronen et al. 1999; Krajňáková et al. 2011). It has also been reported that cryopreservation of embryonic tissues followed by long-term exposure to picloram or BA most likely increases the sensitivity of that tissue to stressful conditions and ultimately increases the chance of somaclonal variation in conifers (Hazubska-Przybył and Dering 2017). Xiaoyi et al. (2021) also reported that the recovery of embryonic calliclones of *Larix gmelinii* var. *principis-rupprechtii* can be influenced by the specific culture conditions. Moreover, pretreatment of the embryogenic cell lines before being cryopreserved was effective in mitigating any damage or morphological changes observed in *Pinus radiata* after regeneration (Lineros et al. 2018).

Based on the aforementioned studies, the genetic variation after cryopreservation appears to largely depend on the cryoprotectant utilized. Therefore, utilizing a combination of different cryoprotectants might be advantageous in mitigating the cryoprotectant-mediated genetic variation of embryogenic cell lines (Gulyaeva et al. 2020).

4 Detection of Somaclonal Variation

Somaclonal variation detection methods were outlined in detail by Bairu et al. (2011). In this study, we primarily focused on coniferous species. Firstly, the detection of somaclonal variation was executed following non-molecular approaches, such as phenotypic and cytological analyses. Soon thereafter, other methods, such as analyses of DNA, isozyme and protein patterns and epigenetic modifications became available and began being utilized worldwide (Fourré 2000).

The unreliability of time-dependent somaclonal variation-induced morphological changes (leaf morphology, growth habit, etc.), make them less likely to be used for the evaluation of somaclonal variation. This is especially true in the very beginnings of growth, since they are affected by numerous environmental variables and growth stages. Furthermore, the long juvenile phase of conifer species obscures the elucidation of somaclonal variation, particularly when growth habit is a factor of interest. In contrast, variation in gene encoding plastid enzymes and the associated transcription factors often results in a change to chloroplastic pigments (Bairu et al. 2011), which is visually detectable. In the following section we will delineate some of the techniques utilized for the detection of unseen variation during in vitro regeneration of conifers.

4.1 Cytogenetic Survey

Structural abnormalities and ploidy level changes (Hao and Deng 2002), chromosomal breakage across the centromeric and heterochromatin regions (Kaeppler et al. 2000), and chromosomal rearrangement (Krutovsky et al. 2014) are some of the mechanisms which have been previously proposed to be involved in somaclonal variation.

Indeed, the occurrence of chromosomal changes in in vitro cultures has been reported in several conifer species (Salajova and Salaj 1992; von Aderkas and Anderson 1993; Pattanavibool et al. 1995; O'Brien et al. 1996; Roth et al. 1997; von Aderkas et al. 2003). In this line, Loureiro et al. (2007) reported that axillary buds developed from microcutting explants of *J. phoenicea* L. kept their ploidy stability when compared with their corresponding mother plant by flow cytometry techniques after 4 subcultures. The same techniques were used by Marum et al. (2009b) to evaluate the genetic stability during the several steps of somatic embryogenesis in *Pinus pinaster* Ait. No significant differences in DNA ploidy were detected among the most frequently observed abnormal phenotypes, embryogenic cell lines, zygotic and normal somatic embryos, and somatic embryogenesis-derived plantlets. Furthermore, in vitro–mediated aneuploidy has been reported in *Pinus nigra* Arn. (Salajova and Salaj 1992), *Picea abies* (Fourré et al. 1997), *Abies alba* (Roth et al. 1997), *Picea glauca* (Tremblay et al. 1999), *Picea mariana* (Tremblay et al. 1999), *Larix decidua* (Von Aderkas et al. 2003), *Larix sibirica* Ledeb. (Krutovsky et al. 2014), and *Larix sibirica* (Goryachkina et al. 2017). Many chromosomal rearrangements and high levels of aneuploidy were also found in *Larix sibirica* Ledeb. embryogenic cell lines initiated from megagametophyte explants (Krutovsky et al. 2014). In addition to morphological changes, *Picea glauca* and *Picea mariana* plants derived from somatic embryogenesis showed different rates of aneuploidy and polyploidy variation after 5 years of in vitro culture and acclimatization. The regenerated aneuploid plants were morphologically dwarf and had stems with a larger diameter (Tremblay et al. 1999). Another study utilizing 5-year-old embryogenic cell lines of *Picea abies* showed variation in chromosome number, including monoploidy, trisomy, double trisomy, and tetraploidy (Fourré et al. 1997).

Possible mechanisms underlying somaclonal variation have been addressed by several researchers. Some reports found that in some observed abnormal dwarf *Picea* spp., the chromosome number (aneuploidy) had changed (Fourré et al. 1997; Tremblay et al. 1999).

4.2 Epigenetic and Methylation Response

Epigenetics includes heritable and non-heritable changes not due to modifications in the DNA sequence, but which that can affect the pattern of gene expression (Amiri Roudbar et al. 2015). The major epigenetic changes include DNA methylation, nucleosome remodeling, histone modification, and RNA-mediated targeting (Amiri Roudbar et al. 2015).

DNA methylation is one of the events associated with epigenetic diversity that occurs abundantly in eukaryotes (Ford 2013; Piccolo and Fisher 2014). DNA methylation is an important epigenetic change that is highly dependent on cytosine- and guanine–rich regions; which form so-called CpG islands (Du et al. 2012). Several previously published studies have shown that DNA methylation in plants occurs in these CpG islands, and that most methylation happens in the CG coding regions, as compared to the CHG and CHH sequences (Karim et al. 2016; Law and Jacobsen 2010). This statement was also confirmed in *Arabidopsis*, where the most commonly methylated sequence is CG, and this was found to be true 24% of times (Cokus et al. 2008). DNA methylation also affects transposable element (TE) silencing and the regulation of gene expression (Wu and Zhang 2014).

DNA methylation changes can be one of the several observable problems within regenerated plants (Kaeppler and Phillips 1993). DNA methylation has an important effect on the developmental processes of somatic embryogenesis (Nic-Can et al. 2013). The level of DNA methylation in *Pinus nigra* Arn. embryogenic cell lines was investigated, and lines without embryogenic capacity showed the highest level of methylation, whereas lines with high embryogenic potential exhibited the lowest level of methylation. This observation hints at an inverse relationship between 5-methylcytosine (5-mdC) and embryogenic potential. DNA methylation was also assessed during micropropagation of *Cedrus libani* and *Cedrus atlantica* (Renau-Morata et al. 2005a). In vitro culture caused hypomethylation in genomic DNA of the tissue culture progenies of one of the adult *C. libani* genotypes. This DNA demethylation event accompanied an enhancement of the regrowth capacity of this genotype. However, in *C. libani* and *C. atlantica*, the relative amounts of 5-methylcytosine remained stable after the in vitro establishment of seedling-originated shoot stocks (Renau-Morata et al. 2005a). During callus-mediated somatic embryogenesis, DNA methylation occurred due to multiple subcultures and differentiation that may take place over a long period of time (Klimaszewska et al. 2009; Valledor et al. 2007). In the research of Klimaszewska et al. (2009), which was conducted on *Pinus pinaster* (Ait.) somatic embryogenesis, it was observed that the production of mature somatic embryos was stopped as cultures age increased. The aging of embryogenic masses was associated with net DNA demethylation.

Different factors have been proven to change DNA methylation patterns in *in-vitro* culture, and PGRs are one of them (Ghosh et al. 2021). The global DNA methylation in embryogenic cultures of *A. angustifolia* was evaluated by Fraga et al. (2016). They observed that, during long-term subcultures, PGR-supplementation provoked a gradual increase in global DNA methylation, which may compromise genomic stability and induce gene expression modifications. Meanwhile, hypermethylation in embryogenic cultures of *Taxus media* and *Larix* × *eurolepis* was found to be associated with somatic embryo differentiation (Teyssier et al. 2014; Fu et al. 2012). In *Arabidopsis* callus culture, some epigenetic genes, such as *MET1*, *KYP*, *JM14*, and *HAC1* were evaluated to investigate the rate of shoot regeneration (Li et al. 2011).

Bisulfite sequencing (BS-seq) and chromatin immunoprecipitation demonstrated the regulatory role of DNA methylation and histone modification on the indirect organogenesis of *Arabidopsis* via the modulation of *WUS* gene and auxin signaling (Li et al. 2011). A loss-of-function mutation strategy was applied to five DNA methyltransferase genes in *Arabidopsis* to decipher the synergistic and additional roles of CG and non-CG DNA methylation in plant growth and silencing of transposable elements. It has been shown that CG and non-CG DNA methylation synergistically and redundantly regulate plant reproductive development, vegetative development, and TE silencing in *Arabidopsis* (Liang et al. 2022).

In vitro culture-induced epigenetic changes have also been extensively reviewed in previous studies (Us-Camas et al. 2014; Mahdavi-Darvari et al. 2015; Miguel et al. 2016; Sarmast 2016). Deciphering the key regulatory epigenetic mechanisms underlying somatic variation in conifers will greatly aid in avoiding unwanted somaclonal variations, and these parameters can be applied to future plant breeding programs. The auxin-dependent hypermethylation role observed in somaclonal variation also deserves more attention. A precise understanding of different tissue culture-induced modifications, such as DNA methylation and histone modification pattern on targeted genes during micropropagation will impact our ability to begin to more precisely control all aspects of plant tissue culture.

4.3 Molecular Markers

PCR-mediated techniques with the help of various specific and random molecular markers can identify somaclonal variation precisely and quickly, based on the presence or absence of a DNA fragment in gel electrophoresis in a limited period, regardless of the tissue age or plant growth stage.

In *C. libani* Loudon and *A. excelsa* R. Br., DNA fingerprinting of in vitro regenerated plants by using RAPD molecular markers showed a high degree of genetic stability when compared with their corresponding mother plants (Piola et al. 1999; Renau-Morata et al. 2005b; Sarmast et al. 2012a). Furthermore, despite not observing RAPD intraclonal variation in embryogenic masses of *Picea abies* (L.) Karst., certain degrees of trisomy, double trisomy and tetraploidy were observed in the same embryogenic masses (Fourré et al. 1997). In another investigation in embryogenic cultures of *Picea glauca* (Moench) Voss, De Verno et al. (1999) reported somaclonal variation in 2 and 12-month-old cultures following cryopreservation for 3–4 years. Intriguingly, this variation was not observed immediately after thawing. However, in another relevant study, Renau-Morata et al. (2005a) stated that no significant differences were found between *Cedrus libani* and *C. atlantica* mother plants and their corresponding in vitro progenies using RAPD markers after 12 months of culture. Despite confirming the usefulness of this molecular marker in screening the genetic fidelity of axillary bud proliferation progenies, they stated that the degree of stability observed in *Cedrus libani* was higher than that observed in *C. atlantica.* In accordance with Renau-Morata et al. (2005a), who found slight, insignificant variability between in vitro regenerated cedar plants and their corresponding mother plants, Sarmast et al. (2012a) also found a very low variation rate in micropropagated plants of *Araucaria excelsa* R. Br. established from orthotropic stems, compared with their *ex vitro* grown mother plants. In *Chamaecyparis pisifera* Sieb. et Zucc., no genetic differences were found among somatic embryos when using RAPD markers (Hosoi et al. 2015). However, and although no phenotypic differences were observed, the somatic embryo-derived trees showed lower growth compared with seedlings (Hosoi et al. 2015). These observations strengthen the importance of a visual inspection of in vitro-generated progenies after genetic analysis. Five nuclear microsatellites were successfully used to detect somaclonal variation in embryogenic cultures of *Picea abies* (L.) Karst. and *P. omorika* (Pančić) Purk (Hazubska-Przybył and Dering 2017). The study of the relationship between microsatellite sequences and the observed abnormal phenotypes within in vitro regenerated *Pinus pinaster* did not find a correlation between plagiotropism and loss of apical dominance and the analyzed SSR loci (Marum et al. 2009a).

In addition to DNA markers, protein markers were also used to evaluate the genetic fidelity of conifers based on the diversity of selected enzymes. A 2.5-year-old embryogenic cell line of *P. glauca* did not show any somaclonal variation based on an evaluation of 25 loci controlling the diversity of 15 different enzymes (Eastman et al. 1991). As in the previous report, no somaclonal variation was detected in *Pinus resinosa* embryogenic tissues using isoenzyme evaluation (Noh et al. 1988).

However, in their analysis of 8 enzymes, Richard et al. (1995) detected somaclonal variation in in vitro shoots and acclimatized plants of *Larix* × *urolepis* Henry. However, they concluded that this variation could constitute a physiological response to particular culture conditions.

The inefficiency of some molecular markers regarding screening precise genetic variations makes necessary the combined use of the aforementioned markers with other available screening methods. The detection of variations in DNA sequences induced by in vitro culture does not necessarily mean that those variations are responsible for all the changes observed in the offspring phenotype.

There are additional causes of variation, including ploidy change, chromosomal breakage, translocation with inversion, insertions/deletions (point mutations), tissue culture-induced transposition activities, modification of organelle genome, and/or gene activation or silencing (changes in DNA methylation and core histone multi-acetylation) (Kaeppler et al. 1998, 2000). Thus, the molecular markers alone cannot precisely detect somaclonal variation. Hence, by combining different methods in concert, such as visual inspection, the utilization of molecular markers, determining the degree of methylation and acetylation, and examination of the genes involved in methylation, we might begin to recognize the tissue culture–derived variation in a more efficient manner.

5 Conclusion

This paper compiled a great number of previously published reports that were deemed to be relevant to conifer somaclonal variation, to build a more complete understanding of this phenomenon and offer a consumable synopsis for researchers pursuing concise information about conifer micropropagation–derived variations. The tissue culture procedure, the type and origin of explants, and the number of subculture passages were all found to potentially contribute to induce somaclonal variation. Different approaches can be used to detect somaclonal variation. However, only by utilizing a combination of visual inspection, molecular markers, cytogenetic surveys, and analysis of hypo/hypermethylation and acetylation rates, can researchers reasonably expect for a mitigation in the frequency of tissue culture-induced variations.

References

Ahmed EU, Hayashi T, Yazawa S (2004) Auxins increase the occurrence of leaf-colour variants in Caladium regenerated from leaf explants. Sci Hortic 100:153–159

Aitken-Christie J, Thorpe TA (1984) Clonal propagation: gymnosperms. Cell culture and somatic cell genetics of plants, 1st edn. pp 82–95

Alizadeh M, Singh SK (2009) Molecular assessment of clonal fidelity in micropropagated grape (Vitis spp.) rootstock genotypes using RAPD and ISSR markers. Iranian J Biotech 7:37–44

Alizadeh M, Singh SK, Patel VB (2010) Comparative performance of *in vitro* multiplication in four grape (*Vitis* spp.) rootstock genotypes. Int J Plant Prod 4:41–50

Amiri Roudbar M, Mohammadabadi MR, Salmani V (2015) Epigenetics: a new challenge in animal breeding. Genet 3rd millennium 12:3900–3914 (In Persian)

Aquea F, Poupin MJ, Matus JT, Gebauer M, Medina C, Arce-Johnson P (2008) Synthetic Seed production from somatic embryos of *Pinus radiata*. Biotechnol Lett 30:1847–1852. https://doi.org/10.1007/s10529-008-9754-x

Aronen TS, Krajnakova J, Häggman HM et al (1999) Genetic fidelity of cryopreserved embryogenic cultures of open-pollinated Abies cephalonica. Plant Sci 142:163–172. https://doi.org/10.1016/s0168-9452(98)00244-1

Bairu MW, Aremu AO, Staden J (2011) Somaclonal variation in plants: causes and detection methods. Plant Growth Regul 63:147–173. https://doi.org/10.1007/s10725-010-9554-x

Bhardwaj K, Islam MT, Jayasena V, Sharma B, Sharma S, Sharma P, Kuˇca K, Bhardwaj P (2020) Review on essential oils, chemical composition, extraction, and utilization of some conifers in Northwestern Himalayas. Phyther Res 34:2889–2910

Bhardwaj K, Silva AS, Atanassova M, Sharma R, Nepovimova E, Musilek K, Sharma R, Alghuthaymi MA, Dhanjal DS, Nicoletti M (2021) Conifers phytochemicals: a valuable forest with therapeutic potential. Molecules 26(10):3005. https://doi.org/10.3390/molecules26103005

Bhojwani SS, Dantu PK (2013) Plant tissue culture: an introductory text. Springer, India

Bonga J, Park YS, Ding Ch (2018) What technical improvements are needed to achieve industrial application of conifer somatic embryogenesis. In: Proceeding of 5th international conference of the IUFRO unit 2.09.02, 10–15 September, Coimbra, Portugal

Boulay M (1976) Recherches sur la propagation du douglas par culture *in vitro*. Ann AFOCEL 1975:84–145

Boulay M (1977) Multiplication rapide du *Sequoia sempervirens* en culture *in vitro*. Ann AFOCEL pp 38–65

Boulay M (1987) Conifer micropropagation: applied research and commercial aspects. Cell and tissue culture in forestry: case histories: gymnosperms. Angiosperms and Palms. Springer, Netherlands, pp 185–206

Burrows GE, Doley DD, Haines RJ, Nikles DG (1988) *In vitro* propagation of *Araucaria cunninghamii* and other species of the Araucariaceae via axillary meristems. Aust J Bot 36:665–676

Cokus SJ, Feng S, Zhang X, Chen Z, Merriman B, Haudenschild CD, Pradhan S, Nelson SF, Pellegrini M, Jacobsen SE (2008) Shotgun bisulphite sequencing of the Arabidopsis genome reveals DNA methylation patterning. Nature 452:215–219

Cooper RJ (2003) World markets for coniferous forest products: recent trends and future prospects. Acta Hort 615:349–353

D'Amato F (1985) Cytogenetics of plant cell and tissue culture and their regenerates. CRC Crit Rev Plant Sci 3:73–112

David H, Isemukali K, David A (1978) Obtention de pin maritime (*Pinus pinaster* Sol.) a partir de brachyblastes ou d'apexcaulinaires de tres jeunes sujets cultivés *in vitro*. C R Acad Sci, Paris 287:245–247

De Verno LL, Park YS, Bonga JM, Barrett JD (1999) Somaclonal variation in cryopreserved embryogenic clones of white spruce (*Picea glauca* (Moench)) Voss. Plant Cell Rep 18:948–953

Du X, Han L, Guo AY, Zhao Z (2012) Features of methylation and gene expression in the promoter-associated CpG islands using human methylome data. Comp Funct Genom 59:8987–5994

Eastman PA, Webster FB, Pitel JA et al (1991) Evaluation of somaclonal variation during somatic embryogenesis of interior spruce (*Picea glauca* engelmanii complex) using culture morphology and isozyme analysis. Plant Cell Rep 10(8):425–430. https://doi.org/10.1007/bf00232617

Farjon A, Page CN (1999) Conifers. Status survey and conservation action plan. IUCN/SSC conifer specialist group. IUCN, Gland, Switzerland and Cambridge, UK, p 121

Flanagan M (2003) History of conifers in horticulture: an overview. Acta Hortic 615:241–245

Ford D (2013) Honeybees and cell lines as models of DNA methylation and aging in response to diet. Exp Ger 48(7):614–619. https://doi.org/10.1016/j.exger.2012.07.010

Fourré JL (2000) Somaclonal variation and genetic molecular markers in woody plants. In: Jain SM, Minocha SC (eds) Molecular biology of woody plants. Kluwer Academic publisher, Netherlands, p 299

Fourré JL, Berger P, Niquet L, Andre P (1997) Somatic embryogenesis and somaclonal variation in Norway spruce: morphogenetic, cytogenetic, and molecular approaches. Theor Appl Genet 94:159–169

Fraga HPF, Vieira LN, Heringer AS, Puttkammer CC, Silveira V, Guerra MP (2016) DNA methylation and proteome profiles of *Araucaria angustifolia* (Bertol.) Kuntze embryogenic cultures as affected by plant growth regulators supplementation. Plant Cell Tissue Organ Cult 125:353–374

Fu C, Li L, Wu W, Li M, Yu X, Yu L (2012) Assessment of genetic and epigenetic variation during long-term *Taxus* cell culture. Plant Cell Rep 31:1321–1331

George EF, Debergh, PC (2008) Micropropagation: uses and methods. In: George EF, Hall MA, Klerk GJD (eds) Plant propagation by tissue culture, vol 1. Springer, Dordrecht, pp 29–64. https://doi.org/10.1007/978-1-4020-5005-3_2

Ghorbanzadeh A, Ghasemnezhad A, Sarmast MK, Ebrahimi SN (2021) An analysis of variations in morphological characteristics, essential oil content, and genetic sequencing among and within major Iranian Juniper (*Juniperus* spp.) populations. Phytochemistry 186:112737. https://doi.org/10.1016/j.phytochem

Ghosh A, Igamberdiev AU, Debnath SC (2021) Tissue culture-induced DNA methylation in crop plants: a review. Mol Biol Rep 48(1):823–841

Goryachkina OV, Pak ME, Tret'yakova IN, (2017) Cytogenetic features of Larix sibirica (Ledeb.) embryogenic cell lines in in vitro culture. Vestn Tomsk Gos Univ, Ser Biol 39:140–153

Goryachkina OV, Park ME, Tretyakova IN, Badaeva ED, Muratova EN (2018) Cytogenetic stability of young and long-term embryogenic cultures of *Larix sibirica*. Cytologia 83:323–329

Gulyaeva EN, Ignatenko RV, Galibina NA (2020) Somaclonal variability of conifers in culture in vitro. Ecol Genet 18:301–315. https://doi.org/10.17816/ecogen19143

Häggman H, Ryynänen L, Aronen T et al (1998) Cryopreservation of embryogenic culture of Scots pine. Plant 11 Tissue Organ Cult. 54(1):45–53. https://doi.org/10.1023/a:1006104325426

Haisel D, Hofman P, Vágner M, Lipavska H, Ticha I, Schäfer C, Capkova V (2001) Ex vitro phenotype stability if affected by *in vitro* cultivation. Biol Plant 44:321–324

Hao YJ, Deng XX (2002) Occurrence of chromosomal variations and plant regeneration from long-term-cultured citrus callus. In Vitro Cell Dev Biol Plant 38:472–476

Hazubska-Przybył T, Dering M (2017) Somaclonal variation during *Picea abies* and *P. omorika* somatic embryogenesis and cryopreservation. Acta Biol Cracoviensia Bot 59:93–103. https://doi.org/10.1515/abcsb-2017-0003

Hosoi Y, Kuramoto N, Maruyama TE (2015) Screening RAPD primers to assess clonal fidelity in somatic embryos of Sawara cypress (*Chamaecyparis pisifera* Sieb. et Zucc.) and field performance of somatic embryo-derived trees. Plant Biotechnol 32:149–155

Isabel N, Boivin R, Levasseur C et al (1996) Occurrence of somaclonal variation among somatic embryo-derived white spruces (*Picea glauca*, Pinaceae). Am J Bot 83:1121–1130. https://doi.org/10.2307/2446195

Jinks RL (2003) Propagation of conifers. Acta Hortic. 615:303–307. https://doi.org/10.17660/ActaHortic.2003.615.32

Kaeppler S, Phillips R (1993) DNA methylation and tissue culture induced variation in plants. *in vitro* Cell Dev Biol Plant 29:125–130

Kaeppler SM, Phillips RL, Olhoft P (1998) Molecular basis of heritable tissue culture-induced variation in plants. In: Jain et al (eds) Somaclonal variation and induced mutations in crop improvement. Current plant science and biotechnology in agriculture, vol 32. Kluwer Academic Publishers, Dordrecht, Netherlands, pp 465–484

Kaeppler SM, Kaeppler HF, Rhee Y (2000) Epigenetic aspects of somaclonal variation in plants. Plant Mol Biol 43:179–188

Karim R, Nuruzzaman M, Khalid N, Harikrishna JA (2016) Importance of DNA and histone methylation in *in vitro* plant propagation for crop improvement: a review. Ann Appl Biol 169:1–16. https://doi.org/10.1111/aab.12280
Karp A (1994) Origins, causes and uses of variation in plant tissue cultures. In: Vasil IK, Thorpe TA (eds) Plant cell and tissue culture. Kluwer Academic Publishers, Dordrecht, pp 139–152
Katiyar RK, Chopra VL (1995) A somaclone of Brassica juncea is processed into a variety and is released for commercial cultivation in India. Cruciferae Newsl 17:92–93
Klimaszewska K, Noceda C, Pelletier G, Label P, Rodriguez R, Lelu-Walter MA (2009) Biological characterization of young and aged embryogenic cultures of *Pinus pinaster* (Ait). In Vitro Cell Dev Biol Plant 45:20–33
Klimaszewska K, Hargreaves C, Lelu-Walter MA, Trontin JF (2015) Advances in conifer somatic embryogenesis since year 2000. In: Germanà MA, Lambardi M (eds) *In vitro* plant embryogenesis in higher plants. Methods in molecular biology, Chapter 7. Springer Science+Business Media, New York, in press. https://doi.org/10.1007/978-1-4939-3061-6_8
Krajňáková J, Sutela S, Aronen T et al (2011) Longterm cryopreservation of Greek fir embryogenic cell lines: recovery, maturation and genetic fidelity. Cryobiology 63:17–25. https://doi.org/10.1016/j.cryobiol.2011.04.004
Krutovsky KV, Tretyakova IN, Oreshkova NV, Pak ME, Kvitko OV, Vaganov EA (2014) Somaclonal variation of haploid *in vitro* tissue culture obtained from Siberian larch (*Larix sibirica* Ledeb.) megagametophytes for whole genome de novo sequencing. In Vitro Cell Dev Biol Plant 50:655–664
La Goublaye NH (1981) Vieillissement et rajeunissement chez Ie Sequoia sempervirens (Endl.) en relation avec la propagation vegetative. Th~se 3~me Cycle Univ Paris VI, p 170
Larkin PJ, Scowcroft WR (1981) Somaclonal variation—a novel source of variability from cell cultures for plant improvement. Theor Appl Genet 60(4):197–214
Law JA, Jacobsen SE (2010) Establishing, maintaining, and modifying DNA methylation patterns in plants and animals. Nat Rev Genet 11:204–220
Lee M, Phillips RL (1988) The chromosomal basis of somaclonal variation. Annu Rev Plant Physiol Plant Mol Biol 39:413–437
Lelu-Walter M-A, Thompson D, Harvengt L, Sanchez L, Toribio M, Pâques LE (2013) Somatic embryogenesis in forestry with a focus on Europe: state-of the-art, benefits, challenges, and future direction. Tree Genet Genomes 9:883–899
Lelu-Walter M-A, Gautier F, Elišov K et al (2018) High gellan gum concentration and secondary somatic embryogenesis: two key factors to improve somatic embryo development in *Pseudotsuga menziesii* [Mirb.]. Plant Cell Tissue Organ Cult 132:137–155. https://doi.org/10.1007/s11240-017-1318-0
Li W, Liu H, Cheng ZJ, Su YH, Han HN, Zhang Y, Zhang XS (2011) DNA methylation and histone modifications regulate de novo shoot regeneration in Arabidopsis by modulating WUSCHEL expression and auxin signaling. PloS Genet 7(8):1002243. https://doi.org/10.1371/journal.pgen.1002243
Liang W, Li J, Sun L, Liu Y, Lan Z, Qian W (2022) Deciphering the synergistic and redundant roles of CG and non-CG DNA methylation in plant development and transposable element silencing. New Phytol 233(2):722–737
Lineros Y, Balocchi C, Munoz X, Sanchez M, Rios D (2018) Cryopreservation of *Pinus radiata* embryogenic tissue: effects of cryoprotective pretreatments on maturation ability. Plant Cell Tissue Org Cult 135:357–366
Loureiro J, Capelo A, Brito G, Rodriguez E, Silva S, Pinto G, Santos C (2007) Micropropagation of *Juniperus phoenicea* from adult plant explants and analysis of ploidy stability using flow cytometry. Biol Plant 51:7–14
Mahdavi-Darvari F, Mohd Noor N, Ismanizan I (2015) Epigenetic regulation and gene markers as signals of early somatic embryogenesis. Plant Cell Tissue Org Cult 120:407–422

Marum L, Rocheta M, Maroco J, Oliveira M, Miguel C (2009) Analysis of genetic stability at SSR loci during embryogenesis in maritime pine (*Pinus pinaster*). Plant Cell Rep 28(4):673–682. https://doi.org/10.1007/s00299-008-0668-9

Marum L, Loureiro J, Rodriguez E, Santos C, Oliveira MM, Miguel C (2009) Flow cytometric and morphological analyses of *Pinus pinaster* somatic embryogenesis. J Biotechnol 143:288–295

Miguel CM, Rupps A, Raschke J (2016) Impact of molecular studies on somatic embryogenesis development for implementation in conifer multi-varietal forestry. In: Park Y-S, Bonga JM, Moon H-K (eds) Vegetative propagation of forest trees. National Institute of Forest Science (NIFoS), Seoul, pp 373–421

Mohanty S, Panda M, Subudhi E, Nayak S (2008) Plant regeneration from callus culture of *Curcuma aromatica* and *in vitro* detection of somaclonal variation through cytophotometric analysis. Biol Plant 52:783–786

Nawrot-Chorabik K (2009) Somaclonal variation in embryogenic cultures of silver fir (*Abies alba* Mill.). Plant Biosyst 43(2):377–385. https://doi.org/10.1080/11263500902722717

Neelakandan AK, Wang K (2012) Recent progress in the understanding of tissue culture-induced genome level changes in plants and potential applications. Plant Cell Rep 31:597–620

Nic-Can GI, López-Torres A, Barredo-Pool F, Wrobel K, Loyola-Vargas VM, Rojas-Herrera R, De-la-Peña C (2013) New insights into somatic embryogenesis: LEAFY COTYLEDON1, BABY BOOM1 and WUSCHEL-RELATED HOMEOBOX4 are epigenetically regulated in Coffea canephora. PLoS ONE 8(8):72160

Noceda C, Salaj T, Perez M et al (2009) DNA demethylation and decrease on free polyamines is associated with the embryogenic capacity of Pinus nigra Arn. cell culture. Trees 23:1285–1293. https://doi.org/10.1007/s00468-009-0370-8

Noh EW, Minocha SC, Riemenschneider DE. (1988) Adventitious shoot formation from embryonic explants of red pine (*Pinus resinosa*). Physiol Plant 74(1):119–124. https://doi.org/10.1111/j.1399-3054.1988.tb04951.x.61

O'Brien IEW, Smith DR, Gardner RC, Murray BG (1996) Flow cytometric determination of genome size in *Pinus*. Plant Sci 115:91–99

Pattanavibool R, von Aderkas P, Hanhijarvi A, Simola LK, Bonga JM (1995) Diploidization in megagametophyte-derived cultures of the gymnosperm *Larix decidua*. Theor Appl Genet 90:671–674

Piccolo FM, Fisher AG (2014) Getting rid of DNA methylation. Trends Cell Biol 24:136–143

Piola F, Rohr R, Heizmann P (1999) Rapid detection of genetic variation within and among *in vitro* propagated cedar (*Cedrus libani* Loudon) clones. Plant Sci 141:159–163

Renau-Morata B, Nebauer SG, Arrillaga I, Segura J (2005) Assessment of somaclonal variation in micropropagated shoots of Cedrus: consequences of axillary bud breaking. Tree Genet Genom 1:3–10

Renau-Morata B, Ollero J, Arrillagam I, Segura J (2005) Factor influencing axillary proliferation and adventitious budding in cedar. Tree Physiol 25:477–486

Richard S, Gauthier S, Laliberté S (1995) Isozyme assessment of the genetic stability of micropropagated hybrid larch (Larix × eurolepis Henry). Can j of for Res 25(7):1103–1112

Roth R, Ebert I, Schmidt J (1997) Trisomy associated with loss of maturation capacity in a long-term embryogenic culture of Abies alba. Theor Appl Genet 95:353–358

Sahavacharin O (1982) Rapid propagation of caladium through tissue culture. In: Plant tissue culture 1982: proceedings, 5th international congress of plant tissue and cell culture held at Tokyo and Lake Yamanake, Japan, 11–16 July 1982

Salajova T, Salaj J (1992) Somatic embryogenesis in European black pine (*Pinus nigra* Arn.). Biol Plant 4:213–218

Sarmast MK (2016) Genetic transformation and somaclonal variation in conifers. Plant Biotechnol Rep 10(6):309–325

Sarmast MK, Salehi H, Ramzani Abolimoghadam AA, Niazi A, Khosh-Khui M (2012a) RAPD fingerprint to appraise the genetic fidelity of *in vitro* propagated *Araucaria excelsa* R Br Var. Glauca Plantlets. Mol Biotechnol 50:181–188

Sarmast MK, Salehi H, Khosh-Khui M (2012b) *In vitro* rooting of Araucaria excelsa R. Br. Var. glauca using *Agrobacterium rhizogenes*. J Cent Eur Agric 13:123–130

Sarmast MK, Salehi H, Khosh-Kui M (2012c) Micropropagation of *Araucaria excelsa* R. Br. Var. Glauca Carrière from orthotropic stem explants. Physiol Mol Biol of Plants 18:265–271

Shemer O, Landau U, Candela H, Zemach A, Williams LE (2015) Competency for shoot regeneration from Arabidopsis root explants is regulated by DNA methylation. Plant Sci 238:251–261

Skirvin RM, McPheeters KD, Norton M (1994) Sources and frequency of somaclonal variation. Hort Science 29:1232–1237

Sutton BCS, Atree SM, El-Kassaby YA, Grossnickle SC, Polonenko DR (2004) Commercialization of somatic embryogenesis for plantation forestry. In: Walter C and Carson M (eds) Plantation forest biotechnology for the 21st century. Research Signpost 37/661 (2), Fort PO, Trivandrum695 023, Kerala

Teyssier C, Maury S, Beaufour M, Grondin C, Delaunay A, Le Mette´ C, et al (2014) In search of markers for somatic embryo maturation in hybrid larch (*Larix* × *eurolepis*): global DNA methylation and proteomic analyses. Physiol Plant 150:271–291

Tremblay L, Levasseur C, Tremblay FM (1999) Frequency of somaclonal variation in plants of black spruce (*Picea mariana*, Pinaceae) and white spruce (*P. glauca*, Pinaceae) derived from somatic embryogenesis and identification of some factors involved in genetic instability. Am J Bot 86:1373–1381

Us-Camas R, Rivera-Solıs G, Duarte-Ake´ F, De la Peña C (2014) *In vitro* culture: an epigenetic challenge for plants. Plant Cell Tiss Organ Cult 118:187–201. https://doi.org/10.1007/s11240-014-0482-8

Valledor L, Hasbún R, Meijón M, Rodrıguez JL, Santamarıa E, Viejo M, Berdasco M, Feito I, Fraga MF, Can˜ al MJ, Rodrıguez R, (2007) Involvement of DNA methylation in tree development and micropropagation. Plant Cell Tissue Organ Cult 91:75–86

Venkatachalam L, Sreedhar RV, Bhagyalakshmi N (2007) Genetic analyses of micropropagated and regenerated plantlets of banana as assessed by RAPD and ISSR markers. In Vitro Cell Dev Biol Plant 43:267–274

Von Aderkas P, Anderson P (1993) Aneuploidy and polyploidization in haploid tissue cultures of Larix decidua. Physiol Plant 88(1):73–77

Von Aderkas P, Pattanavibool R, Hristoforoglu K, Ma Y (2003) Embryogenesis and genetic stability in long term megagametophyte- derived cultures of larch. Plant Cell Tissue Org Cult 75:27–34

Wendling I, Trueman SJ, Xavier A (2014) Maturation and related aspects in clonal forestry—part II: reinvigoration, rejuvenation, and juvenility maintenance. New for 45:437–486

Wu H, Zhang Y (2014) Reversing DNA methylation: mechanisms, genomics, and biological functions. Cell 156:45–68

Xiaoyi C, Shuaifei J, Dai J, Deshui Y, Kong L, Jinfeng Z, Jian Z (2021) Cryopreservation of embryogenic callus for Larix gmelinii var. principis-rupprechtii. Beijing Forestry University 43:47–53

Tissue Culture of Cacao (*Theobroma cacao* L.): Between Somaclonal Variation and Mass Propagation

Ana M. Henao, Tatiana Osorio, and Aura I. Urrea

Abstract Cacao (*Theobroma cacao* L.) is an important commercial and agricultural crop. It has been the subject of varied research aimed at optimizing the propagation of materials selected for different traits of interest. Tissue culture is one of the promising tools for this purpose, as well as for plant breeding and other basic research. Somatic embryogenesis (SE) in cacao has been shown to be a viable and cost-effective alternative; the process, standardized by different authors, is associated with significant advantages. However, difficulties related to somaclonal variation have also been identified, which are the main focus of this work. Of the two pathways for inducing SE, direct (DSE) and indirect somatic embryogenesis (ISE), cacao primarily responds to ISE, where an intermediate callus phase is present. Large-scale multiplication has been effectively achieved from primary somatic embryos (PSE) in hormone-supplemented culture media, which has been shown to generate somaclonal variation. It was detected through changes in embryos development and conversion rates, plant morphology and architecture, and supported at the genetic and molecular levels. This work provides an overview of the state of development of plant production technology by SE for cacao and the challenges in terms of scaling, efficiency, and final production costs. The most prominent factors inducing somaclonal variation are also addressed, as well as the anomalies in somatic embryos and cacao seedlings found in different research studies.

Keywords Cacao (*Theobroma cacao* L.) · Tissue culture · Somatic embryogenesis · Micropropagation · Somaclonal variation · Abnormality

---

Ana M. Henao and Aura I. Urrea are contributed equally to this work and share first authorship

---

A. M. Henao (✉) · T. Osorio · A. I. Urrea
Biotechnology Group, Faculty of Exact and Natural Sciences, University of Antioquia, Medellín, Colombia
e-mail: amaria.henao@udea.edu.co

C. Sánchez-Romero (ed.), *Somaclonal Variation: Basic and Practical Aspects*,
https://doi.org/10.1007/978-3-031-51626-9_8

1 Introduction

Cacao (*Theobroma cacao* L.), known worldwide as the raw material of chocolate, belongs to the tribe Theobromae and subfamily Byttnerioideae of family Malvaceae, genus *Theobroma*. It is the primary fruit of the cultivated genus due to the value and importance of the seeds (Richardson et al. 2015). This evergreen crop is native to the South American rainforest, with its center of origin in the upper Amazon regions of Brazil, Peru, and Southern Colombia, where the greatest diversity was observed (Osorio-Guarín et al. 2017).

The traditional classification of cacao under domestication, which is applied in the cacao trade, recognizes some main categories: Criollo, Forastero, and Trinitario, along with Nacional, Amelonado, and other classes (Bekele et al. 2020). This traditional nomenclature is based on botanical traits, particularly those of the fruit (pod) and seed (bean), although recently it has been shown that it also has a genetic basis (Zarrillo et al. 2018). In the wild, Criollo and Forastero cacao are found in the Amazon Basin, in South America, at the putative center of the genetic diversity of cacao. In contrast, Trinitario cacao comprises many hybrids between Criollo and Forastero cacao (Souza et al. 2018).

The lucrative global cacao and chocolate market has a significant economic and social importance worldwide (Cadby and Araki 2021). Cacao grows in up to 50 countries throughout the humid tropical region (Boza et al. 2013). The major producing countries are Ivory Coast, Ghana, Indonesia, Brazil, and Nigeria, which contribute 82% of the world's production. African countries produced 68% of the total, while Central and South American countries produced 14%. Asian countries produced the remaining 18% (Nair 2021).

Cacao processing at origin includes cacao seed cultivation, harvesting, fermenting, and drying (Lafargue-Molina 2021). Fermented and dried cacao seeds, hereby referred to as "beans", are the majority of traded processed cacao used for making finished chocolate. Commodity cacao, also referred to as bulk or industrial cacao, constitutes most of the global cacao produced nowadays, estimated at around 93–95% of global production, mainly out of West Africa (Cadby and Araki 2021). Most of this cacao is produced from small and family-run farms, with an average size of two to four hectares. More massive plantations for cacao production are also found in countries such as in Brazil and Malaysia (Cadby and Araki 2021).

The dark chocolate segment of the market is expected to have the highest growth rate (5.2%) in the next five years, perhaps due to the acclaimed health benefits resulting from the high content of cocoa solids (60% and over) of dark chocolate (Cádiz-Gurrea et al. 2019; Swamy 2020). Most consumers are currently purchasing dark chocolate over milk and white chocolate (Clark 2022). Dark and gourmet chocolates are mainly manufactured from cacao beans of different varieties like Criollo, Trinitario, Nacional, and Piura Porcelana, which are designated as "fine flavor" cocoa (Arevalo-Gardini et al. 2019).

The cocoa sector is projected to grow, driven primarily by its extensive appeal, popularity, and wide use in the food and beverage industry. The research in genetics and plant breeding continues producing new varieties of cacao and improving propagation systems to attend to the growing demand. There is already a large demand for plantlets, which clearly indicates the need for large-scale propagation. This demand has been estimated at millions of plantlets annually. Thus, the production of plantlets will be increasingly necessary.

Cacao can be propagated both by seed and vegetative means. Seed propagation is the easier strategy, but when this method is followed, the obtained seedlings will be highly variable genetically. All cacao-producing countries have practiced vegetative propagation, ensuring greater uniformity. The methods are grafting (top grafting and budding grafting), rooting of cuttings, and tissue culture (Sodré and Gomes 2019). Top grafting is the method most used by producers to propagate genotypes in regions of South America, Central America, and the Caribbean. Top grafting achieved survival rates from 55 to 92%, varying based on the clone, place of grafting, and season of the year. Tissue culture deserves special attention among the advanced technologies applied for cacao propagation. The main advantages of tissue culture methods include rapid production of asexually propagated, uniform plants with highly valued genetic traits. Another benefit is avoiding the dissemination of diseases that conventional cuttings and grafting can transmit. Moreover, tissue culture can accelerate the multiplication of trees with interesting characteristics related to disease resistance, growth pattern, or oil composition. Tissue culture is also an important tool for regenerating plantlets modified by modern genetic techniques, such as the CRISPR/Cas 9 system (Fister et al. 2018; Etaware 2021), which will become very important for creating novel genetic variants in the future.

However, somaclonal variation, defined as genetic and phenotypic differences in in vitro regenerated plants, may occur due to tissue culture per se. These differences may be caused by genomic changes, such as deletions, changes in chromosome structure and number, and directed and undirected point mutations. Thus, passing cells through in vitro culture can induce genomic changes in the regenerated plants (Fang et al. 2009).

Different factors have been reported to affect the level of both genetic and epigenetic somaclonal variation induced during in vitro culture, including (1) Medium composition (Wang and Wang 2012); (2) In vitro culture technique used, with methods involving a dedifferentiated callus phase expected to show higher levels of variability (Azizi et al. 2020); (3) The source of the donor tissue (Bhojwani and Dantu 2013); (4) Time in culture and number of regeneration events (Sarmah et al. 2017). Therefore, somaclonal variation generally increases with a prolonged callus phase and higher number of multiplication cycles (Duta-Cornescu et al. 2023).

In this context, this chapter aims to provide a complete summary of the level of maturity of SE technology in cacao, successful experiences, and bottlenecks in the scaling-up process for producing plantlets by this way. It also presents the factors that may influence the appearance of somaclonal variation and contains extensive information on the most recent strategies to estimate these variations in cacao plants regenerated via SE. The main morphological and developmental malformations that can be found in somatic embryos at different stages and during the plantlet conversion phase are also included. Finally, some suggestions to reduce somaclonal variation and general conclusions on the subject are addressed.

2 SE for Scaling Up Cacao Plant Production

SE is considered a next-generation propagation method with additional practical advantages compared to micropropagation by organogenesis and conventional methods (Wickramasuriya and Dunwell 2018) such as, greater efficiency, plant uniformity, reduced space requirements, and year-round plant production (Peña-López et al. 2016).

SE progresses through a developmental pathway that is similar to that of its zygotic counterpart (Noah et al. 2013; Ikeuchi et al. 2016; Winkelmann 2016). Normal development in synchrony of shoot and root apical meristems results in the growth of plants with a desirable orthotropic architecture, avoiding problems with plagiotropic growth and the formation of adventitious roots obtained in most cases through propagation via rooting of cuttings (Entuni et al. 2018). Furthermore, SE maintains the genetic identity of the produced cacao plants, while conventional grafting may show variability (Henao-Ramírez et al. 2021).

The possibility of scaling up somatic embryo production is one of the most significant advantages of SE once the technology has been optimized, as demonstrated in various species (Suárez-Castellá et al. 2012). However, large-scale cacao plant production using SE has not yet been established. Guillou and Verdier (2022) reported the evaluation of the SE process in 170 cacao genotypes: around 80% of genotypes responded positively to SE, and 95% of the genotypes that formed PSE also developed secondary somatic embryos (SSE). To date, the highest production efficiency was reported by Henao et al. (2022b), with 70% conversion to plantlets, i.e., 70% of embryos develop both photosynthetic tissues and roots. These results are similar to those of reported by Fang et al. (2009), with 65% conversion rate and multiplication coefficients of 5–58 embryos per explant, depending on the genotype (Garcia et al. 2016). Scaling up the process from the laboratory to a production scale remains challenging for SE culture systems. Pilot plant data and operating know-how are critical to improve large-scale plant process (Fig. 1). As a productive approach, SE is a complex process involving many mechanisms in each one of its stages: initiation, multiplication, maturation, germination, and acclimatization (Egertsdotter et al. 2019; Henao et al. 2022b).

Explant disinfection for in vitro introduction and callogenesis are included in the initiation stage (Henao et al. 2022b). At this stage, the composition of the culture medium is decisive for promoting dedifferentiation and redifferentiation processes. One of the most important factors is selecting a potentially embryogenic tissue. Petals and the staminodes are the explants with the best response, depending on the cacao genotype (Macias-Naranjo et al. 2023; Osorio et al. 2022). In cacao, the multiplication stage includes the induction of PSE (expression I), the induction of recurrent or secondary embryogenesis (expression II), and a second proliferation

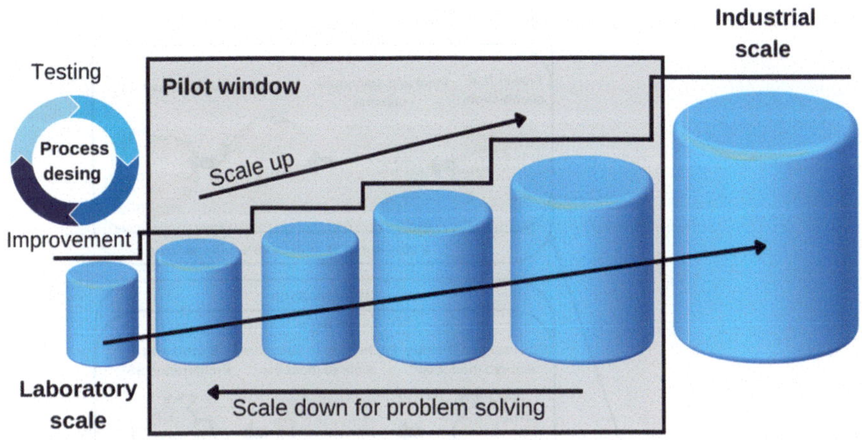

Fig. 1 Industrial process scale-up

cycle of SSE (expression III). During expression I, once the tissues become calli, they are transferred to a medium for the induction of PSE. During expression II, once PSE have been obtained, the embryos are chopped with a scalpel and pieces of somatic embryos are placed in the secondary or recurrent embryogenesis medium. Since somatic embryos develop asynchronously, they are harvested at the globular and early cotyledonary stages to prevent embryogenic axes from fusing during growth. Subsequently, during expression III, the embryos are transferred to a medium for somatic embryo multiplication. At this point, SE allows unlimited multiplication of the original plant material (Fehér 2019).

Adjusting the culture conditions, such as lighting and the medium used, enhances the maturation and germination of somatic embryos (Bustami and Werbrouck 2017, 2018). Subsequently, the obtained plantlets develop true leaves. At this moment, they can be transferred to *ex vitro* conditions for hardening or acclimatization (Henao et al. 2022b).

A pilot production scale of some genotypes demonstrated that the minimal time for production was 30 days for initiation, 70 days for multiplication (30 days for PSE, 20 days for SSE, and 20 days for expression III, 50 days for somatic embryos maturation, 70 days for embryos germination, and 70 days for plants acclimatization to ex vitro culture conditions (Fig. 2). The decrease in the time of the production process directly influences the cost of the process, and represents an important issue in the development of the technology.

However, despite having some protocols with adequate times, reductions in efficiency occur due to the presence of malformations, which impact each stage (Fig. 2). It has been found that the most critical stages of the process are embryo maturation and germination, and plant adaptation to greenhouse conditions. The response does not exceed 60–70% in these stages, mainly due to malformations that do not allow the plant to develop normally. Despite extensive in vitro research in various

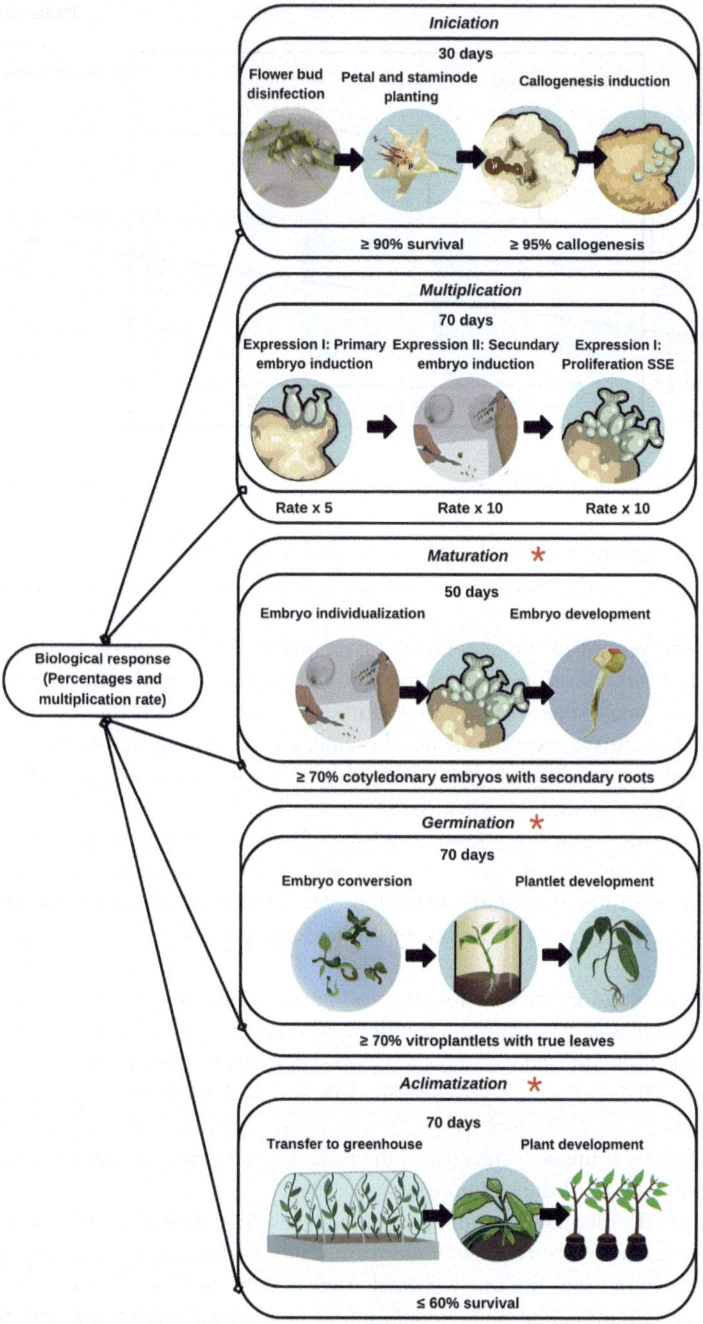

Fig. 2 Representation of cacao plant production through SE from floral explants: efficiency ranges and documented timeframes for each stage. * Represents the most critical stages of the process with high reduction of biological responses by malformations due to somaclonal variation, as described in Sect. 3.1

cacao genotypes, the high incidence of malformations during the process still leads to reductions in the efficiency.

In scaling in vitro culture processes, liquid culture medium has some advantages over semisolid medium. It has been considered more appropriate in some crops because it has been shown to improve the efficiency of the process, increasing the multiplication rate, rooting percentage, and survival of the plants during the acclimatization stage (Peña-López et al. 2016). Additionally, it reduces labour and input costs, with a notable improvement in the profitability of the propagation process (Carvalho et al. 2019).

In cacao, some studies on SE have been carried out using not only liquid media, but also temporary immersion systems, in which explant regeneration and embryo synchronization to the same development stage increase. Moreover, recurrent morphological abnormalities in the embryos and tissue oxidation are reduced, compared to the semisolid medium. Additionally, a decrease in the time necessary to renew the culture medium and execute the process has been observed (Minyaka et al. 2008; Solano 2008; Soraia 2013). According to Guillou and Verdier (2022), 9000 plants could be produced by one person per year using solid culture media, while 20,000 plants could be produced using liquid culture media. The multiplication rate of embryogenic calli is about 1.5 to 2, and the most efficient cell lines can produce between 1500 and 2000 embryos per gram of embryogenic calli. Once embryos are obtained, embryo conversion into plants is about 20–40%, with large batch-to-batch variability by epigenetic modifications, such as DNA methylation and chromatin remodeling. These alterations may cause morphological abnormalities that can reduce the germination capacity, thereby impacting the success rate of embryo conversion to plants (Osorio-Montalvo et al. 2018; Sivanesan et al. 2022; Garcia et al. 2022). However, in ongoing studies with the *IMC67* cacao genotype, embryo multiplication rates recorded in RITA temporary immersion systems were up to 4 times higher than those obtained in semisolid media. Nevertheless, normal germination and conversion to plants have not been still possible by using this system (results not yet published).

3 Somaclonal Variation in Cacao SE: Genetic and Epigenetic Changes

Each of the cases in which somaclonal variation has been reported in cacao due to SE is analyzed below (Table 1), detailing the possible factors or causes involved in the appearance of variation. The methods used to detect somaclonal variation and the results obtained in each case are shown in Table 2.

The first study on somaclonal variation in cacao was performed by Rodríguez López et al. (2004), (Table 1). The aim of this investigation was to present a simple protocol that enables a provisional diagnosis of both homogenous and chimeric mutants among large regenerant populations, together with a semi-quantitative means

Table 1 Methodological details of the main works on somaclonal variation in cacao

References	Culture protocol	Type of explant	Phase	PGR	Explant for analysis	Detection method
Rodríguez López et al. (2004)	Li et al. (1998)	Staminodes	Primary callus growth (PCG)	1.98 mg L^{-1} 2,4-Dichlorophenoxyacetic acid (2,4-D) + 0.005 mg L^{-1} Thidiazuron (TDZ)	Cotyledons and leaves of in vitro plantlets	SSR markers
			Secondary callus growth (SCG)	2 mg L^{-1} 2,4-D + 0.30 mg L^{-1} Kinetin (KIN)		
			Embryo development (ED)	Without PGR		
			Plant regeneration medium (PR)			
Fang et al. (2009)	Maximova et al. (2002)	Staminodes	PCG	1.98 mg L^{-1} 2,4-D + 0.005 mg L^{-1} TDZ	Somatic embryos	SSR markers
			SCG (Li et al. 1998 modified)	0.53 mg L^{-1} 2,4-D + 0.315 mg L^{-1} BA		
			ED (Li et al. 1998)	Without PGR		
Rodríguez López et al. (2010a)	Li et al. (1998)	Staminodes	PCG	1.98 mg L^{-1} 2,4-D + 0.005 mg L^{-1} TDZ	In vitro plantlets and plantlets from recently expanded leaves of trees used as the source for explant material	CAPS
			SCG	2 mg L^{-1} 2,4-D + 0.30 mg L^{-1} KIN		
			ED	Without PGR		
			PR			
Rodríguez López et al. (2010b)	Li et al. (1998)	Staminodes	PCG	1.98 mg L^{-1} 2,4-D + 0.005 mg L^{-1} TDZ	Source trees, regenerated plants, and staminodes	SSR and MSAP markers
			SCG	2 mg L^{-1} 2,4-D + 0.30 mg L^{-1} KIN		
			ED	Without PGR		
			PR			

(continued)

Table 1 (continued)

References	Culture protocol	Type of explant	Phase	PGR	Explant for analysis	Detection method
Ajijah et al. (2016)	Ajijah et al. (2016)	Basal petals and staminodes	Callus initiation (CI)	1.04 mg L^{-1} 2,4-D and either 0.07, 0.13 or 0.27 mg L^{-1} KIN. Themolar ratios of those KIN: 2,4-D were 1.0:15.5, 1.0:7.8 or 1.0:3.9	Fresh leaves of in vitro plantlets and the mother plant	SSR markers
			SCG	1.04 mg L^{-1} 2,4-D and either 0.13 or 0.27 mg L^{-1} KIN. Themolar ratios of those KIN: 2,4-D were 1.0:7.8 or 1.0:3.9. The KIN: 2,4-D ratio of 1.0:7.8		
			ED	Without PGR		
Adu-Gyamfi et al. (2016)	Maximova et al. (2002)	Staminodes	PCG (Li et al. 1998 modified)	1.98 mg L^{-1} 2,4-D + 0.005 mg L^{-1} TDZ	Leaves collected from tree and somatic embryos; somatic embryos cryopreserved in liquid nitrogen	MSAP markers
			SCG (Li et al. 1998 modified)	0.53 mg L^{-1} 2,4-D + 0.315 mg L^{-1} benzyl adenine (BA)		
			ED (Li et al. 1998)	Without PGR		
Garcia et al. (2021)	Li et al. (1998)	Petals	PCG (Li et al. 1998 modified)	2 mg L^{-1} 2,4-D and 5 μg L^{-1} TDZ	PSE and SSE	Global DNA methylation by whole-genome bisulfite sequencing
			SCG (Li et al. 1998 modified)	2 mg L^{-1} 2,4-D		
			ED (Li et al. 1998)	Without PGR		
Henao et al. (2022a)	Fontanel et al. (2002) with minor changes	Staminodes and petals	Induction medium (INDI) (Fontanel et al. 2002)	1 mg L^{-1} 2,4-D + 0.25 mg L^{-1} KIN	Leaves	SSR markers

(continued)

Table 1 (continued)

References	Culture protocol	Type of explant	Phase	PGR	Explant for analysis	Detection method
			Expression medium (INDexp) (Fontanel et al. 2002)	Without PGR		
			Callus multiplication (CM2) (Fontanel et al. 2002)	1 mg L^{-1} 2,4,5-T (2,4,5-Trichlorophenoxyacetic acid)		
			Maturation medium (MM6) (Fontanel et al. 2002)	0.01 mg L^{-1} of 1-naphthaleneacetic acid (NAA) + 0.02 mg L^{-1} gibberellic acid (GA_3)		
			In vitro development: (Fontanel et al. 2002)	0.02 mg L^{-1} GA_3		

Abbreviations 2,4-D: 2,4-Dichlorophenoxyacetic acid, 2,4,5-T: 2,4,5-Trichlorophenoxyacetic acid, *BA* Benzyl adenine, *CAPS* Cleaved amplified polymorphic sequence, *CI* Callus initiation, *ED* Embryo development, GA_3 Gibberellic acid, *INDI* Induction medium, *Kin* Kinetin, *MSAP* Methylation sensitive amplification polymorphism, *NAA* 1-naphthaleneacetic acid, *PCG* Primary callus growth, *PGR* Plant growth regulators, *PR* Plant regeneration medium, *PSE* Primary somatic embryos, *SCG* Secondary callus growth, *SSR* Simple sequence repeat, *SSE* Secondary somatic embryos, *TDZ* Thidiazuron

Table 2 Genotype, source of variation and main results obtained in works on somaclonal variation in cacao

References	Genotype	Source of variation	Results
Rodríguez López et al. (2004)	*LCT EEN 37/I, LCT EEN162/S-1010, SC3* and *SIAL 93*	Embryogenic culture	• The assay was able to predict chimeric composition with a reasonable level of confidence (± 5%), provided the rare allele exceeded 15% of the template • The study analysed 233 plants developed from SE • The study identified 72 (31%) putative chimeric mutants resulting from slippage mutations or allele loss across two loci
Fang et al. (2009)	*LCTEEN162/ S-10101, LCTEEN 37/I, SC3, SIAL93* and *AMAZ 12*	Cryopreservation methodology	• The study found that PSE for a range of cacao genotypes showed high levels of somaclonal variation compared to their donor trees, with a polymorphic profile frequency of 38.1% • SSE showed lower mutation frequencies, with a polymorphic profile frequency of 23.3% • Cacao somatic embryos cryopreserved through encapsulation-dehydration had no polymorphism, indicating that this cryopreservation method does not impose any significant extra mutational load on the susceptible cacao SE
Rodríguez López et al. (2010a)	*LCTEEN162/S-1010* and *SIAL93*	Embryogenic culture	• The study found 49 polymorphisms, indicating a relatively high rate of mutations in the clonal population • The high rate of mutations suggests that caution should be taken when using SE for plant germplasm conservation

(continued)

Table 2 (continued)

References	Genotype	Source of variation	Results
Rodríguez López et al. (2010b)	*LCTEEN37/1, LCTEEN162/ S-1010, SC3,* and *SIAL93*	Embryogenic culture	• Plant regenerants from calli exhibited 35% polymorphism, with genetic variation accumulating with culture age initially, but subsequently declining • The epigenetic patterns varied between samples taken from leaves and staminodes of the original trees. Leaves from plants regenerated over an extended period showed fewer genetic and epigenetic changes compared to those regenerated after a short callus phase • The study contradicts the common belief that genetic variations accumulate with exposure to callus culture • Embryos initiated late in the lifespan of a callus yielded regenerants with significantly fewer aberrant alleles
Ajijah et al. (2016)	*Sca6, Pa300, UIT1, ICS13, GC7, DR2, ICCRI2, CIMANGGU1* and *CIMANGGU2*	Embryogenic culture	• The developed protocol resulted in a high percentage of somatic embryo formation, germination, and plantlet conversion, and a medium percentage of somaclonal variations based on SSR marker analysis
Adu-Gyamfi et al.(2016)	*AMAZ 15*	In vitro conservation, cryopreservation and post cryopreservation generation of somatic embryos	• Somatic embryos accumulated epigenetic changes, but the changes were less severe compared to embryos regenerated after being stored in liquid nitrogen • Passing cryopreserved embryos through another embryogenic stage increased somaclonal variation, which was partially reversible • Cryopreservation-induced phenotypic variability could be due to DNA methylation changes. It was partially reversible, indicating that the cryopreservation-induced phenotypic variability could be due to DNA methylation changes

(continued)

Table 2 (continued)

References	Genotype	Source of variation	Results
Garcia et al. (2021)	*CCN10* and *CCN51*	Embryogenic culture	• The authors identified DMGs in the CpG context between zygotic embryos and normal and abnormal somatic embryos, which can help explain the morphological abnormalities • They also identified the role of ethylene and its precursor in several biological processes associated with abnormal somatic embryos
Henao et al. (2022a)	*TSH565* and *CCN51*	Embryogenic culture	• The leaves of the IVL and FPL seedlings were compared, and their genetic composition was found to be similar with no significant differences • The Jaccard's coefficient showed high similarity between the two genotypes, with more than 91% for *TSH565* and 92% for *CCN51* • The study concludes that the protocol used for SE is efficient and suitable for large-scale propagation of these two genotypes for commercial purposes

Abbreviations CpG Cytosine-phosphate-guanine sequence, *DMGs* Differentially methylated genes, *FPL* Leaves of field plants, *IVL* Leaves in SE plants, *PSE* Primary somatic embryos, *SE* Somatic embryogenesis, *SSE* Secondary somatic embryos

of estimating the proportion of mutant cells. The assay exploited consistent differential amplification of alternate, simple sequence repeat (SSR) alleles at heterozygous loci. Calibration of the relative amplification of alleles from two different genotypes and the synthetic chimeras created from them revealed a strong linear relationship between "peak heights" representing alternate alleles following capillary electrophoresis. The assay predicts chimeric composition to a reasonable confidence level (±5%) so long as the rare allele exceeds 15% of the template. The system was applied to 233 cocoa regenerants from SE and identified 72 (31%) putative chimeric mutants for slippage mutation or allele loss across two loci.

To evaluate genetic fidelity during the SE process, Fang et al. (2009) used microsatellites as molecular markers. Ten cocoa SSR were used to screen a population of PSE and SSE. They found 38.1% of polymorphic profiles in the PSE, while the frequency was 23.3% in SSE. PSE for a range of cocoa genotypes exhibited pronounced levels of somaclonal variation, as evidenced by microsatellite profiles divergent from those of their donor trees. Their descendant SSE showed lower mutation frequencies. An additional population of SSE produced after a comparable time in culture and then taken through cryopreservation, exhibited no mutations when screened with the same microsatellite markers.

In the next year, Rodríguez López et al. (2010b) surveyed both genetic and epigenetic changes in plants regenerated from calli of various ages and between tissues from the source trees. They used 15 SSR and 386 MSAP markers to study genetic and epigenetic modifications, respectively. In relation to genetic alterations, regenerated plants exhibited 35% polymorphism. Genetic variation initially accumulated with culture age but subsequently declined. Epigenetic profiles diverged between leaf and staminode samples from source trees. Leaves of late regenerants exhibited significantly less genetic and epigenetic divergence than those exposed to short periods of callus growth. Contrarily to the widespread belief that somaclonal genetic variants accumulate with exposure to callus culture, embryos initiated late in the lifespan of a callus yielded regenerants with significantly fewer aberrant alleles.

Ajijah et al. (2016) optimize a SE method and assessed somaclonal variation using SSR markers. The developed SE protocol resulted in a high percentage of somatic embryo formation (5.6–66.7%), germination (50%), and plantlet conversion (65%), and a medium percentage of somaclonal variations (16%) based on SSR marker analysis. In relation to somaclonal variation evaluation, 19 SSR markers were used to compare mother plant and plantlets regenerated from cacao somatic embryos of *CIMANGGU2*. Out of 19 regenerated plantlets, 3 exhibited different SSR allele profiles than the mother plant. The frequency of observed mutations was 6.2×10^{-3} (4 out of 646 events). Three out of f 4 events were allele loss, while the remaining one was new allele formation. Cluster analysis among regenerated plantlets based on 19 SSR loci resulted in a 0.97–100% level of similarity.

The frequency of somaclonal variation in this study was found to be lower than that described by Rodríguez López et al. (2010b), where it ranged from 35 to 50% depending on the genotype. Furthermore, findings in this research regarding regenerated cacao seedlings with a high frequency of allele loss and multiple mutations align with those reported by Rodríguez López et al. (2010b).

Likewise, Henao-Ramírez et al. (2021) studied the genetic stability of cacao plantlets propagated by SE and conventional grafting of genotypes *CCN51* and *TSH565* using 13 SSR. They found that the 13 analyzed loci revealed 25 alleles in genotype *CCN51* and 24 alleles in genotype *TSH565*. Higher polymorphic information contact (PIC) values were observed for all SSR. Only *MTCCIR8* and *MTCUNICAMP09* were intermediate, with PIC values of less than 0.250. In vitro SE and *ex vitro* field graft plant populations were genetically equal. According to the results, no differences in allelic composition were observed between leaves of field plants (FPL) and leaves of SE plants (IVL) in each genotype. This indicates that plants propagated by SE did not show perceptible detriment to their genome with the used SSR. In addition, Jaccard's coefficient showed more than 91% similarity for *TSH565* and 92% for *CCN51*. The unweighted pair group method with arithmetic average (UPGMA) and principal component analysis (PCA) showed that the populations tended to group within two genotypes. The SSR results obtained do not exclude the occurrence of other changes in the nuclear genome. However, considering the morphological stability of in vitro propagated plants, the results indicate that the protocol is suitable and efficient for large scale, true-to-type propagation of genotypes *CCN51* and *TSH565* for commercial purposes.

Considering the relevance of SE for germplasm conservation, several works have been carried out, seeking to determine the genetic stability of the regenerants after the conservation treatments have been applied. Rodríguez López et al. (2010a) characterized new mutations among a clonal population of cocoa plants regenerated via SE after cryopreservation by cleaved amplified polymorphic sequence (CAPS). They found 49 polymorphisms, comprising 26.3% of the screened samples, with an inferred rate of 2.89×10^{-3} substitutions/screened base. This elevated rate is of a similar order of magnitude than previous reports of de novo microsatellite length mutations arising in the crop. It suggests that caution should be exercised when applying SE to conserve plant germplasm. In another study carried out by Fang et al. (2009), SSE cryopreserved through the encapsulation-dehydration method showed no signs of polymorphism. This underscores that while cocoa SE might be prone to genetic variations, the encapsulation-dehydration cryopreservation technique does not contribute to any additional mutational load.

Adu-Gyamfi et al. (2016) evaluated the effect of in vitro conservation, cryopreservation and generation of somatic embryos post-cryopreservation on the appearance of epigenetic somaclonal variation using methylation sensitive amplification polymorphism (MSAP). Somatic embryos accumulated epigenetic changes, but in less extension than after storage in liquid nitrogen. Furthermore, the passage of cryopreserved embryos through another embryogenic stage led to further increases in variation. Interestingly, this detected variability appears to be, in some measure, reversible. In the context of the study, the term "reversible" refers to the fact that some of the epigenetic changes observed in somatic embryos after cryopreservation can be partially restored or reversed during the subsequent rounds of SE. The authors observed that some epialleles (epigenetic variants) that showed changes in frequency associated with SE/cryopreservation returned to donor tree allele frequency levels after cryopreservation or after tertiary embryogenesis. This suggests that some of the epigenetic changes induced by cryopreservation and in vitro culture are not permanent and can be modified through subsequent developmental stages. However, it is important to note that not all of the observed variations were reversible, and some of them may have lasting effects on the phenotype of the plant. In addition, the outcome of this study indicates that the cryopreservation-induced phenotypic variability could be, at least partially, due to DNA methylation changes.

Finally, the methylome landscape of cacao SE was examined by Garcia et al. (2021), using a whole-genome bisulfite sequencing technique, with the aim to understand the epigenetic basis of somatic embryo abnormalities. As a result, they identified 873 differentially methylated genes (DMGs) in the cytosine-phosphate-guanine sequence (CpG) context between zygotic embryos and normal and abnormal somatic embryos, with important roles in development, programmed cell death, oxidative stress, and hypoxia induction. This can help explain the morphological abnormalities of somatic embryos. They also identified the role of ethylene and its precursor 1-aminocyclopropane-1-carboxylate in several biological processes, such as hypoxia induction, cell differentiation and cell polarity, which could be associated with the development of abnormal somatic embryo.

3.1 *Abnormalities in the Development of Somatic Embryos of Cacao*

The production of cacao via SE is divided into five well-defined steps as follows: initiation, multiplication, maturation, germination, and acclimatization (Henao et al. 2022a). All these steps are mediated by a complex regulatory network where different genes are regulated at the transcriptional level by both genetic and epigenetic mechanisms (Maximova et al. 2014).

Several factors can influence the level of genetic and epigenetic somaclonal variation that occurs during in vitro culture, including the composition of the culture medium, regeneration technique, origin of the donor tissue, duration of culture, and number of regeneration events (Adu-Gyamfi et al. 2016) This process can lead to a number of problems, including low survival rates, abnormal plant growth, malformed leaves, root malformation, stunted growth, or poor growth. Abnormalities in the somatic embryos may cause these problems, in addition to other factors, such as contamination, nutrient deficiencies, and hormonal imbalances (El-Sherif 2019). All these issues could threaten the commercial viability of cacao SE for clonal multiplication purposes. Therefore, it is very important to know and detail them.

Plant growth regulators (PGRs) like auxins as 2,4-D and 2,4,5-T, play a crucial role in inducing SE in many plant species. Auxin is a highly potent signaling molecule that controls multiple aspects in plants, including cell division and elongation, and development of root, leaf, flower, zygotic embryo, and fruit (Sivanesan et al. 2022). A basic mechanism by which auxin triggers diverse developmental processes includes the regulation of gene expression at the transcriptional level (Wójcikowska et al. 2020). However, excessive use or high concentrations of 2,4-D can lead to genetic and epigenetic changes in cells, such as DNA mutation and methylation. These changes can disrupt normal somatic embryo development, leading to abnormal formation. Long-term exposure or accumulation of 2,4-D can also disrupt the normal balance of auxins in the tissue, interfering with the polarization of cells and leading to abnormal bilateral symmetry during early plant embryogenesis (Osorio-Montalvo et al. 2018). Therefore, it is important to use auxins at appropriate concentrations and exposure times to avoid disrupting normal somatic embryo development. Moreover, the crosstalk of auxin with epigenetic processes needs also to be considered (Fehér 2015; Garcia et al. 2019).

Morphological deformation in somatic embryos commonly occurs as a result of physiological disorders or somaclonal variation. Understanding the causes of somaclonal variation is important to establish reliable protocols, alternatives, and solutions for plant propagation through in vitro culture.

In general, the main phenotypic abnormalities in cacao somatic embryos are observed during the cotyledonary stages and during the conversion of the embryo into a plantlet. The most recognized abnormalities are malformed or absent cotyledons, absence of apical and radical meristems, and loss of bipolarity, among others (Hussain et al. 2009; el Dawayati et al. 2012).

Considering the results obtained in the evaluation of the regeneration potential via SE of different cacao genotypes, it was possible to identify different abnormalities based on the morphology and development of somatic embryos. They can be classified in three categories based on the structures affected, similar to the abnormal types reported by Garcia et al. (2022): (1) The number and shape of the cotyledons, such as multicotyledonary, monocotyledons, cotyledon with overgrowth, bulbiform cotyledons, or poorly developed cotyledons (Fig. 3; Table 3); (2) The shape and development of the axes, such as fused axes, atrophy of the hypocotyl, coiled hypocotyl, or axes without development (Fig. 4; Table 3) and (3) The morphology and development of the seedling and its radicle system, such as polyradiculation, lacking apical and/or radical meristems, abnormal plant growth, variegated and malformed leaves, root malformation, or stunted growth, among others (Fig. 5; Table 3).

3.2 Aspects to Consider in Order to Reduce Somaclonal Variation in Cocoa Tissue Culture

Some of the changes produced by somaclonal variation have been utilized in plant breeding to give rise to genetic diversity through tissue culture-derived plant material. However, the production through SE of a high number of mutant regenerants may negatively impact the economic viability of this approach, as the generated variations are not always desirable or useful. It is a great biological barrier to the propagation of plants, but its risks can be increasingly reduced with the knowledge of the causes that originate it. It is important to minimize the frequency of somaclonal variation to below 10% to ensure the production of high-quality cocoa plants (Ajijah and Hartati 2019).

It has been sufficiently documented that the genotype has a significant effect on somaclonal variation and the percentage of regeneration (Garcia et al. 2018). From a genetic perspective, the frequency at which mutants appear varies by gene. Specifically, some genes, located in chromosomal "hot spots", tend to mutate more frequently. In studies in our laboratory, we have identified malformations in different cacao genotypes, such as *CCN51*, *TSH565*, *CNCh12*, *CNCh13*, *IMC67*, and *PA46*, when propagated in vitro, as reflected in Figs. 3, 4 and 5. Specifically, the genotypes *IMC67* and *PA46* stand out for having a high percentage of malformations during seedling germination (data not shown). Adopting genetically stable genotypes in tissue culture is a key strategy to mitigate the risk of genetic alterations (Bairu et al. 2011; Weckx et al. 2019).

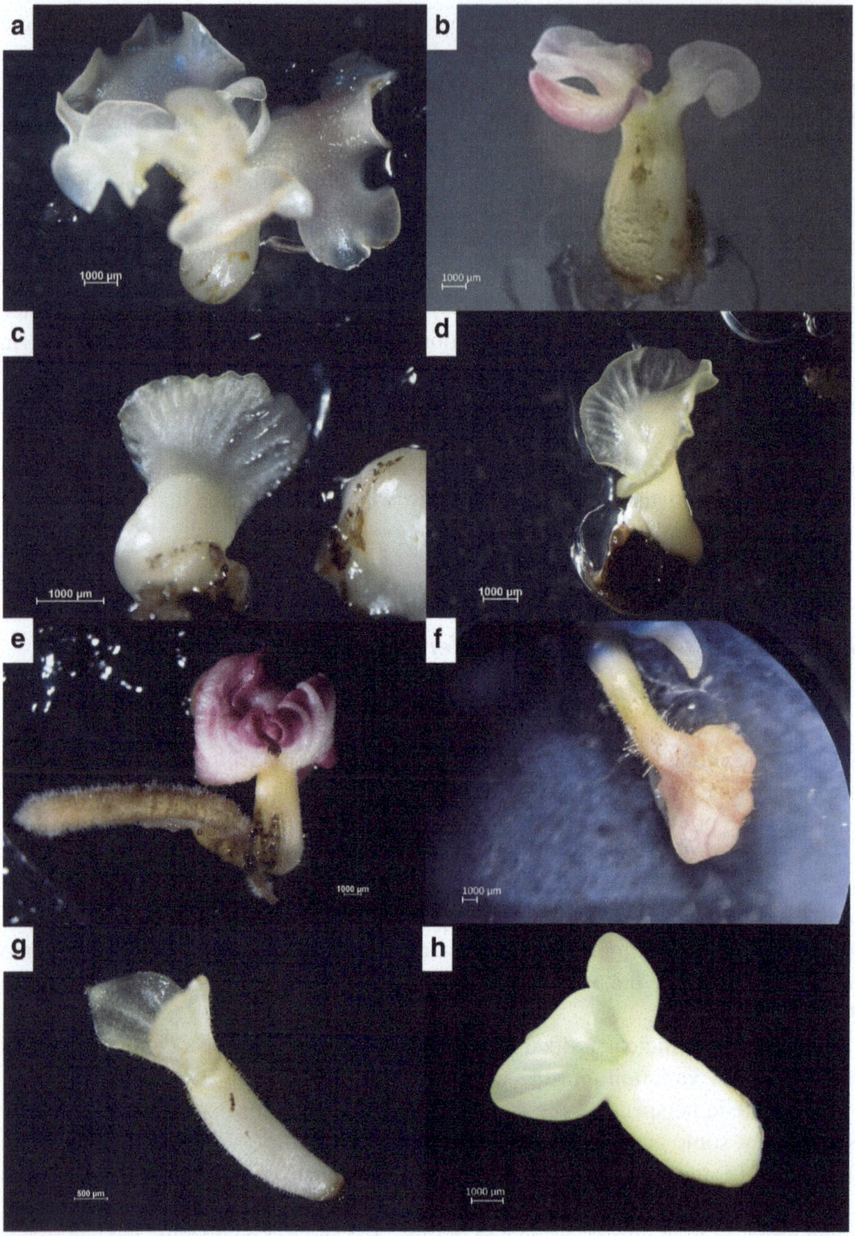

Fig. 3 Main abnormalities in cotyledonary somatic embryos observed in different cocoa genotypes. **a**, **b** Multicotyledonary; **c**, **d** monocotyledons; **e** cotyledon with overgrowth; **f** bulbiform cotyledon; **g** poorly developed cotyledons; **h** normal development

Table 3 Main abnormalities in cacao cotyledonary somatic embryos and regenerated plantlets

Number and shape of the cotyledons	Shape and development of the axis	Shape and development of the root
Multicotyledonary: embryos with multiple cotyledonary structures (3 or more cotyledons) emerging from the apex **Monocotyledons**: embryos that only develop one cotyledon, usually a centralized structure, can have a webbed shape **Cotyledon with overgrowth**: embryos with an excessive and evident development of cotyledons, that may present coiling **Bulbiform cotyledons**: embryos in which the cotyledons fuse and manifest as a compact, fleshy mass **Poorly developed cotyledons**: embryos in which the size of the hypocotyl is larger, and the cotyledons show very incipient or undeveloped growth	**Fused axes**: the process by which two or more embryos are completely united at their axial base **Atrophy of the hypocotyl**: embryos in which a hypocotyl with insufficient development is observed, which is imperceptible or merely incipient **Coiled hypocotyl**: embryos with an elongated axis that grows in a spiral or coils on itself **Axes without development**: embryos that lack an apical and root axis and whose manifestation is reduced to a developed tissue	**Polyradiculation**: plantlets in which two primary roots have developed **Stunted growth:** normal plantlets with a reduced size, also known as dwarf plants **Abnormal apical meristems**: plants with a partial or complete lack of a shoot apical meristem **Variegated and malformed leaves:** leaves bearing more than one color that develop distorted foliage as they grow, such as curled or cupped leaves **Root malformation**: irregular or abnormal structural development due to the organization of a quiescent center and root cap in the root apex **Total abnormal plant growth**: when growth is not properly regulated, anomalies and tumors may develop

Additionally, based on the genotype, specific in vitro culture conditions can enhance embryo quality by reducing abnormalities. One of the pivotal PGRs for inducing SE is 2,4-D. High doses and extended exposure to this PGR might lead to abnormal embryo development (Garcia et al. 2019). When applied to embryos in vitro, 2,4-D disrupts the natural auxin balance. Both 2,4-D and 2,4,5-T are PGRs, and their usage can lead to alterations, such as methylation or mutations, resulting in both genetic and epigenetic changes. For many genotypes, it is common to use these compounds, particularly 2,4-D and 2,4,5-T, as stress-inducing agents in plants. In fact, various studies, such as those by Kouassi et al. (2017), López-Baez et al. (2000), Maximova et al. (2002), pinpoint 2,4-D and 2,4,5-T as primary activators of the SE process. It is known that stress in plants, caused by chemical or physical factors, can trigger the SE process. Thus, there could be consideration for exploring other alternatives that do not involve the use of 2,4-D or 2,4,5-T to stimulate embryonic development. It is worth noting that these auxins, once introduced, are challenging to remove and tend to linger in tissues for extended periods.

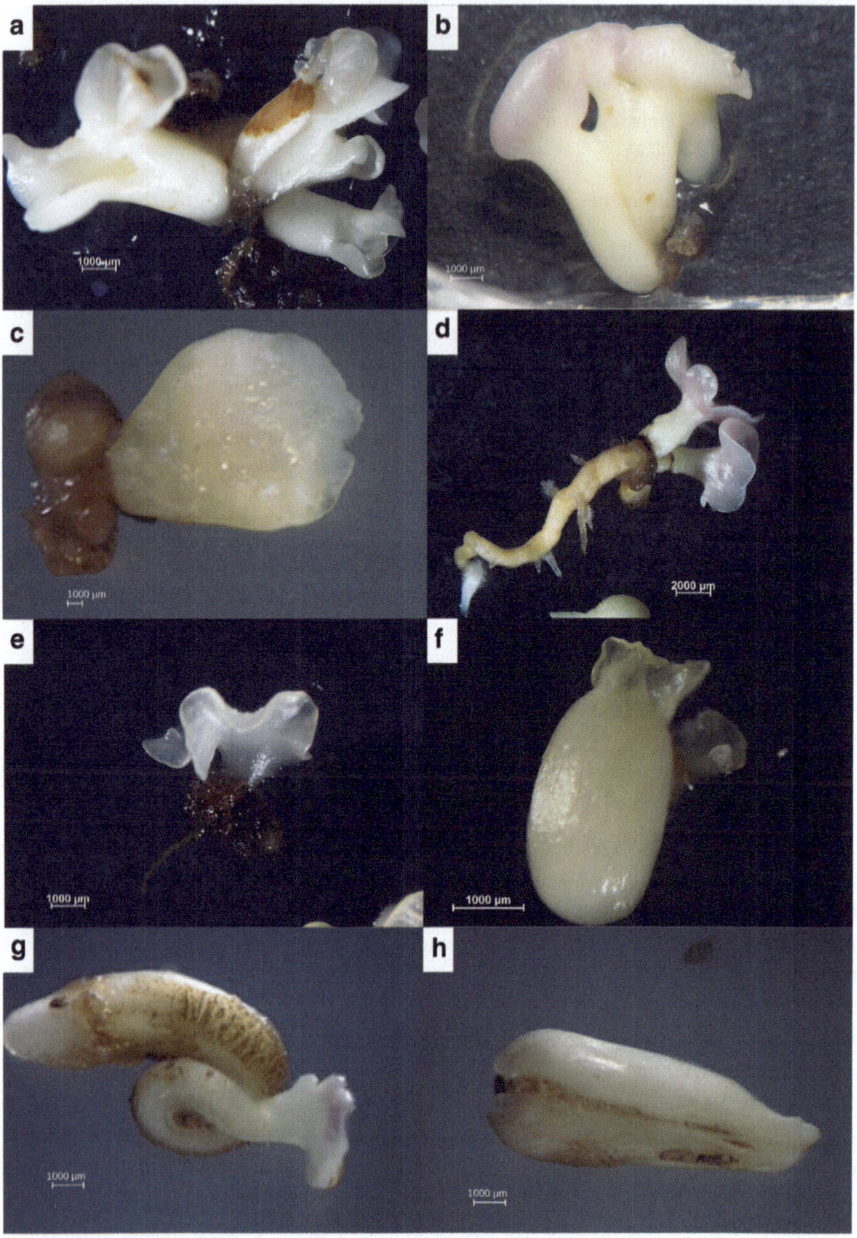

Fig. 4 Main abnormalities in cocoa cotyledonary somatic embryos related to shape and development of the axis. **a**, **b**, **d** Fused axes; **c** axes without development; **e** atrophy of the hypocotyl; **f**, **h** axes without development; **g** coiled hypocotyl

Fig. 5 Main abnormalities in the shape and development of cacao in vitro regenerated plantlets and roots. **a**, **f** Abnormal apical meristems; **b** polyradiculation; **c** variegated and malformed leaves; **d** stunted growth; **e** total abnormal plant growth; **g**, **h** normal plant growth

4 Conclusions

The SE process presents significant challenges for the large-scale production of economically valuable cocoa genotypes, for instance, in relation to somaclonal variation. However, with ongoing research, it is possible to overcome these challenges and achieve effective commercial propagation. To address these issues, it is essential to delve deeper into areas such as physiology, molecular biology, and genetics. This comprehensive approach will allow to better understand problems like somaclonal variation and to develop tailored solutions, such as creating adjusted protocols for each genotype and cultivation stage. The SSE technique, which has shown greater efficacy and lower mutation risk than PSE, stands out as a crucial tool for future research aimed at large-scale production. As the SE process progresses, tools like bioreactors and temporary immersion systems can be incorporated, or even combined with solid culture media, with the aim of maximizing productivity and minimizing costs.

Acknowledgements We would like to thank the Laboratory of Plant Physiology and Plant Tissue Culture of Universidad de Antioquia. A special acknowledgement to Universidad de Antioquia's Research Development Committee (CODI).

References

Adu-Gyamfi R, Wetten A, Rodríguez C, Marcelino (2016) Effect of cryopreservation and post-cryopreservation somatic embryogenesis on the epigenetic fidelity of cocoa (*Theobroma cacao* L.). PLoS One 11(7):1–13. https://doi.org/10.1371/journal.pone.0158857

Ajijah N, Hartati R (2019) Primary and secondary somatic embryogenesis of cacao: the effect of explant types and plant growth regulators. Indonesian J Agric Sci 20(2):69–76. https://doi.org/10.21082/ijas.v.20.n2.2019.p69-76

Ajijah N, Hartati R, Rubiyo R, Sukma D, Sudarsono D (2016) Effective cacao somatic embryo regeneration on kinetin supplemented DKW medium and somaclonal variation assessment using SSRs markers. Agrivita 38(1):80–92. https://doi.org/10.17503/agrivita.v38i1.619

Arevalo-Gardini E, Meinhardt LW, Zuñiga LC, Arévalo-Gardni J, Motilal L, Zhang D (2019) Genetic identity and origin of "Piura Porcelana"—a fine-flavored traditional variety of cacao (*Theoborma cacao*) from the Peruvian Amazon. Tree Genet Genomes 15(1). https://doi.org/10.1007/s11295-019-1316-y

Azizi P, Hanafi MM, Sahebi M, Harikrishna JA, Taheri S, Yassoralipour A, Nasehi A (2020) Epigenetic changes and their relationship to somaclonal variation: a need to monitor the micropropagation of plantation crops. Funct Plant Biol 47(6):508–523

Bairu MW, Aremu AO, van Staden J (2011) Somaclonal variation in plants: causes and detection methods. Plant Growth Regul 63(2):147–173. https://doi.org/10.1007/s10725-010-9554-x

Bekele FL, Bidaisee GG, Singh H, Saravanakumar D (2020) Morphological characterisation and evaluation of cacao (*Theobroma cacao* L.) in Trinidad to facilitate utilisation of Trinitario cacao globally. Genet Resour Crop Evol 67(3):621–643. https://doi.org/10.1007/s10722-019-00793-7

Bhojwani SS, Dantu PK (2013) Somaclonal variation. In: Plant tissue culture: an introductory text [place unknown]: Springer, pp 141–154

Boza EJ, Irish BM, Meerow AW, Tondo CL, Rodríguez OA, Ventura-López M, Gómez JA, Moore JM, Zhang D, Motamayor JC, Schnell RJ (2013) Genetic diversity, conservation, and utilization of *Theobroma cacao* L.: genetic resources in the Dominican Republic. Genet Resour Crop Evol [Internet] 60(2):605–619. https://doi.org/10.1007/s10722-012-9860-4

Bustami M, Werbrouck S (2018) Somatic Embryogenesis in Elite Indonesian cacao *Theobroma cacao* L. In: Jain SM, Gupta P (eds) Step wise protocols for somatic embryogenesis of important woody plants [Internet] [place unknown]. Springer International Publishing AG, pp 73–81. https://doi.org/10.1007/978-1-4939-8594-4_15

Bustami MU, Werbrouck S (2017) Comparison of two protocols for somatic embryo induction in a Sulawesi elite *Theobroma cacao* L. clone. Acta Hortic 1155:71–76. https://doi.org/10.17660/ActaHortic.2017.1155.9

Cadby J, Araki T (2021) Towards ethical chocolate: multicriterial identifiers, pricing structures, and the role of the specialty cacao industry in sustainable development. SN Bus Econ 1(3). https://doi.org/10.1007/s43546-021-00051-y

Cádiz-Gurrea M, Fernández I, Aguilera LM, Fernández S, Legeai L, Bouaziz M, Segura A (2019) Bioactive compounds from *Theobroma cacao*: effect of isolation and safety evaluation. Plant Foods Hum Nutr 74(1):40–46. https://doi.org/10.1007/s11130-018-0694-x

Carvalho LSO, Ozudogru EA, Lambardi M, Paiva L (2019) Temporary immersion system for micropropagation of tree species: a bibliographic and systematic review. Not Bot Horti Agrobot Cluj Napoca 47(2):269–277

Clark C (2022) Chocolate. In: Miller J, Van C (eds) Food and health. Springer, Cham

el Dawayati MM, Abd EL Bar OH, Zaid ZE, Zein El Din AFM (2012) *In vitro* morpho-histological studies of newly developed embryos from abnormal malformed embryos of date palm cv. Gundila under desiccation effect of polyethelyne glycol treatments. Ann Agric Sci [Internet]. 57(2):117–128. https://doi.org/10.1016/j.aoas.2012.08.005

Duta-Cornescu G, Constantin N, Pojoga D-M, Nicuta D, Simon-Gruita A (2023) Somaclonal variation—advantage or disadvantage in micropropagation of the medicinal plants. Int J Mol Sci 24(1):838

Egertsdotter U, Ahmad I, Clapham D (2019) Automation and scale up of somatic embryogenesis for commercial plant production, with emphasis on conifers. Front Plant Sci 10(February):1–13. https://doi.org/10.3389/fpls.2019.00109

El-Sherif NA (2019) Impact of plant tissue culture on agricultural sustainability. In: Handbook of environmental chemistry, vol 77. [place unknown], Springer, pp 93–107. https://doi.org/10.1007/698_2017_160

Entuni G, Edward R, Nori H, Ahmad A (2018) Field performance of selected Malaysian cocoa clones regenerated through somatic embryogenesis cultures. Malays Appl Biol 47(1):97–102

Etaware PM (2021) The effects of the phytochemistry of cocoa on the food chemistry of chocolate(s) and how disease resistance in cocoa can be improved using CRISPR/Cas9 technology. Food Chem Mol Sci 3.https://doi.org/10.1016/j.fochms.2021.100043

Fang JY, Wetten A, Adu-Gyamfi R, Wilkinson M, Rodríguez-López C (2009) Use of secondary somatic embryos promotes genetic fidelity in cryopreservation of cocoa (*Theobroma cacao* L.). Agric Food Sci 18(2):152–159. https://doi.org/10.2137/145960609789267579

Fehér A (2015) Somatic embryogenesis—stress-induced remodeling of plant cell fate. Biochim Biophys Acta Gene Regul Mech 1849(4):385–402. https://doi.org/10.1016/j.bbagrm.2014.07.005

Fehér A (2019) Callus, dedifferentiation, totipotency, somatic embryogenesis: what these terms mean in the era of molecular plant biology? Front Plant Sci 10(April):1–11. https://doi.org/10.3389/fpls.2019.00536

Fister AS, Landherr L, Maximova SN, Guiltinan MJ (2018) Transient expression of CRISPR/Cas9 machinery targeting TcNPR3 enhances defense response in *Theobroma cacao*. Front Plant Sci 9.https://doi.org/10.3389/fpls.2018.00268

Fontanel A, Gire S, Labbe G, Favereau P, Alvarez M, Von S, Petiard V (2002) *In vitro* multiplication and plant regeneration of *Theobroma cacao* L. via stable embryogenic calli. IAPTC Congress, Plant Biotechnology 2002 and beyond:23–28

Garcia C, Corrêa F, Seth F, Alex A, Marcio C, Motamayor C, Schnell R, Jean M (2016) Optimization of somatic embryogenesis procedure for commercial clones of *Theobroma cacao* L. Afr J Biotechnol 15(36):1936–1951. https://doi.org/10.5897/AJB2016.15513

Garcia C, Britto D, Marelli JP (2018) Transcription factors: their role in the regulation of somatic embryogenesis in *Theobroma cacao* L. and other species. In: Methods in molecular biology, vol 1815. [place unknown]: Humana Press Inc., pp 385–396. https://doi.org/10.1007/978-1-4939-8594-4_27

Garcia C, Furtado de Almeida A, Costa M, Britto D, Valle R, Royaert S, Marelli J (2019) Abnormalities in somatic embryogenesis caused by 2,4-D: an overview. Plant Cell Tissue Organ Cult [internet]. 137(2):193–212. https://doi.org/10.1007/s11240-019-01569-8

Garcia C, Almeida A-AF de, Costa M, Britto D, Mangabeira P, Silva L, Silva J, Royaert S, Marelli J-P (2021) Single-base resolution methylomes of somatic embryogenesis in *Theobroma cacao* L. Reveal epigenome modifications associated with somatic embryo abnormalities. Sci Rep.: under review

Garcia C, Furtado de Almeida AA, Costa M, Britto D, Correa F, Mangabeira P, Silva L, Silva J, Royaert S, Marelli JP (2022) Single-base resolution methylomes of somatic embryogenesis in *Theobroma cacao* L. reveal epigenome modifications associated with somatic embryo abnormalities. Sci Rep 12(1). https://doi.org/10.1038/s41598-022-18035-9

Guillou C, Verdier D (2022) *Theobroma cacao*: somatic embryogenesis. In: Ramírez-Mosqueda MA (ed) Somatic embryogenesis methods in molecular biology, vol 2527. New York, pp 69–81

Henao AM, Hajduk D, Cano D, Trujillo A (2022a) Indicator framework for large-scale cacao (*Theobroma cacao* L.) *in vitro* plant production planning and controlling. Bionatura [Internet] 8(1):1–7. https://doi.org/10.21931/RB/2023.08.01.8

Henao A, Palacio D, Urrea A (2022b) Cost analysis of cacao (*Theobroma cacao* L.) plant propagation through the somatic embryogenesis method. Bionatura 7(2):1–13. https://doi.org/10.21931/rb/2022.07.02.2

Henao-Ramírez AM, Salazar Duque HJ, Calle Tobón AF, Urrea Trujillo AI (2021) Determination of genetic stability in cacao plants (*Theobroma cacao* L.) derived from somatic embryogenesis using microsatellite molecular markers (SSR). Int J Fruit Sci 21(1):284–298. https://www.tandfonline.com/, https://doi.org/10.1080/15538362.2021.1873219

Hussain SS, Rao AQ, Husnain T, Riazuddin S (2009) Cotton somatic embryo morphology affects its conversion to plant. Biol Plant [internet] 53(2):307–311. https://doi.org/10.1007/s10535-009-0055-6

Ikeuchi M, Ogawa Y, Iwase A, Sugimoto K (2016) Plant regeneration: cellular origins and molecular mechanisms. Development (cambridge). 143(9):1442–1451. https://doi.org/10.1242/dev.134668

Kouassi M, Kahia J, Kouame C, Tahi M, Koffi K (2017) Comparing the effect of plant growth regulators on callus and somatic embryogenesis induction in four elite *Theobroma cacao* L. genotypes. HortScience 52:142–145. https://doi.org/10.21273/HORTSCI11092-16

Lafargue-Molina P (2021) Marker development for the traceability of certified sustainably produced cacao (*Theobroma cacao*) in the chocolate industry [place unknown]

Li Z, Traore A, Maximova S, Guiltinan M (1998) Somatic embryogenesis and plant regeneration from floral explants of cacao (*Theobroma cacao* L.) using thidiazuron. In vitro Cell Dev Biol 34:293–299. https://doi.org/10.1007/BF02822737

López-Baez O, Moreno J, Pacheco S (2000) Avances en Propagación de cacao *Theobroma cacao* por embriogenesis Somática en Mexico. In: Cocoa biotech. Malaysia

Macias-Naranjo S, Henao-Ramírez A, Urrea-Trujillo A (2023) Propagation of the Colombian genotype of cacao (*Theobroma cacao* L.) *CNCh12* by somatic embryogenesis. Bionatura 8(1)

Maximova S, Alemanno L, Young A, Ferriere N, Traore A, Guiltinan M (2002) Efficiency, genotypic variability, and cellular origin of primary and secondary somatic embryogenesis of *Theobroma cacao* L. In Vitro Cell Develop Biol Plant 38(3):252–259. https://doi.org/10.1079/IVP2001257

Maximova SN, Florez S, Shen X, Niemenak N, Zhang Y, Curtis W, Guiltinan MJ (2014) Genome-wide analysis reveals divergent patterns of gene expression during zygotic and somatic embryo maturation of *Theobroma cacao* L., the chocolate tree. BMC Plant Biol 14(1):1–18

Minyaka E, Niemenak N, Sangare A, Ndoumou D (2008) Effect of $MgSO_4$ and K_2SO_4 on somatic embryo differentiation in *Theobroma cacao* L. Plant Cell Tissue Organ Cult 94(2):149–160. https://doi.org/10.1007/s11240-008-9398-5

Nair K (2021) Cocoa (*Theobroma cacao* L.). In: Tree crops. [place unknown]. Springer International Publishing. https://doi.org/10.1007/978-3-030-62140-7

Noah AM, Niemenak N, Sunderhaus S, Haase C, Omokolo DN, Winkelmann T, Braun H-P (2013) Comparative proteomic analysis of early somatic and zygotic embryogenesis in *Theobroma cacao* L. J Proteomics [Internet] 78:123–133. https://doi.org/10.1016/j.jprot.2012.11.007

Osorio-Guarín JA, Berdugo-Cely J, Coronado RA, Zapata YP, Quintero C, Gallego-Sánchez G, Yockteng R (2017) Colombia a source of cacao genetic diversity as revealed by the population structure analysis of germplasm bank of *Theobroma cacao* l. Front Plant Sci 8.https://doi.org/10.3389/fpls.2017.01994

Osorio-Montalvo P, Sáenz-Carbonell L, De-la-Peña C (2018) 5-azacytidine: a promoter of epigenetic changes in the quest to improve plant somatic embryogenesis. Int J Mol Sci 19(10). https://doi.org/10.3390/ijms19103182

Osorio T, Henao A, de la Hoz T, Urrea A (2022) Propagation of *IMC67* plants, universal cacao (*Theobroma cacao* L.) rootstock via somatic embryogenesis. Int J Fruit Sci 22(1):78–94. https://doi.org/10.1080/15538362.2021.2023067

Peña-López JL, Azpeitia-Morales A, Mirafuentes-Hernández F, Ruíz-Carrera V, Sáenz-Carbonell L (2016) Embriogénesis somática en cacao (*Theobroma cacao*) en sistemas de inmersión automatico. Ecosistemas y Recursos Agropecuarios [Internet] 3(8):215–224. www.ujat.mx/era215

Richardson JE, Whitlock BA, Meerow AW, Madriñán S (2015) The age of chocolate: a diversification history of Theobroma and Malvaceae. Front Ecol Evol 3(NOV). https://doi.org/10.3389/fevo.2015.00120

Rodríguez López C, Wetten A, Wilkinson M (2004) Detection and quantification of *in vitro*-culture induced chimerism using simple sequence repeat (SSR) analysis in *Theobroma cacao* (L.). Theor Appl Genet 110(1):157–166. https://doi.org/10.1007/s00122-004-1823-5

Rodríguez López C, Sicilia H, Wetten A, Wilkinson M (2010) Detection of somaclonal variation during cocoa somatic embryogenesis characterised using cleaved amplified polymorphic sequence and the new freeware Artbio. Mol Breeding 25(3):501–516. https://doi.org/10.1007/s11032-009-9348-x

Rodríguez López C, Sicilia H, Wetten A, Wilkinson M (2010) Progressive erosion of genetic and epigenetic variation in callus-derived cocoa (*Theobroma cacao*) plants. New Phytol 186(4):856–868

Sarmah D, Sutradhar M, Singh BK (2017) Somaclonal variation and its' application in ornamentals plants. Int J Pure App Biosci 5:396–406

Sivanesan I, Nayeem S, Venkidasamy B, Kuppuraj SP, Rn C, Samynathan R (2022) Genetic and epigenetic modes of the regulation of somatic embryogenesis: a review. Biol Futur 73(3):259–277. https://doi.org/10.1007/s42977-022-00126-3

Sodré GA, Gomes ARS (2019) Cocoa propagation, technologies for production of seedlings. Rev Bras Frutic 41(2):1–22. https://doi.org/10.1590/0100-29452019782

Solano W (2008) Embriogénesis Somática en Clones Superiores de Cacao (*Theobroma cacao* L.) Obtenidos en el Programa de Mejoramiento Genético CATIE [place unknown]: Centro Agronómico Tropical de Investigación y Enseñanza, Costa Rica

Soraia T (2013) Somatic Embryogenesis of *Theobroma cacao* L.: developmental physiology of the embryo and improvement of culture conditions. University of Hamburg, Hamburg

Souza PA, Moreira LF, Sarmento D, da Costa FB (2018) Cacao—*Theobroma cacao*. In: Exotic fruits [place unknown]. Elsevier, pp 69–76. https://doi.org/10.1016/b978-0-12-803138-4.00010-1

Suárez-castellá M, Kosky RG, Chong-pérez B, Reyes M, García- L, Sarría Z, Orellana P, Rodríguez A, Triana R, Pérez Z et al (2012) Estrategia de innovación tecnológica para el empleo de embriogénesis somática en medios de cultivo semisólido en *Musa spp.* y su impacto económico. Biotecnología Vegetal 12(1):41–48

Swamy MK (2020) Plant-derived bioactives: Production, properties and therapeutic applications [place unknown]. Springer, Singapore. https://doi.org/10.1007/978-981-15-1761-7

Wang QM, Wang L (2012) An evolutionary view of plant tissue culture: Somaclonal variation and selection. Plant Cell Rep 31(9):1535–1547. https://doi.org/10.1007/s00299-012-1281-5

Weckx S, Inzé D, Maene L (2019) Tissue culture of oil palm: finding the balance between mass propagation and somaclonal variation. Front Plant Sci 10.https://doi.org/10.3389/fpls.2019.00722

Wickramasuriya A, Dunwell J (2018) Cacao biotechnology: current status and future prospects. Plant Biotechnol J 16(1):4–17. https://doi.org/10.1111/pbi.12848

Winkelmann T (2016) In: Germana MA, Lambardi M (eds) Somatic versus zygotic embryogenesis: learning from seeds BT—*In vitro* embryogenesis in higher plants [internet]. Springer, New York, pp 25–46. https://doi.org/10.1007/978-1-4939-3061-6_2

Wójcikowska B, Wójcik AM, Gaj MD (2020) Epigenetic regulation of auxin-induced somatic embryogenesis in plants. Int J Mol Sci 21(7). https://doi.org/10.3390/ijms21072307

Zarrillo S, Gaikwad N, Lanaud C, Powis T, Viot C, Lesur I, Fouet O, Argout X, Guichoux E, Salin F et al (2018) The use and domestication of *Theobroma cacao* during the mid-Holocene in the upper Amazon. Nat Ecol Evol 2(12):1879–1888. https://doi.org/10.1038/s41559-018-0697-x

Factors Influencing Somaclonal Variation in Date Palm, Detection and Selection for Application in the Plantation

Abdul Aziz Mirani, Jennifer Ann Harikrishna, Chee How Teo, and Najamuddin Solangi

Abstract The use of somaclonal variations may suppose an advantage over conventional plant breeding aiming at introducing new traits for increased fertility, high yield, low stature (dwarfness) and tolerance to abiotic and biotic stresses such as salt, drought, water logging or diseases. Date palm is a dioecious plant with a long life cycle that makes its breeding more challenging. The length of the commercial life cycle of date palms also means that it is important to have the best breeding outcomes for longer-term plant health and sustainable yields in the plantation. Currently, tissue culture techniques for date palm focus on micropropagation and the occurrence of somaclonal variations during the micropropagation process is deemed useless. This chapter covers the current research progress in date palm somaclonal variation in terms of sources of variation, its effects and methods of detection; and identifies some gaps in knowledge and recommendations for future research. The comprehensive knowledge of somaclonal variations will help date palm researchers in both micropropagating true-to-type plants and developing new genotypes.

Keywords Date palm · Micropropagation · Somaclonal variations · Transposable elements

---

A. A. Mirani (✉) · N. Solangi
Date Palm Research Institute (DPRI), Shah Abdul Latif University, Khairpur 66020, Sindh, Pakistan
e-mail: abdulazizmirani@gmail.com

N. Solangi
e-mail: najamsolangi@gmail.com

J. A. Harikrishna · C. H. Teo
Centre for Research in Biotechnology for Agriculture (CEBAR), Universiti Malaya, 50603 Kuala Lumpur, Malaysia
e-mail: jennihari@um.edu.my

C. H. Teo
e-mail: cheehow.teo@um.edu.my

C. Sánchez-Romero (ed.), *Somaclonal Variation: Basic and Practical Aspects*,
https://doi.org/10.1007/978-3-031-51626-9_9

1 Introduction

Date palm (*Phoenix dactylifera* L.) is a dioecious, monocotyledonous plant with an estimated genome size of 1.59–1.84 pg $2C^{-1}$ (Jatt et al. 2019a) or 550–658 Mbp in length (Racchi et al. 2014; Al-Faifi et al. 2017). It is a socioeconomically important horticultural crop for sustainable agriculture in tropical and sub-tropical regions of the world that involves 45% of the labor force directly or indirectly in date-producing countries (Raza et al. 2012). Date palm is propagated through different means including seeds, but the most genetically stable and homogenous propagation is based on transplantation of offshoots from a donor plant. The development of new varieties with increased yield and resistance to biotic and abiotic stresses is challenging due to the long-life cycle and dioecious nature of the date palm. The application of tissue culture technology to date palm has brought two major benefits: Firstly, tissue culture has revolutionized rapid, disease-free and true-to-type propagation, which was not possible with offshoots due to their low availability per plant (about 5–15) and low survivability caused by the poorly developed root system. Secondly, tissue culture has opened the gate for the introduction of new traits in date palm through somaclonal variation (SV).

SV is defined as the variation originated in cell or tissue culture (Larkin and Scowcroft 1981). Generally, the factors contributing to SV during tissue culture include nutrient stress, osmotic stress, dehydration, wounding, and toxic effects of high concentrations of plant growth regulators (PGR) (Filipecki and Malepszy 2006). Other factors including the type of regeneration method (e.g. via organogenesis or embryogenesis), age of culture (Etienne and Bertrand 2003), plant genotype (Etienne and Bertrand 2003), and types of PGR used (Bordallo et al. 2004) also contribute to the mutagenic impact on the cultured cells or tissues.

Studies in various crops have shown SV to be useful in producing new varieties (Katiyar and Chopra 1995; Baer et al. 2007). However, in date palm such studies are limited (Baklouti et al. 2022) and no positive use case of SV has been reported yet. One of the challenges for the application of SV in in vitro tissue culture grown date palm plants is that it is generally manifested as phenotypic abnormalities in economically important characters or undesirable phenotypes, which are not deemed useful in the market. These undesirable phenotypes are a drawback for the commercial production of planting materials through plant tissue culture that has limited the use of this technology for commercial micropropagation and germplasm conservation for most crop species. In general, with the exception of the banana crop (10%), the acceptable range of SV in tissue culture plants for commercial micropropagation industries is 3% to 5% (Côte et al. 1993). It can be noted, however, that relatively high levels of SV can be manageable in plant tissue culture and regeneration programs that already engage a high degree of selection.

2 Somaclonal Variation and Its Sources

The variations in tissue culture-derived date palms could be either generated during in vitro tissue culture conditions or pre-existing in the source material (explant) used for culture establishment (Peschke and Phillips 1992). The possibility of these variations in tissue culture plants may increase with the use as explant source of chimeric plants (Gurevich et al. 2005), plants with an altered mechanism of cell cycle regulation (Beemster et al. 2003), and plants with activated cryptic transposable elements (TEs) (Banks et al. 1988; Mujib et al. 2007). In date palms, the existence of chimeric plants has been reported as SV (Gurevich et al. 2005).

Variations observed in date palm tissue cultured plants are either due to genetic mutations or epigenetic variations in the genome. Genetic mutations in a genome are more stable and difficult to reverse, whereas epigenetic variations can be unstable and reversible. Epigenetic variations have been associated with in vitro oxidative stress (Cassells and Curry 2001; Joyce et al. 2003) and altered levels of DNA methylation (Jaligot et al. 2002, 2000; Bender 2004; Xu et al. 2004). Such epigenetic variations are generally unstable, but may sometimes remain stable for many generations (Johannes et al. 2009) being transferred to progeny through the sexual life cycle as epimutations (Brettell and Dennis 1991). Epigenetic variations in date palms, caused by various factors discussed below, are commonly reported as abnormal phenotypes that revert to their normal phenotypes within a few years of field plantation (Mirani et al. 2022). For example, the complete reversion of inability to form inflorescence (delayed flowering), low level of fruit set, production of deformed offshoots, and abnormal growth of frond in date palm cultivars (cv.) Gulistan and Kashuwari were reported by Mirani et al. (2019). The reversion of dwarf plants to their normal stature has also been reported in cv. Medjool (Gurevich et al. 2005). Similar behavior was observed in other date palm cultivars (Mirani et al. 2022; Cohen et al. 2004, 2003; Cohen 2011; Varughese 2000). These variations have been mostly linked to PGR or physiological imbalance in the plants (Varughese 2000). Abnormal morphologies, which may later revert to their natural state, can also be found during the in vitro differentiation, proliferation or regeneration stages (Miguel and Marum 2011), For example, in Fig. 1 abnormal embryo development and embryonic leaf formation can be observed, which later reverted to their normal shape in the subsequent subculture. Development of abnormal embryos during micropropagation in date palm cv. Barhi (Bhaskaran and Smith 1992) and unproportionate ratio of root and shoot, sheath restriction on leaf emergence and development of blind meristems were also reported as in vitro abnormalities in date palm cultures.

Fig. 1 In vitro abnormal embryo development and embryonic leaf formation in date palm cultures: **a** normal embryo (left) and abnormal embryo (right), **b** abnormal thick leaf (Photo by Najamuddin Solangi)

3 Factors Inducing Somaclonal Variations in Date Palm

It is well established that several factors contribute to the occurrence of SV in date palm cultured tissues, namely the explant source, genotype of the mother plant, culture age, subculturing frequency, proliferation rate of the cultured cells or tissues, type of regeneration method, as well as type and the concentration of PGRs (Alkhateeb 2008; Karp 1994). Explant sterilization, wounding, illumination, humidity and misbalance in culture media components also contribute to the occurrence of SV in plants (Bidabadi and Jain 2020).

In date palm, the application of tissue culture has focused on commercial micropropagation and uses a wide range of explant sources, genotypes, regeneration methods, PGR types and concentrations, and in vitro environmental conditions that sometimes trigger the SV found in cultured tissues. A brief account of these factors associated with the possible occurrence of SV in date palm tissue cultured plants is documented below.

3.1 Explant Type and Source

The degree of SV in tissue cultured plants may vary with the type of explant (Skirvin et al. 1994). The use of explants from different parts of the date palm such as zygote, leaves, apical meristem of a shoot tip and spikelets of an inflorescence can result in different levels of SV. In general, explants from shoot tips produce plants that are more true-to-type and have a lower incidence of SV compared to explants from leaves. For example, Kumar et al. (2010a, b), Othmani et al. (2010) and Javouhey et al. (1999) reported that the use of shoot tip explants produced no SV in tissue culture-derived date palms.

The association of the type of explant with the occurrence of SV in cultured tissues may be due to the maturity or differentiation stage of the cells in the explant. Thus, the chances of SV are lower in meristematic cells such as those from shoot or root tips, where no dedifferentiation is required, than in mature cells from leaves, roots, and stems (Sharma et al. 2007). This generalization may be supported by the findings reported by Bregitzer et al. (2002) in barley, where a high level of SV was found in plant lines derived from embryogenic callus generated from mature cells compared to plants derived from meristematic tissues. The induction of dedifferentiation of mature cells into meristemoid cells and their redifferentiation into organs or somatic embryos generally requires the use of high hormonal concentrations, which eventually imposes several abiotic stresses on the cultured cells. These overall physiological imbalances in the culture medium or abiotic stresses trigger various genetic or epigenetic malfunctioning in the genome, which results in the generation of stable or unstable SV in the cultured tissues. In date palm, meristematic tissue from shoot tips of an offshoot (Jatoi et al. 2019; Samiei et al. 2022) and flowers from an immature inflorescence (Solangi 2021; Abul-Soad and Al-Khayri 2018; Jatoi et al. 2015; Abul-Soad 2011; Abul-Soad and Mahdi 2010; Zaid and Al Kaabi 2006) are preferred for micropropagation to minimize the risk of SV. For example, Fki et al. (2003) reported the true-to-type micropropagation of date palm plants of cv. Deglet Noor, using juvenile leaves and inflorescences as explants. Similarly, Mirani et al. (2022) used inflorescence explants for producing true-to-type date palm plants of cv. Dedhi.

3.2 Genotype

The chances and degree of SV in tissue culture-derived plants, not only differ between species but also within the species, with certain cultivars being relatively stable, while others show a high level of variation. This was also observed for date palm (El Hadrami et al. 2011), with each cultivar showing its own responses to a specific stress condition, including the in vitro abiotic stresses that trigger SV and result in variation in the stability level (Bairu et al. 2011). This differential behavior of cultivars of the same species under in vitro culture conditions can be explained by variations in repetitive DNA sequence content of the genomes (Lee and Phillips 1988; Biscotti et al. 2015).

For instance, while working with date palm plants of different cultivars produced through somatic embryogenesis, Saker et al. (2006) found that cv. Sakkoty showed the most Amplified Fragment Length Polymorphism (AFLP) variation (2.6%) compared to cv. Bertamoda (1%) and Gandila (0.79%). Similarly, Mirani et al. (2020) reported the occurrence of more variation in date palm cv. Kashuwari (14%) than in cv. Gulistan (4.26%), although in both cases, embryogenic cultures derived from inflorescence explants under similar culture conditions. While working with date palm plants of five cultivars derived through organogenesis and somatic embryogenesis, Al-Mazroui et al. (2007) noted the occurrence of SV in plants regenerated via somatic embryogenesis. They also found that cvs. Khalas and Barhi were comparatively more sensitive, showing a high rate of SV.

3.2.1 Transposable Elements

TEs are the predominant part of the interspersed repetitive sequences in the genome of almost all eukaryotes. In plants, they comprise up to 80% of the nuclear genome (Alzohairy et al. 2012). In date palm, they account for about 22.95% of the genome, of which 21.99% are retrotransposons and only 0.96% are DNA transposons (Al-Mssallem et al. 2013). Al-Dous et al. (2011) identified 45 families of long terminal repeat (LTR) retrotransposons and 35 families of Miniature Inverted-repeat Transposable Elements (MITE) transposable elements in date palm cv. Khalas. The mobilization of TEs under in vitro (stressful) conditions has also been reported as a cause of genome sequence variation in plants. In tissue culture-derived date palms, the mobilization of TEs was confirmed through Inter-Retrotransposon Amplified Polymorphism (IRAP) and MITE molecular markers (Mirani et al. 2020). The mobilization of plant retrotransposons into coding sequences in cell, protoplast or tissue cultured plants has also been demonstrated in tobacco (Grandbastien et al. 1989), rice (Hirochika et al. 1996), maize (Peschke et al. 1987), African violet plant (Masuta et al. 2017), lotus (Madsen et al. 2005) and orchid (Hsu et al. 2019).

In plants, TEs can be mobilized when the plant is subjected to any abiotic or biotic stress including radiation (Peterson 1960), ultraviolet light (Neuffer 1966), and virus infection (Dellaporta et al. 1984). The in vitro stress conditions such as explant sterilization, wounding, high PGR concentrations, malnutrition, salt type and concentration, microbial contamination, heat and drought stress, and light quality and intensity have mutagenic effects on cultured tissues, thus being capable of activating TEs (Alzohairy et al. 2012; McClintock 1978), with strong evidence describing the loss of epigenetic regulation (Smulders and De Klerk 2011). This loss of epigenetic regulation or change in DNA methylation pattern during the tissue culture process triggers an enhanced mobilization of TEs (Lisch and Bennetzen 2011), which leads to the production of various SV in cultured tissues (Chiu et al. 2010; Hirochika et al. 1996; Kashkush et al. 2002). For example, in tissue culture-derived date palm cv. Barhi a significant increase (13.8–18.93%) in DNA methylation rate was recorded, which not only produced epigenetic changes but also genetic variations detected through High-Performance Liquid Chromatography (HPLC) analysis of nucleosides and Inter Simple Sequence Repeat (ISSR) analysis of genomic DNA, respectively (Baklouti et al. 2022). While working on molecular characterization of tissue culture-derived off-types (dwarfism and reduced fruit set) plants of date palm cvs. Barhi and Medjool, Cohen (2011) also detected several alterations in DNA methylation pattern using Methylation-Sensitive Amplified Fragment Length Polymorphism (MS-AFLP) and changes in chloroplast-derived sequences and transposable elements sequences using Representation Difference Analyses (RDA).

3.3 Method of Regeneration

The method of regeneration used in tissue culture can also have an effect on SV in date palm. Different in vitro regeneration methods, such as direct or indirect organogenesis and direct or indirect somatic embryogenesis can result in different levels of SV. Somatic embryogenesis is the most efficient protocol for mass propagation of endangered species or commercial crops (Bidabadi and Jain 2020). However, indirect somatic embryogenesis, which requires the dedifferentiation of cultured cells to produce callus and the subsequent differentiation of the callus cells into embryos, is more likely to generate SV compared to direct embryogenesis or the direct formation of organs (organogenesis). The regeneration of plants following indirect organogenesis (Bader et al. 2007) or somatic embryogenesis (Al-Mazroui et al. 2007) has been associated with SV in date palm. However, comparing the two methods, Al Kaabi et al. (2005b) reported higher SV (0.6%) in tissue culture-derived date palms produced through somatic embryogenesis than in those regenerated via organogenesis (0.038%). As stated above, Al-Mazroui et al. (2007) used both, organogenesis and somatic embryogenesis, for producing plants of five date palm cultivars and found phenotypic variation only in plants derived through somatic embryogenesis.

From the above discussion, it can be stated that irrespective of the genotype and the explant type used, SV have occurred in date palm during production of plants via somatic embryogenesis. However, compared to somatic embryogenesis, very little or no variation has been reported in organogenesis-derived date palms (Al-Mazroui et al. 2007).

3.4 Culture Age

The chances and degree of variation in tissue-cultured plants increase with the age of culture, generally meaning a higher number of in vitro subcultures on nutrient medium (Bairu et al. 2006). In date palm cv. Zaghloul, plants derived from 3-, 6- and 12-month old embryogenic cultures initiated from shoot tip cultures showed respectively 0%, 1.5% and 2.5% variation, assessed by Random Amplified Polymorphism DNA (RAPD) marker (Saker et al. 2000). Similarly, date palm plants developed from materials that underwent more than 10 subcultures revealed SV, while no SV was found in plants developed after less than 10 subcultures (Mirani et al. 2019, 2020). According to Rival et al. (2013), prolonged subculturing of tissues has a significant impact on the DNA methylation pattern. It is also believed that the variation rate increases over each generation, giving rise to an accumulation of mutations in the cultures (Saker et al. 2000; Bobadilla Landey et al. 2015). Hence, reducing the culture age or subculture period may reduce the chances and rate of SV. To maintain the genomic stability of tissue culture plants, most commercial laboratories limit the number of subcultures per explant and regularly shift their plants to the greenhouse.

3.5 *Type and Concentration of Plant Growth Regulators*

PGRs play an important role in the dedifferentiation and redifferentiation of cultured cells and tissues to produce organs or somatic embryos (Solangi et al. 2020). Generally, the regeneration of date palm plants through somatic embryogenesis requires media with a high concentration of the synthetic auxin 2,4-dichlorophenoxyacetic acid (2,4-D), often in combination with other PGRs, including 6-benzyladenine (BA) (Moghaieb et al. 2011), 1-naphthaleneacetic acid (NAA), gibberellic acid (GA) and 2-isopentenyl adenine (2-iP) (McCubbin et al. 2004), to induce callus formation and embryo development. This is followed by abscisic acid (ABA) treatment for the final maturation of the developed embryos (Al-Mazroui et al. 2007). These high doses of PGRs, especially of synthetic auxins (2,4-D) (Baklouti et al. 2022) and their cumulative effects on cells, due to their non-degradable nature, produce a long-lasting impact on cell metabolism and eventually become the cause of SV in the cultured tissues of date palm (McCubbin et al. 2004). The use of synthetic PGRs has been reported to cause increased cytological, epigenetic and genetic variations not only in date palm (Baklouti et al. 2022; Saker et al. 2006), but in several other species such as barley (Ziauddin and Kasha 1990), cotton (Jin et al. 2008), potato (Bordallo et al. 2004) and garlic (Al-Zahim et al. 1999). For example, Baklouti et al. (2022) used 1, 5, 10, 20 and 40 mg/L of 2,4-D to induce SV in date palm cv. Barhi and found significant physiological and genetic variations in the cultures as well as in the produced plants.

4 Somaclonal Variations Observed in Date Palm

SV in date palm tissue culture-derived materials may be the result of stable (genetic) or unstable (epigenetic) events in the genome. Naturally, the mutation rate for a given locus is about 1 in 100,000–1,000,000 (0.001–0.0001%) (Skirvin et al. 2000); however, the rate of mutation in tissue culture plants commonly varies from 15 to 20% (Evans and Sharp 1983) to as high as 100% in celery (Orton 1983). In date palm, this mutation rate in tissue culture-derived plants has reached up to 25% in cv. Bertamoda, 23% in cv. Sakkoty (El-Assar et al. 2005), 17% in cv. Khanzi (Eshraghi et al. 2005) and 14% in cv. Kashuwari (Mirani et al. 2020). SV in tissue culture plants can manifest as stable genetic events that include point mutations, excision and insertions of TEs, somatic recombination, gene amplification, chromosomal rearrangements, and chromosomal aberrations that may result in polyploidy or aneuploidy (Madlung and Comai 2004; Skirvin et al. 1994, 2000). Added to this, epigenetic variations, which are sometimes unstable and transitory, can be found in tissue cultured plants as the result of alteration in DNA methylation pattern and/or histone modification (Kaeppler et al. 2000; Smulders and De Klerk 2011). Altogether, these variations can phenotypically be seen as abnormalities of various kinds, which are not deemed useful in the date palm tissue culture industry.

4.1 Phenotypic Effects

The impact of SV on plant phenotypic traits came into account in 1969 when researchers observed significant morphological changes in sugarcane plants derived from cell cultures (Heinz and Mee 1971). Phenotypic expressions of the SV in date palm (Fig. 2; Table 1) may include plant dwarfism (Mirani et al. 2019; Al-Wasel 2001); abnormal growth of leaves or fruit strands and inability to form inflorescences (Mirani et al. 2019; McCubbin et al. 2004); leaf albinism and dryness of apical bud (Alkhateeb 2008); change in fruit quality and delayed flowering time (Alkhateeb and Ali-Dinar 2002); production of deformed offshoots and twisted inflorescences (Mirani et al. 2019; Zaid and Al-Kaabi 2003; Hassanpour-Estahbanati and Hamidian 2007); twisted leaves, necrosis on midrib and abnormal buds (Hassanpour-Estahbanati and Hamidian 2007), low fruit set (Mirani et al. 2019; McCubbin et al. 2004) and production of abnormal parthenocarpic fruits (Mirani et al. 2019; Cohen et al. 2004).

A wide range of such variations in phenotypic traits has also been reported in other plant species, for example, the development of albino plants and change in leaf morphology and vigor in *Lolium* spp. (Ahloowalia 1983); change in plant height, biomass, grain yield and agronomic performance in *Zea* spp. (Kaeppler et al. 2000), and change in vine morphology, tuber yield, tuber composition, maturity and flowering in *Solanum* spp. (Secor and Shepard 1981).

Mirani et al. (2019) reported six different types of abnormal phenotypes in 2–3-year-old tissue culture derived date palms of cvs. Kashuwari and Gulistan. The plants were regenerated from inflorescence explant via somatic embryogenesis and were grouped according to the age of the culture from which they were obtained, i.e., 1–10 and 11–25 in vitro subcultures in the multiplication stage. The effects of culture age (1–10 subculture and 11–25 subculture), cultivar (Kashuwari and Gulistan), year of study (2016, 2017 and 2018) and their interactions on phenotypic variations revealed that increasing the period of in vitro culture on nutrient medium caused significant SV. Among the six different abnormal phenotypes detected in the open field trials, "inflorescence-absent" was the most frequent. The overall phenotypic variation percentage in tissue culture plants derived from subcultures 11–25 was higher (64.19%) than that of the tissue culture plants derived from up to 10 subcultures (2.42%). The total percentage of variant plants was higher in cv. Gulistan (44.29%) than in cv. Kashuwari (22.32%) in both blocks at the end of the first season in the year 2016. However, a significant reduction was observed in the overall phenotypic variation for plants derived from block II cultures (from 61.19% to 13.76%) and block I cultures (from 2.42% to 0.69%), as well as for both cvs. Gulistan (from 44.29% to 8.13%) and Kashuwari (from 22.32% to 6.23%) at the end of the third season in the year 2018. This reversion of abnormal phenotypes to normal phenotypes within 10 years of planting has also been observed in other date palm cultivars (Cohen et al. 2004).

Fig. 2 Phenotypic expression of SV in *P. dactylifera*: **a** normal plant, **b** excessive vegetative growth without inflorescence, **c** dwarf/inflorescence absent, **d** high number of offshoots without inflorescence, **e** excessive vegetative growth of leaf, **f** fertilization failure, **g** deformed offshoots, **h** abnormal frond/leaf growth and **i** twisted inflorescence

4.2 Cytological Effects

SV can be manifested cytologically in tissue culture plants as a change in chromosome structure or number (Kharabian and Darabi 2005). The change in chromosome structure through chromosomal rearrangements and/or breakage is more frequent than the change in chromosome number (Lee and Phillips 1988). A karyotypic study of celery cell suspension cultures revealed that about 68% of the protoplasts had a change in the number of chromosomes and more than 95% showed a change in the structure of the chromosomes (Murata and Orton 1983). Increase or decrease in nuclear DNA content, missing chromosomes, presence of extra chromosomes and polyploidy have been observed in different plant species (Cao and Deng 2020; Azizi

Table 1 Phenotypic expression of SV in different date palm cultivars derived through tissue culture

Phenotypic expression	Cultivar name	Reference
Apical bud bending	Barhi, Sukkary	Soliman et al. (2009), Al-Wasel (2005)
Abnormal frond/leaflet growth	Gulistan	Mirani et al. (2019)
Dwarfism	Medjool, Barhi, Khlass, Sukkary, Sultana, Oum Dahn, Bou-feggous, Kashuwari, Gulistan	McCubbin et al. (2004), Soliman et al. (2009), Gurevich et al. (2005), Al-Mazroui et al. (2007), Al Kaabi et al. (2005b), Oihabi and De Wet (2005), Cohen et al. (2005), Mirani et al. (2019)
Deformed offshoot production	Khlass, Barhi, Gulistan	Al-Mazroui et al. (2007), Al Kaabi et al. (2005b), Mirani et al. (2019)
Delayed flowering/inability to form inflorescence	Khlass, Barhi, Kashuwari, Gulistan	Al-Mazroui et al. (2007), Al Kaabi et al. (2005b), Mirani et al. (2019)
Delayed ripening	Medjool, Khlass	Gurevich et al. (2005), Al-Mazroui et al. (2007)
Excessive vegetative growth (broad leaf and leaflets)	Barhi, Khlass, Gulistan	Soliman et al. (2009), Gurevich et al. (2005), Al-Mazroui et al. (2007), Al Kaabi et al. (2005b), Mirani et al. (2019)
Fertilization and fruit set failure	Khlass, Barhi, Medjool, Sukkary, Deglat-noor, Nabtet seif, Kashuwari, Gulistan	Al-Mazroui et al. (2007), Al-Wasel (2005), Mirani et al. (2019), Al Kaabi et al. (2005b), McCubbin et al. (2004)
Leaf and spine structure variation	Medjool	McCubbin et al. (2004)
Leaf variegation or albinism	Medjool, Khlass, Barhi, Nabtet seif, Sultana	McCubbin et al. (2004), Cohen et al. (2005), Al-Mazroui et al. (2007), Al Kaabi et al. (2005b)
Low level of fruit setting	Barhi, Medjool, Hallas	Gurevich et al. (2005), Cohen et al. (2005)
Multiple carpel production	Barhi, Medjool	Gurevich et al. (2005), Al Kaabi et al. (2005b)
Seedless fruit production	Medjool	McCubbin et al. (2004)
Susceptibility to disease	Medjool	Cohen et al. (2005)
Twisted inflorescence development	Gulistan	Mirani et al. (2019)

et al. 2022). The possible mechanisms responsible for chromosomal breakage or rearrangements in cultured cells may be the late replication of heterochromatin (Wear et al. 2017), nucleotide pool imbalance (Donne et al. 2022), mitotic recombination or somatic crossing overs (Gisler et al. 2002; Jia et al. 2021).

In date palm, varied levels of ploidy (2n = 28, 32, 36 and 64) (Jatt et al. 2019a), variation in genome size (780.11–903.06 Mbp) (Ouarda et al. 2012; Jatt et al. 2019a) and variation in total chromosome length, arm ratio, relative length and centromeric index (Jatt et al. 2019b) have been reported in non-cultured materials (offshoots). However, such cytological events or variations are yet to be observed in tissue culture-derived date palm plants.

4.3 Genomic Effects

The most frequent genomic effects of SV reported in cultured tissues of date palm are sequence variations that have been analyzed through molecular marker systems. A number of genomic markers have been used, including RAPD (Adawy et al. 2002; Kumar et al. 2010a; Othmani et al. 2010; Saker et al. 2006, 2000), AFLP (Al Kaabi et al. 2005a; Othmani et al. 2010; Saker et al. 2006), ISSR (Baklouti et al. 2022; Kumar et al. 2010a), IRAP (Mirani et al. 2020) and MITE (Mirani et al. 2020) markers. These markers have proved useful in detecting a number of genomic variations in tissue culture derived plants, including point mutations, insertions and deletions. These genomic variations can lead to biochemical, physiological or phenotypic abnormalities in tissue culture-derived date palm plants.

Depending upon the specific changes that occur in the genome, the SV generated during tissue culture process can be either beneficial or detrimental (unwanted). The beneficial SV can be utilized to develop new date palm cultivars with improved traits. Some examples of released cultivars in other plant species are celery cv. UC-T3 (*Fusarium* resistance) (Heath-Pagliuso and Rappaport 1990), tomato cv. DNAP-17 (*Fusarium* resistance) (Evans 1989) and *Capsicum* cv. A-D4 (early maturing and high yielding) (Xu and Chen 1995). However, the detrimental SV cause a great loss to tissue culture industries, especially when true-to-type regeneration of a plant is required for micropropagation or transgenesis. The detrimental SV may include change in growth habit, change in fruit quality, reduced fruit production or fertility, and increased susceptibility to disease or stress etc. (Mirani 2019).

4.4 Epigenetic Effects

Epigenetic regulation has several functions including as a defense system against retrotransposon and TE activities (Hosaka and Kakutani 2018). This system includes DNA methylation and/or modification of amino acids in the tails of histones, which interact and reinforce one another. The mobilization of retrotransposons (a class of

TEs) has been reported in *P. dactylifera* (Cohen 2011), *Oryza sativa* (Hirochika et al. 1996), *Nicotiana tabacum* (Grandbastien et al. 1989), *Ipomoea batatas* (L.) Lam (Tahara et al. 2004) and *Nicotiana plumbaginifolia* (Leprince et al. 2001).

Changes in DNA methylation during tissue culture may sometimes result in phenotypic changes, i.e. SV (Brown 1989). The occurrence of hypomethylation in abnormal calli, leaves and immature inflorescence from mantled somaclones of oil palm has been reported (Jaligot et al. 2000). In carrot cultures, an increase in the concentration of 2,4-D in the culture medium corresponded with increased DNA methylation, but methylation levels decreased when the concentration of kinetin was increased (LoSchiavo et al. 1989). Abnormal rejuvenation, habituation, flower abnormalities, bushiness, tumorous outgrowths, and delay in flowering time are some of the non-permanent variations triggered by epigenetic mechanisms in tissue cultured plants (Smulders and De Klerk 2011). Such temporary variations can also be noticed in plant tissue cultures during in vitro differentiation, proliferation, and regeneration stages (Miguel and Marum 2011), including in date palm (Fig. 1). Epigenetic variations reported in date palm tissue cultured plants include delayed flowering, production of parthenocarpic fruit, growth-retarded and low level of fruit setting (Gurevich et al. 2005). In our study on date palm cvs. Kashuwari and Gulistan (Mirani et al. 2019, 2020), and Dedhi (Mirani et al. 2022), we also found a similar behavior in the plants, where most of the abnormal phenotypes reverted to their normal phenotypes.

5 Methods Used to Detect Somaclonal Variations in Date Palm

The screening of variations at the genomic, cytological, or phenotypic levels is an essential step in the application of biotechnology for the optimization of micropropagation protocols to minimize the risks of SV in clonal plants (Eshraghi et al. 2005). Generally, the date palm tissue culture industry exploits phenotypic, cytological, biochemical, and molecular markers for the screening of SV in micropropagated plants (Table 2).

5.1 Phenotypic Markers

The phenotypic markers can be documented through visual observations of the plant phenotypic traits (both vegetative and reproductive) in the field or laboratory. These markers are valuable due to their low cost, simplicity and ease of use, with no requirement for complicated equipment or procedures. However, this approach requires a large set of morphological data and visualization may only be apparent in mature or sometimes fruit-bearing trees. In date palm, the phenotypic traits may include

Table 2 Methods for detecting variation in date palm cultivars

Cultivar name	Explant source	Method of regeneration	Variation detected	Method of detection	References
Karama	–	SE	No	ISSR	Abd-Alla (2010)
Khlass	–	Organogenesis, SE	Yes	Phenotyping, AFLP	Al Kaabi et al. (2006)
Khlass, Barhi, Nabtet seif, Sultana, Sukkary	–	Organogenesis, SE	Yes	Phenotyping	Al-Mazroui et al. (2007)
Bou-feggous, Jihel, Bou-skri	Shoot tip	Organogenesis	Yes	MS-AFLP, bisulfate sequencing, phenotyping	Azeqour et al. (2002)
Barhi	Leaf	Callogenesis	Yes	ISSR, GMR	Baklouti et al. (2022)
Barhi, Medjool	–	–	Yes	MS-AFLP, Bisulfate sequencing, phenotyping	Cohen (2011)
Barhi, Deglat-noor, Khlass, Khadrawy, Medjool	Shoot tip	SE	Yes	RFLP, RAPD	Corniquel and Mercier (1994)
Khanizi	Shoot tip	SE	Yes	RAPD	Eshraghi et al. (2005)
Deglat-noor	Leaf and inflorescence	SE	No	Flow cytometry	Fki et al. (2003)
Barhi, Medjool	–	–	Yes	AFLP, Phenotyping	Gurevich et al. 2005
–*	Shoot tip	SE	No	RAPD	Javouhey et al. (1999)
–	Shoot tip	SE	No	RAPD, ISSR	Kumar et al. (2010a)
–	Shoot tip	SE	No	SSR	Kumar et al. (2010b)
Barhi, Medjool	Shoot tip	SE	Yes	Phenotyping	McCubbin et al. (2004)
Dedhi	Inflorescence	SE	No	Phenotyping, IRAP	Mirani et al. (2022)
Kashuwari, Gulistan	Inflorescence	SE	Yes	Phenotyping, IRAP, MITE	Mirani et al. (2020)
Kashuwari, Gulistan	Inflorescence	SE	Yes	Phenotyping	Mirani et al. (2019)

(continued)

Table 2 (continued)

Cultivar name	Explant source	Method of regeneration	Variation detected	Method of detection	References
Ferhi, unknown	Shoot tip	SE	Yes	RAPD	Moghaieb et al. (2011)
Deglat-noor	Juvenile leaf	SE	No	RAPD, AFLP	Othmani et al. (2010)
Zaghloul	Shoot tip	SE	Yes	RAPD, isozyme	Saker et al. (2000)
Sakkoty, Gandila, Bertamoda	Shoot tip	SE	Yes	AFLP, RAPD	Saker et al. (2006)
Barhi	–	SE	Yes	Phenotyping, isozymes, ISSR, RAPD	Soliman et al. (2009)
Medjool, Barhi, Sukkary, Toory, Deglat-noor, Khalas, and Nabtet-seif	–	–	Yes	Phenotyping	Al-Wasel (2005)
Barhi, Medjool, Sukkary, Sultana, Khlass	–	SE	Yes	Phenotyping	Al Kaabi et al. (2005b)
Medjool, Bou-feggous, Deglat-noor, Barhi, Hilali, Ashrasi	Shoot tip	Organogenesis, SE	Yes	Phenotyping	Oihabi and De Wet (2005)
Barhi, Medjool	–	–	Yes	Phenotyping, AFLP, DNA methylation pattern analysis	Cohen et al. (2005)

SE somatic embryogenesis, *ISSR* inter simple sequence repeat, *AFLP* amplified fragment length polymorphism, *MS-AFLP* methylation-sensitive amplified fragment length polymorphism, *GMR* global DNA methylation rates, *RFLP* restriction fragment length polymorphism, *RAPD* random amplified polymorphism DNA, *SSR* simple sequence repeats, *IRAP* inter-retrotransposon amplified polymorphism, * information not found

leaf characteristics, pollen morphology, fruit quantitative and qualitative characters (Mirani et al. 2019), palm height, number of offshoots per palm and total fruit production per palm (Faissal et al. 2013; Hammadi et al. 2009). However, phenotypic markers are yet to be identified for early detection of SV in date palm under in vitro conditions or at the nursery stage to minimize the cost of plant production, similar to those for identifying dwarf plants in banana at the nursery stage (Reuveni and Israeli 1989; Smith and Hamill 1993).

Phenotypic markers have been employed by McCubbin et al. (2004) to screen off-type tissue cultured date palm plants produced via somatic embryogenesis. A wide range of variant traits was documented, including leaf variegation (0.005%), compact growth (4%), seedless fruit (0.1%), low pollination (100%), and leaf and spine variations (1.4%). The variations observed during the study adhered to three main factors i.e., human error (mislabeling, improper pollination, watering, and fertilization), environmental stress (disease, nutrition, and climate) and physiological factors (in vitro PGR imbalance). A phenotypic survey in four groups of plants of *P. dactylifera* cvs. Khlass, Barhi, Nabtet seif, Sultana and Sukkary was carried out by Al-Mazroui et al. (2007) to assess the effect of the propagation method: offshoot and three different tissue culture techniques, including organogenesis, indirect somatic embryogenesis, and direct somatic embryogenesis. According to the report, 18% plant dwarfism was observed in cv. Sukkary propagated through indirect embryogenesis. The presence of deformed offshoots, fertilization failure and low fruit set was less than 5% among all plants produced using the three tissue culture techniques. Uneven ripening, delay in flowering, excessive vegetative growth and abnormal morphology and structure were also observed in the plants, but they did not exceed 1%. Specifically, no variation was observed in offshoot-grown plants and in the tissue cultured plants produced via in vitro organogenesis as compared to the tissue cultured plants derived through the somatic embryogenesis method. The variation in leaf whitening that appeared in tissue-cultured plants derived through the organogenesis method was temporary and reverted to the green leaf color. Comparing the five cultivars, most of the observed variations (8 out of 9) were found in cvs. Khlass and Barhi.

Phenotypic detection of off-types in tissue culture derived plants is still being used successfully. For example, recent reports published by Mirani et al. (2019, 2022) and Mirani (2019) on the detection of SV in three micropropagated date palm cultivars derived from inflorescence explants through somatic embryogenesis shows the successful use of phenotypic markers in screening SV and abnormal phenotypes.

5.2 *Cytological Markers*

Cytological markers, which include variations in chromosome number, size, shape, and position of centromere, are useful for detecting changes in ploidy level and genome size, and their relationship with phenotypic diversity in date palm cultivars (Ouarda et al. 2012; Jatt et al. 2019b). The cytological analysis of an organism's chromosomes is direct and gives strong evidence of genetic variation (Sahara et al. 2019); hence, it is useful in evaluating the genetic stability of tissue culture-derived date palm plants (Salman et al. 1988). The cytological variations in tissue culture plants may be linked to chromosomal breakage or heterochromatin aggregation/disaggregation (Jain 2007). These markers have also been used in screening SV in different plants species, including date palm. For example, Fki et al. (2003) screened tissue culture-derived plants of date palm cv. Deglet Nour for nuclear genome size and ploidy level; however, no cytological variation could be found in comparison to offshoot grown plants. In general, the cytological markers that are associated with variations present in chromosome size, shape, position of centromere, and number have been used very rarely in date palm studies.

5.3 Biochemical Markers

Biochemical markers (protein or isozymes markers) have been used for species or cultivar identification (Sammour 2014) and detection of off-types in tissue culture-derived plants (Zivdar et al. 2008). Observing polymorphism in the form of isozyme fingerprints on a gel is a more economical method than using DNA-based markers. Isozyme markers have been applied in date palm for the detection of off-type plants generated through micropropagation (Saker et al. 2000; Zivdar et al. 2008; Salman et al. 1988), cultivar identification (Torres and Tisserat 1980), screening of Bayoud-resistant cultivars (Bendiab et al. 1992), population studies (Bendiab et al. 1998), and in vitro characterization of date palm calli and somatic embryos (Baaziz et al. 1994). The most commonly used enzymes in date palm evaluation are esterase (EST), glutamate oxaloacetate transaminase (GOT), polyphenol oxidases (PPO), endopeptidase (ENP), alcohol dehydrogenase (ADH), peroxidases (POX), phosphoglucose isomerase (PGI), leucine (cytosol) aminopeptidase (LAP) and phosphoglucose mutase (PGM) (Kunert et al. 2003). However, there are only a few pieces of evidence that protein markers can significantly distinguish between individual plants of the same cultivar to detect off-type plants caused by SV during the micropropagation process. For example, while working on ten year old field-grown date palm plants derived from tissue culture, Azeqour et al. (2002) reported phenotypic variations in the number of inflorescences and offshoots produced per plant. The observed variations were confirmed through isozyme polymorphism using five different assays, namely ENP, EST, GOT, PPO and POX. Similarly, Zivdar et al. (2008) screened the off-type plants amongst micropropagated date palm plants of cv. Barhi using POX, shikimate dehydrogenase (SHD) and acid phosphatase (ACP) enzyme systems.

5.4 Molecular Markers

DNA markers show a high degree of non-tissue-specific polymorphism and simple inheritance pattern with minimal influence of environment and epistasis as compared to phenotypic or biochemical markers. A DNA marker can be defined as a DNA sequence with an identifiable location on a chromosome or in the genome (Jonah et al. 2011). The marker may be derived from a single base-pair change (single nucleotide polymorphism; SNP), or a repeat sequence such as Simple Sequence Repeats (SSR) or transposable elements (IRAP). DNA markers are also described as the differences found among the individuals of a cultivar/species or, among the individuals of different cultivars/species which occur naturally within a genome sequence as a result of base pair insertion, deletion, translocation, duplication or transposition of a transposable element (Mondini et al. 2009).

In date palm, molecular markers like ISSR (Hamza et al. 2012, 2013), RAPD (Sedra et al. 1998; Soliman et al. 2003; Williams et al. 1990), AFLP (Al-Khalifah and Askari 2003; Cao and Chao 2002; El-Assar et al. 2003, 2005), SSR (Ahmed and Al-Qaradawi 2009; Al-Najm et al. 2017; Al-Ruqaishi et al. 2008; Elshibli and Korpelainen 2008; Zehdi-Azouzi et al. 2015, 2016; Zehdi et al. 2004, 2012), Inter Primer Binding Sequence (iPBS) (Al-Najm et al. 2016) and chloroplast DNA analysis (Rhouma-Chatti et al. 2014; Sakka et al. 2013) have been used to assess the genetic diversity of date palm germplasm. However, only a few studies reported the use of molecular markers in screening the off-types among tissue culture date palm plants, such as IRAP (Mirani et al. 2020), MITE (Mirani et al. 2020), RAPD (Eshraghi et al. 2005; Kumar et al. 2010a; Othmani et al. 2010; Saker et al. 2006), ISSR (Abd-Alla 2010; Kumar et al. 2010a), AFLP (Gurevich et al. 2005; Othmani et al. 2010; Saker et al. 2006) and SSR markers (Kumar et al. 2010b).

A report published by Gurevich et al. (2005) showed the use of AFLP markers for the characterization of SV in tissue cultured *P. dactylifera* cvs. Barhi, Medjool and male palms. Phenotypically, the tissue cultured plants exhibit various abnormal traits including low level of fruit set, leaf variegation, inability to form inflorescences, and abnormal growth of leaves. AFLP markers produced an approximately 13.56% polymorphic bands (5.97 average gene diversity) among plants of cv. Medjool and 1.26% polymorphic bands (0.79 average gene diversity) among plants of cv. Barhi. The regenerated plants of cv. Medjool with variegated leaves were found to be chimeric when root and leaf samples of the plant were screened using AFLP markers. However, the phenotypic variation of low level of fruit set in cv. Barhi and dwarf phenotype in cv. Medjool could not be correlated to any genetic variation detected using AFLP markers and 50% of the plants with these two phenotypic abnormalities returned to normal phenotype after 10 years of field planting. The authors suggested the occurrence of these two off-types as an epigenetic effect.

Saker et al. (2006) used RAPD and AFLP markers to test genomic instabilities in plants of three date palm cultivars (Sakkoty, Bertamoda, and Gandila) derived from tissue culture through somatic embryogenesis using shoot tip explant. A sum of 259 RAPD fragments and 380 AFLP fragments were generated from 37 RAPD primers and 13 AFLP primer combinations, respectively. The results revealed that no polymorphic fragments were produced with RAPD markers. The total polymorphism detected in tissue culture-derived *P. dactylifera* cvs. Sakkoty, Bertamoda and Gandila were 2.6%, 1.0% and 0.79%, respectively. These low polymorphisms detected in the tissue cultured plants supported the validity of the tissue culture protocol used for micropropagation of these three cultivars.

In an unknown cultivar and cv. Ferhi, Moghaieb et al. (2011) tested the genetic similarity between mother plants and plants regenerated from petiole explants via somatic embryogenesis. A total of 10 random primers were used for RAPD analysis, which revealed 36.2% and 37.8% of polymorphism in the unknown and Ferhi date palm cultivars, respectively, as compared to their respective mother palms. Bader et al.

(2007) also reported the successful use of RAPD markers for detecting SV in date palm cvs. Barhi and Maktoom derived from inflorescence explants. Eshraghi et al. (2005) also used RAPD markers to determine the genetic stability in tissue cultured plants of *P. dactylifera* cv. Khanizi produced via somatic embryogenesis. The results revealed 6% polymorphism between the mother plant and the tissue culture-derived plantlets. In another study, Othmani et al. (2010) used RAPD and AFLP markers to assess the genetic stability in tissue cultured plants of *P. dactylifera* cv. Deglet Noor and revealed no polymorphism compared to the mother plant. Similarly, in a study in date palm plants regenerated through somatic embryogenesis after forty-two subcultures, RAPD and ISSR markers revealed no polymorphism (Kumar et al. 2010a). According to the report, 160 RAPD and 21 ISSR primers were used, which revealed homogenous banding patterns (no variation) across all the tested tissue culture plants as compared to their mother plants.

5.5 Transposable Element-Based Molecular Markers

The transposable elements (TEs), especially the retrotransposons, are an ideal source for developing an effective molecular marker system for screening off-type plants among tissue cultured clones. These are abundantly dispersed in the plant genome (40–70%) (Nadeem et al. 2018). The widely reported association of transposable elements with plant genomic variation supports the use of TE-based molecular marker systems together with phenotypic observations for the screening of SV and understanding their molecular nature.

The molecular markers developed from retrotransposons include IRAP, Retrotransposon Microsatellite Amplified Polymorphism (REMAP), iPBS, Sequence Specific Amplified Polymorphism (SSAP) and Retrotransposon-Based Insertion Polymorphism (RBIP). Other types of TE markers such as MITE are based on the transposable nature of non-autonomous DNA transposons. The use of retrotransposon-based markers for population studies and detection of the off-type plants in tissue culture has several advantages over other molecular markers due to their high heterogenicity, widespread nature on chromosomes and insertional polymorphism within and between plant species. This marker system has been applied to study SV in date palm (Mirani et al. 2020, 2022).

Mirani et al. (2020) used IRAP and MITE markers for detection of SV in 2–3 year old tissue culture derived date palm plants of cvs. Kashuwari and Gulistan, differing in their culture age, i.e., 1–10 and 11–25 in vitro subcultures in the multiplication stage. A total of 27 primer combinations from 9 Hordeum and 4 Phoenix IRAP together with Phoenix MITE markers were tested. Only 6 primer combinations from Hordeum IRAP, 8 Phoenix IRAP and 3 Phoenix MITE successfully produced reproducible fragments of defined size. The percentage of somaclonal variants per population detected by Phoenix IRAP, Hordeum IRAP, and MITE markers individually and in combination was the same, i.e., 14% in cv. Kashuwari and 4.3% in cv. Gulistan. As compared to the IRAP and MITE markers, IRAP from Phoenix revealed the highest average number of scorable bands (29.4 bands), average polymorphic bands (4.75 bands), average monomorphic bands (24.7 bands), total polymorphism (16.1%) and average non-parental bands per primer (2.69 bands). IRAP and MITE marker analysis of tissue cultured plants of cv. Gulistan revealed that among all the primers tested only primer Sukkula and primer PdGy01F distinguished between the tissue cultured plant no. 44 and 45 (4.3%). While in cv. Kashuwari, all the tested markers generated different banding patterns. Almost all the markers distinguished between the tissue cultured plants no. 29 and 23 to 28 except primer PdGyChromo01F and combinations of Sukkula + 3'LTR, PdGy01F + PdGyChromo01R, PdGy01R + PdGyChromo01R and MITE-F + MITE-R. Mirani et al. (2020) reported that out of the total 24% phenotypic variation found in tissue culture plants of cv. Kashuwari, only 14% of plants with inflorescence-absent and dwarf phenotype could be confirmed through molecular analysis with all the primers tested individually and in combinations. In cv. Gulistan, out of the total 21.3% phenotypic variant tissue culture plants, only 4.3% plants with excessive vegetative growth phenotype were confirmed through molecular analysis.

6 Possible Applications of the Somaclonal Variants Obtained

While undesirable in true-to-type propagation of date palm, SV has the potential to be harnessed for introducing desirable genetic variation for improvement, especially for increased fertility, high yield, low stature (dwarfness) and tolerance to abiotic and biotic stresses.

Where true-to-type materials are desired, it is clear that it is better to use meristematic tissues as explants, organogenesis as plant regeneration method and to limit the time in culture and number of subcultures before transferring plantlets to the nursery. In addition, minimal exposure to high concentrations of PGR is preferred. Finally, a number of marker systems can be used to screen for "off-types", which can either be discarded or forwarded to breeding selection programs. Future research can focus on improving the number of markers for fuller coverage of the genome and identifying more cultivar-specific allelic markers.

In contrast, if SV is to be used as a tool to increase genetic diversity, then while some general principles are known, several gaps remain to be addressed, particularly for date palm: Firstly, due to the long productive life of palms, stable mutations are essential and so conditions that induce stable genetic, rather than transient epigenetic changes (that could revert over the life span of a palm) are needed. To support this, it is necessary to know which conditions and treatments can induce a high rate of such changes, for example, a high rate of mobilization of TE. This may include manipulation of culture conditions (levels of PGR) and time (higher number of subcultures), as well as the selection of the explant and the regeneration method (use of somatic embryogenesis via callus). Also, it can be noted that there is still not much in-depth knowledge on the TE in date palm cultivars, so this could be further explored in the future. Secondly, there will be a need for early screening of changes that may only emerge phenotypically in older plant materials, such as abiotic stress resistance. For such screening, various targeted genomic and marker systems could be developed, especially as more knowledge of the genetics of stress and disease resistance is developed. The use of information from related species can also be helpful and as new technology emerges, faster and less expensive screening should become more feasible. For either situation, improved knowledge on SV can support a sustainable date palm industry in the future.

References

Abd-Alla MM (2010) Genetic stability on *Phoenix dactylifera* var. Karama produced in vitro. New York Sci J 3:70–75

Abul-Soad AA (2011) Micropropagation of date palm using inflorescence explants. In: Date palm biotechnology. Springer, pp 91–117

Abul-Soad AA, Mahdi SM (2010) Commercial production of tissue culture date palm (*Phoenix dactylifera* L.) by inflorescence technique. J Genet Eng Biotechnol 8(2):39–44

Abul-Soad AA, Al-Khayri JM (2018) Date palm somatic embryogenesis from inflorescence explant. In: Step wise protocols for somatic embryogenesis of important woody plants. Springer, pp 329–347

Adawy SS, Hussein EHA, El-Khishin D, Saker MM, El-Itriby HA (2002) Genetic variability studies and molecular fingerprinting of some Egyptian date palm (*Phoenix dactylifera* L.) cultivars II-RAPD and ISSR profiling. Arab J Biotechnol 5:225–236

Ahloowalia BS (1983) Spectrum of variation in somaclones of triploid ryegrass 1. Crop Sci 23(6):1141–1147

Ahmed TA, Al-Qaradawi AY (2009) Molecular phylogeny of Qatari date palm genotypes using simple sequence repeats markers. Biotechnol 8(1):126–131

Al-Dous EK, George B, Al-Mahmoud ME, Al-Jaber MY, Wang H, Salameh YM, Al-Azwani EK, Chaluvadi S, Pontaroli AC, DeBarry J (2011) De novo genome sequencing and comparative genomics of date palm (*Phoenix dactylifera*). Nat Biotechnol 29(6):521

Al-Faifi SA, Migdadi HM, Algamdi SS, Khan MA, Al-Obeed RS, Ammar MH, Jakse J (2017) Development of genomic simple sequence repeats (SSR) by enrichment libraries in date palm. In: Date palm biotechnology protocols volume II. Springer, pp 315–337

Al-Khalifah NS, Askari E (2003) Molecular phylogeny of date palm (*Phoenix dactylifera* L.) cultivars from Saudi Arabia by DNA fingerprinting. Theor Appl Genet 107(7):1266–1270

Al-Mazroui HS, Zaid A, Bouhouche N (2007) Morphological abnormalities in tissue culture-derived date palm (*Phoenix dactylifera* L.). In: Zaid A, Kaabi HHSA (eds) International society for horticultural science (ISHS), Leuven, Belgium, pp 329–335. https://doi.org/10.17660/ActaHortic.2007.736.31

Al-Mssallem IS, Hu S, Zhang X, Lin Q, Liu W, Tan J, Yu X, Liu J, Pan L, Zhang T (2013) Genome sequence of the date palm *Phoenix dactylifera* L. Nat Commun 4:2274

Al-Najm A, Luo S, Ahmad NM, Trethowan R (2016) Molecular variability and genetic relationships of date palm (*Phoenix dactylifera* L.) cultivars based on inter-primer binding site (iPBS) markers. Aust J Crop Sci 10(5):732–740

Al-Najm A, Luo S, Ahmad NM, Pourkheirandish M, Trethowan R (2017) Molecular variability and population structure of a core collection of date palm (*Phoenix dactylifera* L.) cultivars from Australia and the Middle East. Aust J Crop Sci 11(09):1106–1115

Al-Ruqaishi IA, Davey M, Alderson P, Mayes S (2008) Genetic relationships and genotype tracing in date palms (*Phoenix dactylifera* L.) in Oman, based on microsatellite markers. Plant Genet Resour 6(1):70–72

Al-Wasel AS (2001) Field performance of somaclonal variants of tissue culture-derived date palm (*Phoenix dactylifera* L.). Plant Tissue Cult 11(2):97–105

Al-Wasel A (2005) Survey study on: somaclonal variations in in-vitro derived date palm trees. In: Proceedings of the international workshop on true-to-typeness of date palm tissue culture-derived plants, pp 23–25

Al-Zahim MA, Ford-Lloyd BV, Newbury HJ (1999) Detection of somaclonal variation in garlic (*Allium sativum* L.) using RAPD and cytological analysis. Plant Cell Rep 18(6):473–477

Al Kaabi HH, Zaid A, Ainsworth C (2005a) AFLP variation in tissue culture-derived date palm plants. In: Zaid A, Hegarty V, Kaabi HHSA (eds) Proceeding of the international workshop on true-to-typeness of date palm tissue culture-derived plants, Morocco, p 93

Al Kaabi HH, Zaid A, Ainsworth C (2005b) Plant-off-types in tissue culture derived date palm (*Phoenix dactylifera* L). In: Zaid A, Hegarty V, Kaabi HHSA (eds) Proceeding of the international workshop on true-to-typeness of date palm tissue culture-derived plants, Morocco, pp 23–25

Al Kaabi H, Zaid A, Ainsworth C (2006) Plant-off-types in tissue culture-derived date palm (*Phoenix dactylifera* L) plants. In: III International date palm conference, vol 736, pp 267–274

Alkhateeb AA (2008) A review the problems facing the use of tissue culture technique in date palm (*Phoenix dactylifera* L.). Sci J King Faisal Univ (Basic Appl Sci) 9:85–104

Alkhateeb AA, Ali-Dinar HM (2002) Date palm in Kingdom of Saudi Arabia: cultivation, production and processing. Translation. Authorship and Publishing Center, King Faisal University, Kingdom of Saudi Arabia 188:281–292

Alzohairy AM, Yousef MA, Edris S, Kerti B, Gyulai G, Bahieldin A (2012) Detection of LTR retrotransposons reactivation induced by in vitro environmental stresses in Barley (*Hordeum vulgare*) via RT-qPCR. Life Sci J 9(4):5019–5026

Azeqour M, Majourhat K, Baaziz M (2002) Morphological variations and isoenzyme polymorphism of date palm clones from in vitro culture acclimatized and established on soil in South Morocco. Euphytica 123(1):57–66

Azizi SA, Farahani F, Sheidai M (2022) Somaclonal variation in pomegranate (Punica granatum L.): ISSR and cytological evidences. Genetika 54(1):207–217

Baaziz M, Aissam F, Brakez Z, Bendiab K, Hadrami I, Cheikh R (1994) Electrophoretic patterns of acid soluble proteins and active isoforms of peroxidase and polyphenoloxidase typifying calli and somatic embryos of two reputed date palm cultivars in Morocco. Euphytica 76(3):159–168

Bader S, Baum M, Khierallah H, Choumane W (2007) The use of RAPDs technique for the detection of genetic stability of date palm plantlets derived from *in vitro* culture of inflorescence. J Educ Sci 20(2):149–159

Baer GY, Yemets A, Stadnichuk N, Rakhmetov D, Blume YB (2007) Somaclonal variability as a source for creation of new varieties of finger millet (*Eleusine coracana* (L.) Gaertn.). Cytol Genet 41:204–208

Bairu MW, Fennell CW, van Staden J (2006) The effect of plant growth regulators on somaclonal variation in Cavendish banana (*Musa* AAA cv.'Zelig'). Sci Hortic 108(4):347–351

Bairu MW, Aremu AO, Van Staden J (2011) Somaclonal variation in plants: causes and detection methods. Plant Growth Regul 63(2):147–173

Baklouti E, Beulé T, Nasri A, Ben Romdhane A, Drira R, Doulbeau S, Rival A, Drira N, Fki L (2022) 2,4-D induction of somaclonal variations in in vitro grown date palm (*Phoenix dactylifera* L. cv Barhee). Plant Cell Tissue Organ Cult 150(1):191–205. https://doi.org/10.1007/s11240-022-02259-8

Banks JA, Masson P, Fedoroff N (1988) Molecular mechanisms in the developmental regulation of the maize Suppressor-mutator transposable element. Genes Dev 2(11):1364–1380

Beemster GT, Fiorani F, Inze D (2003) Cell cycle: the key to plant growth control? Trends Plant Sci 8(4):154–158

Bender J (2004) DNA methylation and epigenetics. Annu Rev Plant Biol 55:41–68
Bendiab K, Baaziz M, Brakez Z, Sedra MH (1992) Correlation of isoenzyme polymorphism and Bayoud-disease resistance in date palm cultivars and progeny. Euphytica 65(1):23–32
Bendiab K, Baaziz M, Majourhat K (1998) Preliminary date palm cultivar composition of Moroccan palm groves as revealed by leaf isoenzyme phenotypes. Biochem Syst Ecol 26(1):71–82
Bhaskaran S, Smith RH (1992) Somatic embryogenesis from shoot tip and immature inflorescence of *Phoenix dactylifera* cv. Barhee. Plant Cell Rep 12(1):22–25
Bidabadi SS, Jain SM (2020) Cellular, molecular, and physiological aspects of in vitro plant regeneration. Plants 9(6):702
Biscotti MA, Olmo E, Heslop-Harrison J (2015) Repetitive DNA in eukaryotic genomes. Chromosome Res 23:415–420
Bobadilla Landey R, Cenci A, Guyot R, Bertrand B, Georget F, Dechamp E, Herrera J-C, Aribi J, Lashermes P, Etienne H (2015) Assessment of genetic and epigenetic changes during cell culture ageing and relations with somaclonal variation in *Coffea arabica*. Plant Cell, Tissue and Organ Culture (PCTOC) 122:517–531
Bordallo PN, Silva DH, Maria J, Cruz CD, Fontes EP (2004) Somaclonal variation on in vitro callus culture potato cultivars. Hortic Bras 22(2):300–304
Bregitzer P, Zhang S, Cho M-J, Lemaux PG (2002) Reduced somaclonal variation in barley is associated with culturing highly differentiated, meristematic tissues. Crop Sci 42(4):1303–1308
Brettell RIS, Dennis ES (1991) Reactivation of a silent Ac following tissue culture is associated with heritable alterations in its methylation pattern. Mol Gen Genet MGG 229(3):365–372
Brown PTH (1989) DNA methylation in plants and its role in tissue culture. Genome 31(2):717–729
Cao BR, Chao CT (2002) Identification of date cultivars in California using AFLP markers. HortScience 37(6):966–968
Cao Z, Deng Z (2020) Morphological, cytological and molecular marker analyses of 'Tapestry' caladium variants reveal diverse genetic changes and enable association of leaf coloration pattern loci with molecular markers. Plant Cell Tissue Organ Cult 143(2):363–375
Cassells AC, Curry RF (2001) Oxidative stress and physiological, epigenetic and genetic variability in plant tissue culture: implications for micropropagators and genetic engineers. Plant Cell Tissue Organ Cult 64(2–3):145–157
Chiu L-W, Zhou X, Burke S, Wu X, Prior RL, Li L (2010) The purple cauliflower arises from activation of a MYB transcription factor. Plant Physiol 154(3):1470–1480
Cohen Y (2011) Molecular detection of somaclonal variation in date palm. In: Date palm biotechnology. Springer, pp 221–235
Cohen Y, Korchinsky R, Tripler E, Ziv G, Cohen R (2003) Fruit setting abnormalities in date palms of the cultivar Barhee originated from tissue culture. Alon Hanotea Hebrew 57:275–279
Cohen Y, Korchinsky R, Tripler E (2004) Flower abnormalities cause abnormal fruit setting in tissue culture-propagated date palm (*Phoenix dactylifera* L.). J Horticult Sci Biotechnol 79(6):1007–1013
Cohen Y, Gurevich V, Korchinsky R, Tripler E, Lavi U (2005) Phenotypic and Molecular Characterization of date palm off-types. In: The international workshop on true-to-typeness of date palm tissue cultured derived plants. Erfoud, Morocco, pp 7–10
Corniquel B, Mercier L (1994) Date palm (*Phoenix dactylifera* L.) cultivar identification by RFLP and RAPD. Plant Sci 101(2):163–172
Côte F-X, Sandoval JA, Marie P, Auboiron E (1993) Variations in micropropagated bananas and plantains: literature survey. Fruits 48(1):15–23
Dellaporta S, Chomet P, Mottinger J, Wood J, Yu S-M, Hicks J (1984) Endogenous transposable elements associated with virus infection in maize. In: Klar AJS, Strathern JN (eds) Cold spring harbor symposia on quantitative biology. Cold Spring Harbor Laboratory Press, pp 321–328
Donne R, Saroul-Ainama M, Cordier P, Hammoutene A, Kabore C, Stadler M, Nemazanyy I, Galy-Fauroux I, Herrag M, Riedl T (2022) Replication stress triggered by nucleotide pool imbalance drives DNA damage and cGAS-STING pathway activation in NAFLD. Dev Cell 57(14):1728–1741. e1726

El-Assar AM, Krueger RR, Devanand PS, Chao CT (2003) Genetic analyses of date palms (*Phoenix dactylifera* L.) from Egypt using fluorescent-AFLP markers. HortScience 38:733–734

El-Assar AM, Krueger RR, Devanand PS, Chao CT (2005) Genetic analysis of Egyptian date (*Phoenix dactylifera* L.) accessions using AFLP markers. Genet Resour Crop Evol 52:601–607

El Hadrami A, Daayf F, Elshibli S, Jain S, El Hadrami I (2011) Somaclonal variation in date palm. In: Date palm biotechnology. Springer, pp 183–203

Elshibli S, Korpelainen H (2008) Microsatellite markers reveal high genetic diversity in date palm (*Phoenix dactylifera* L.) germplasm from Sudan. Genetica 134(2):251–260

Eshraghi P, Zarghami R, Ofoghi H (2005) Genetic stability of micropropagated plantlets in date palm. J Sci Islamic Repub Iran 16(4):311–315

Etienne H, Bertrand B (2003) Somaclonal variation in *Coffea arabica*: effects of genotype and embryogenic cell suspension age on frequency and phenotype of variants. Tree Physiol 23(6):419–426

Evans DA (1989) Somaclonal variation-genetic basis and breeding applications. Trends Genet 5:46–50

Evans DA, Sharp WR (1983) Single gene mutations in tomato plants regenerated from tissue culture. Science 221(4614):949–951

Faissal FA, Mohamed MA, Gobara AA, Ahmed AJSC (2013) Evaluation of some dry date palm varieties propagated through seed and tissue culture technique under Aswan region climatic conditions. Stem Cell 4(3):14–24

Filipecki M, Malepszy S (2006) Unintended consequences of plant transformation: a molecular insight. J Appl Genet 47(4):277–286

Fki L, Masmoudi R, Drira N, Rival A (2003) An optimised protocol for plant regeneration from embryogenic suspension cultures of date palm, *Phoenix dactylifera* L., cv. Deglet Nour. Plant Cell Rep 21(6):517–524

Gisler B, Salomon S, Puchta H (2002) The role of double-strand break-induced allelic homologous recombination in somatic plant cells. Plant J 32(3):277–284

Grandbastien M-A, Spielmann A, Caboche M (1989) Tnt1, a mobile retroviral-like transposable element of tobacco isolated by plant cell genetics. Nature 337(6205):376–380

Gurevich V, Lavi U, Cohen Y (2005) Genetic variation in date palms propagated from offshoots and tissue culture. J Am Soc Hortic Sci 130(1):46–53

Hammadi H, Mokhtar R, Mokhtar E, Ali F (2009) New approach for the morphological identification of date palm (*Phoenix dactylifera* L.) cultivars from Tunisia. Pak J Bot 41(6):2671–2681

Hamza H, Abderrahim MAB, Elbekkay M, Ferdaous G, Triki T, Ferchichi A (2012) Investigation of genetic variation in Tunisian date palm (*Phoenix dactylifera* L.) cultivars using ISSR marker systems and their relation with fruit characteristics. Turk J Biol 36:449–458

Hamza H, Abederrahim MAB, Elbekkay M, Ferchichi A (2013) Comparison of the effectiveness of ISSR and SSR markers in determination of date palm (*Phoenix dactylifera* L.) agronomic traits. Aust J Crop Sci 7(6):763–769

Hassanpour-Estahbanati A, Hamidian S (2007) Morphological abnormalities in tissue culture derived date palm plants (*Phoenix dactylifera* L.) in Southern province of Iran. In: Hassanpour-Estahbanati A, Hamidian S (eds) 4th symposium on date palm in Saudi Arabia, King Faizal University, Al-Hassa, pp 5–8

Heath-Pagliuso S, Rappaport L (1990) Somaclonal variant UC-T3: the expression of Fusarium wilt resistance in progeny arrays of celery, *Apium graveolens* L. Theor Appl Genet 80(3):390–394

Heinz DJ, Mee GWP (1971) Morphologic, cytogenetic, and enzymatic variation in *Saccharum* species hybrid clones derived from callus tissue. Am J Bot 257–262

Hirochika H, Sugimoto K, Otsuki Y, Tsugawa H, Kanda M (1996) Retrotransposons of rice involved in mutations induced by tissue culture. Proc Natl Acad Sci 93(15):7783–7788

Hosaka A, Kakutani T (2018) Transposable elements, genome evolution and transgenerational epigenetic variation. Curr Opin Genet Dev 49:43–48

Hsu C-C, Lai P-H, Chen T-C, Tsai W-C, Hsu J-L, Hsiao Y-Y, Wu W-L, Tsai C-H, Chen W-H, Chen H-H (2019) PePIF1, a P-lineage of PIF-like transposable element identified in protocorm-like bodies of Phalaenopsis orchids. BMC Genomics 20(1):1–13

Jain SM (2007) Recent advances in date palm tissue culture and mutagenesis. Acta Horticulturae

Jaligot E, Beulé T, Rival A (2002) Methylation-sensitive RFLPs: characterisation of two oil palm markers showing somaclonal variation-associated polymorphism. Theor Appl Genet 104(8):1263–1269

Jaligot E, Rival A, Beule T, Dussert S, Verdeil J-L (2000) Somaclonal variation in oil palm (*Elaeis guineensis* Jacq.): the DNA methylation hypothesis. Plant Cell Rep 19(7):684–690

Jatoi MA, Abul-Soad AA, Markhand GS, Solangi N (2015) Establishment of an efficient protocol for micropropagation of some Pakistani cultivars of date palm (*Phoenix dactylifera* L.) using novel inflorescence explants. Pak J Bot 47(5):1921–1927

Jatoi MA, Abul-Soad AA, Markhand GS, Solangi N (2019) Micropropagation of cv. Dhakki a high value date palm cultivar of Pakistan using offshoot and inflorescence explants. Paper presented at the proceedings of sixth international date palm conference, Abu-Dhabi, UAE

Jatt T, Lee M-S, Rayburn AL, Jatoi MA, Mirani AA (2019a) Determination of genome size variations among different date palm cultivars (*Phoenix dactylifera* L.) by flow cytometry. 3 Biotech 9(12):1–10

Jatt T, Markhand GS, Rayburn L, Ming R, Jatoi MA, Mirbahar AA (2019b) Karyotype variations among date palm (*Phoenix dactylifera* L.) cultivars of Sindh, Pakistan. Sarhad J Agric 35:708–718

Javouhey M, Daguin F, Letouze R Somatic embryogenesis, an efficient tool for date palm (*Phoenix dactylifera* L.) industrial micropropagation. Characterization and genetic stability of original offshoots and regenerated plantlets by RAPD markers. In: International symposium on methods and markers for quality assurance in micropropagation, vol 530, pp 237–242

Jia X, Zhang Q, Jiang M, Huang J, Yu L, Traw MB, Tian D, Hurst LD, Yang S (2021) Mitotic gene conversion can be as important as meiotic conversion in driving genetic variability in plants and other species without early germline segregation. PLoS Biol 19(3):e3001164

Jin S, Mushke R, Zhu H, Tu L, Lin Z, Zhang Y, Zhang X (2008) Detection of somaclonal variation of cotton (*Gossypium hirsutum*) using cytogenetics, flow cytometry and molecular markers. Plant Cell Rep 27(8):1303–1316

Johannes F, Porcher E, Teixeira FK, Saliba-Colombani V, Simon M, Agier N, Bulski A, Albuisson J, Heredia F, Audigier P (2009) Assessing the impact of transgenerational epigenetic variation on complex traits. PLoS Genet 5(6):e1000530

Jonah PM, Bello LL, Lucky O, Midau A, Moruppa SM (2011) The importance of molecular markers in plant breeding programmes. Glob J Sci Front Res 11(5):4–12

Joyce SM, Cassells AC, Jain SM (2003) Stress and aberrant phenotypes in vitro culture. Plant Cell Tissue Organ Cult 74(2):103–121

Kaeppler SM, Kaeppler HF, Rhee Y (2000) Epigenetic aspects of somaclonal variation in plants. In: Plant gene silencing. Springer, pp 59–68

Karp A (1994) Origins, causes and uses of variation in plant tissue cultures. In: Plant cell and tissue culture. Springer, pp 139–151

Kashkush K, Feldman M, Levy AA (2002) Gene loss, silencing and activation in a newly synthesized wheat allotetraploid. Genetics 160(4):1651–1659

Katiyar R, Chopra V (1995) A somaclone of *Brassica juncea* is processed into a variety and is released for commercial cultivation in India. Cruciferae Newslett 17:92–93

Kharabian A, Darabi A (2005) Characterization of some chromosomal aberrations in regenerated rice plants (*Oryza sativa*). Plant Cell Tissue Organ Cult 83(2):161–168

Kumar N, Modi AR, Singh AS, Gajera BB, Patel AR, Patel MP, Subhash N (2010a) Assessment of genetic fidelity of micropropagated date palm (*Phoenix dactylifera* L.) plants by RAPD and ISSR markers assay. Physiol Mol Biol Plants 16(2):207–213

Kumar N, Singh AS, Modi AR, Patel AR, Gajera BB, Subhash N (2010b) Genetic stability studies in micropropagated date palm (*Phoenix dactylifera* L.) plants using microsatellite marker. J Forest Environ Sci 26(1):31–36
Kunert KJ, Baaziz M, Cullis CA (2003) Techniques for determination of true-totype date palm plants. Emirates J Agric Sci
Larkin PJ, Scowcroft WR (1981) Somaclonal variation—a novel source of variability from cell cultures for plant improvement. Theor Appl Genet 60(4):197–214
Lee M, Phillips RL (1988) The chromosomal basis of somaclonal variation. Annu Rev Plant Physiol Plant Mol Biol 39(1):413–437
Leprince AS, Grandbastien MA, Meyer C (2001) Retrotransposons of the Tnt1B family are mobile in *Nicotiana plumbaginifolia* and can induce alternative splicing of the host gene upon insertion. Plant Mol Biol 47(4):533–541
Lisch D, Bennetzen JL (2011) Transposable element origins of epigenetic gene regulation. Curr Opin Plant Biol 14(2):156–161
LoSchiavo F, Pitto L, Giuliano G, Torti G, Nuti-Ronchi V, Marazziti D, Vergara R, Orselli S, Terzi M (1989) DNA methylation of embryogenic carrot cell cultures and its variations as caused by mutation, differentiation, hormones and hypomethylating drugs. Theor Appl Genet 77(3):325–331
Madlung A, Comai L (2004) The effect of stress on genome regulation and structure. Ann Bot 94(4):481–495
Madsen LH, Fukai E, Radutoiu S, Yost CK, Sandal N, Schauser L, Stougaard J (2005) LORE1, an active low-copy-number TY3-gypsy retrotransposon family in the model legume Lotus japonicus. Plant J 44(3):372–381
Masuta Y, Nozawa K, Takagi H, Yaegashi H, Tanaka K, Ito T, Saito H, Kobayashi H, Matsunaga W, Masuda S (2017) Inducible transposition of a heat-activated retrotransposon in tissue culture. Plant Cell Physiol 58(2):375–384
McClintock B (1978) Mechanisms that rapidly reorganize the genome. University of Missouri, Columbia, Missouri, Paper presented at the Stadler genetics symposium
McCubbin MJ, Zaid A, Van Stade J (2004) A southern African survey conducted for off-types on date palms produced using somatic embryogenesis. Emirates J Food Agric 16(1):8–14
Miguel C, Marum L (2011) An epigenetic view of plant cells cultured in vitro: somaclonal variation and beyond. J Exp Bot 62(11):3713–3725
Mirani AA (2019) Evaluating genetic stability of micropropagated date palm (*Phoenix dactylifera* L.) varieties using inflorescence. Shah Abdul Latif University, Khairpur, Khairpur, Sindh-Pakistan
Mirani AA, Teo CH, Abul-Soad AA, Markhand GS, Jatt T, Mirbahar AA, Solangi N (2019) Phenotypic reversion of somaclonal variants derived from inflorescence of date palm (*Phoenix dactylifera* L.) in the open field trials. Sarhad J Agric 35(3):719–726
Mirani AA, Teo CH, Markhand GS, Abul-Soad AA, Harikrishna JA (2020) Detection of somaclonal variations in tissue cultured date palm (*Phoenix dactylifera* L.) using transposable element-based markers. Plant Cell Tissue Organ Cult 141(1):119–130
Mirani AA, Jatoi MA, Bux L, Teo CH, Kabiita AI, Harikrishna JA, Markhand GS, Jatt T, Solangi N, Abro S (2022) Genetic stability analysis of tissue culture derived date palm cv. Dedhi plants using IRAP markers. Acta Ecol Sinica 42(1):76–81
Moghaieb RE, Abdel-Hadi A-HA, Ahmed MRA (2011) Genetic stability among date palm plantlets regenerated from petiole explants. Afr J Biotech 10(65):14311–14318
Mondini L, Noorani A, Pagnotta MA (2009) Assessing plant genetic diversity by molecular tools. Diversity 1(1):19–35
Mujib A, Banerjee S, Dev Ghosh P (2007) Callus induction, somatic embryogenesis and chromosomal instability in tissue culture raised hippeastrum (*Hippeastrum hybridum* cv. United Nations). Propag Ornamental Plants 7:169–174
Murata M, Orton TJ (1983) Chromosome structural changes in cultured celery cells. In Vitro 19(2):83–89

Nadeem MA, Nawaz MA, Shahid MQ, Doğan Y, Comertpay G, Yıldız M, Hatipoğlu R, Ahmad F, Alsaleh A, Labhane N (2018) DNA molecular markers in plant breeding: current status and recent advancements in genomic selection and genome editing. Biotechnol Biotechnol Equip 32(2):261–285

Neuffer MG (1966) Stability of the suppressor element in two mutator systems at the A1 locus in maize. Genetics 53(3):541–549

Oihabi A, De Wet P (2005) Date palm abnormalities in Namibia and Niger. In: Proceedings of the international workshop on true-to-typeness of date palm tissue culture-derived plants, p 43

Orton TJ (1983) Experimental approaches to the study of somaclonal variation. Plant Mol Biol Report 1(2):67–76

Othmani A, Rhouma S, Bayoudh C, Mzid R, Drira N, Trifi M (2010) Regeneration and analysis of genetic stability of plantlets as revealed by RAPD and AFLP markers in date palm (*Phoenix dactylifera* L.) cv. Deglet Nour. Int Res J Plant Sci 1(3):48–55

Ouarda HEF, Walker DJ, Khouja ML (2012) Phenotypic and nuclear DNA variation in Tunisian cultivars of date palm (*Phoenix dactylifera* L.). Afr J Biotechnol 11(22):6034–6042

Peschke VM, Phillips RL (1992) Genetic implications of somaclonal variation in plants. In: Adv Genetics, vol 30. Elsevier, pp 41–75

Peschke VM, Phillips RL, Gengenbach BG (1987) Discovery of transposable element activity among progeny of tissue culture-derived maize plants. Science 238(4828):804–808

Peterson PA (1960) The pale green mutable system in maize. Genetics 45(1):115–133

Racchi ML, Bove A, Turchi A, Bashir G, Battaglia M, Camussi A (2014) Genetic characterization of Libyan date palm resources by microsatellite markers. 3 Biotech 4(1):21–32

Raza SA, Ali Y, Mehboob F (2012) Role of agriculture in economic growth of Pakistan. Int Res J Financ Econ 83:180–186

Reuveni O, Israeli Y (1989) Measures to reduce somaclonal variation in in vitro propagated bananas. In: Robinson JC (ed) International symposium on the culture of subtropical and tropical fruits and crops, vol 275, pp 307–314

Rhouma-Chatti S, Choulak S, Zehdi-Azzouzi S, Chatti K, Said K (2014) Molecular polymorphism and phylogenetic relationships withinn Tunisian date palm (*Phoenix dactylifera* L.): evidence of non-coding trnL-trnF regions of chloroplast DNAs. Sci Hortic 170:32–38

Rival A, Ilbert P, Labeyrie A, Torres E, Doulbeau S, Personne A, Dussert S, Beulé T, Durand-Gasselin T, Tregear JW (2013) Variations in genomic DNA methylation during the long-term in vitro proliferation of oil palm embryogenic suspension cultures. Plant Cell Rep 32(3):359–368

Sahara A, Utomo C, Liwang T (2019) Early detection of somaclonal variation in oil palm callus culture through cytological and SDS-PAGE protein analysis. In: Setyobudi RH (ed) IOP conference series: earth and environmental science, vol 1. IOP Publishing, p 012005

Saker MM, Bekheet SA, Taha HS, Fahmy AS, Moursy HA (2000) Detection of somaclonal variations in tissue culture-derived date palm plants using isoenzyme analysis and RAPD fingerprints. Biol Plant 43(3):347–351

Saker MM, Adawy SS, Mohamed AA, El-Itriby HA (2006) Monitoring of cultivar identity in tissue culture-derived date palms using RAPD and AFLP analysis. Biol Plant 50(2):198–204

Sakka H, Baraket G, Dakhlaoui Dkhil S, Zehdi Azouzi S, Salhi-Hannachi A (2013) Chloroplast DNA analysis in Tunisian date palm cultivars (*Phoenix dactylifera* L.): sequence variations and molecular evolution of trnL (UAA) intron and trnL (UAA) trnF (GAA) intergenic spacer. Sci Hortic 164:256–269

Salman RM, Al-Jibouri AAM, Al-Quadhy WK, Omar MS (1988) Isozyme and chromosomal analyses of tissue culture derived date palms. Date Palm J 6:401–411

Samiei S, Tajadod G, Zarghami R, Mirzai M (2022) A 2, 4-D-free combined direct organogenesis-indirect somatic embryogenesis protocol for mass propagation of date palm (*Phoenix dactylifera* L.) cv. Barhi. J Sci Islamic Republic of Iran 33(2):125–132

Sammour RH (2014) Cultivars identification based on biochemical markers. Res Rev BioSci 8(9):347–358

Secor GA, Shepard JF (1981) Variability of protoplast-derived potato clones 1. Crop Sci 21(1):102–105

Sedra MH, Lashermes P, Trouslot P, Combes MC, Hamon S (1998) Identification and genetic diversity analysis of date palm (*Phoenix dactylifera* L.) varieties from Morocco using RAPD markers. Euphytica 103(1):75–82

Sharma SK, Bryan GJ, Winfield MO, Millam S (2007) Stability of potato (*Solanum tuberosum* L.) plants regenerated via somatic embryos, axillary bud proliferated shoots, microtubers and true potato seeds: a comparative phenotypic, cytogenetic and molecular assessment. Planta 226(6):1449–1458

Skirvin RM, McPheeters KD, Norton M (1994) Sources and frequency of somaclonal variation. HortScience 29(11):1232–1237

Skirvin RM, Coyner M, Norton MA, Motoike S, Gorvin D (2000) Somaclonal variation: do we know what causes it? AgBiotechNet 2 (ABN 048):1–4

Smith MK, Hamill SD (1993) Early detection of dwarf off-types from micropropagated Cavendish bananas. Aust J Exp Agric 33(5):639–644

Smulders MJM, De Klerk GJ (2011) Epigenetics in plant tissue culture. Plant Growth Regul 63(2):137–146

Solangi N (2021) Micropropagation of some elite cultivars of date palm (*Phoenix dactylifera* L.) through inflorescence explants. PhD dissertation, Shah Abdul Latif University, Khairpur, Pakistan

Solangi N, Abul-Soad AA, Markhand GS, Jatoi MA, Jatt T, Mirani AA (2020) Comparison among different auxins and cytokinins to induce date palm (*Phoenix dactylifera* L.) somatic embryogenesis from floral buds. Pak J Bot 52(4):1243–1249

Soliman SS, Ali BA, Ahmed MM (2003) Genetic comparisons of Egyptian date palm cultivars (*Phoenix dactylifera* L.) by RAPD-PCR. Afr J Biotechnol 2(4):86–87

Soliman KA, Rizk R, El-Sharabasy S (2009) Genetic analysis of abnormalities in tissue culture-derived date palm (Phoenix dacylifers L.), Barhi cultivars. Egypt J Biotech 33:122–134

Tahara M, Aoki T, Suzuka S, Yamashita H, Tanaka M, Matsunaga S, Kokumai S (2004) Isolation of an active element from a high-copy-number family of retrotransposons in the sweetpotato genome. Mol Genet Genomics 272(1):116–127

Torres AM, Tisserat B (1980) Leaf isozymes as genetic markers in date palms. Am J Bot 162–167

Varughese A (2000) Mass propagation of date palm through tissue culture: an efficient method by SAPAD. In: Varughese A (ed) Procedings of date palm international symposium, Windhoek, Namibia

Wear EE, Song J, Zynda GJ, LeBlanc C, Lee T-J, Mickelson-Young L, Concia L, Mulvaney P, Szymanski ES, Allen GC (2017) Genomic analysis of the DNA replication timing program during mitotic S phase in maize (*Zea mays*) root tips. Plant Cell 29(9):2126–2149

Williams JGK, Kubelik AR, Livak KJ, Rafalski JA, Tingey SV (1990) DNA polymorphisms amplified by arbitrary primers are useful as genetic markers. Nucleic Acids Res 18(22):6531–6535

Xu M, Li X, Korban SS (2004) DNA-methylation alterations and exchanges during in vitro cellular differentiation in rose (*Rosa hybrida* L.). Theor Appl Genet 109(5):899–910

Xu Z, Chen Z (1995) Recent advances of plant tissue culture and biotechnology in China. Plant Tissue Cult Biotech 1:147–153

Zaid A, Al-Kaabi HH (2003) Plant-off types in tissue culture-derived date palm (*Phoenix dactylifera* L.). Emirates J Food Agric 15(1):17–35

Zaid A, Al Kaabi HH (2006) Large scale in vitro propagation of a rare and unique male date palm (*Phoenix dactylifera* L.) using inflorescences technique. In: Zaid A, Hegarty V, Kaabi HHA (eds) III International date palm conference, vol 736, pp 234–254

Zehdi-Azouzi S, Cherif E, Guenni K, Ben Abdelkrim A, Bermil A, Rhouma S, Ben Salah M, Santoni S, Pintaud J, Aberlenc-Bertossi F, Salhi Hannachi A (2016) Endemic insular and coastal Tunisian date palm genetic diversity. Genetica 144(2):181–190

Zehdi S, Sakka H, Rhouma A, Salem AOM, Marrakchi M, Trifi M (2004) Analysis of Tunisian date palm germplasm using simple sequence repeat primers. Afr J Biotech 3(4):215–219

Zehdi S, Cherif E, Rhouma S, Santoni S, HAS, Pintaud J (2012) Molecular polymorphism and genetic relationships in date palm (*Phoenix dactylifera* L.). The utility of nuclear microsatellite markers. Sci Horticult 148:255–263

Zehdi-Azouzi S, Cherif E, Moussouni S, Gros-Balthazard M, Abbas Naqvi S, Ludena B, Castillo K, Chabrillange N, Bouguedoura N, Bennaceur M, Si-Dehbi F, Abdoulkader S, Daher A, Terral J, Santoni S, Ballardini M, Mercuri A, Salah M, Kadri K, Othmani A, Littardi C, Salhi-Hannachi A, Pintaud J, Aberlenc-Bertossi F (2015) Genetic structure of the date palm (*Phoenix dactylifera* L.) in the old world reveals a strong differentiation between eastern and western populations. Ann Bot 116(1):101–112

Ziauddin A, Kasha K (1990) Long-term callus cultures of diploid barley (*Hordeum vulgare*). II. Effect of auxins on chromosomal status of cultures and regeneration of plants. Euphytica 48(3):279–286

Zivdar S, Mousawi M, Ansari NA (2008) Genetic stability in date palm micropropagation. Asian J Plant Sci 7(8):775

Somaclonal Variation in Olive

Carolina Sánchez-Romero

Abstract Tissue culture is commonly used in olive with different purposes. Apart from commercial large-scale plant production, it is applied as a complement for genetic improvement, by both conventional and unconventional methods, with efficient protocols available for genetic transformation, in vitro selection or cryopreservation. However, uncontrolled variations can be observed during tissue culture due to somaclonal variation. These alterations can suppose an important problem in culture systems requiring plant fidelity. However, somaclonal variation can also constitute an alternative source of genetic variability in breeding programs. Therefore, evaluation of somaclonal variation is recommended in order to determine the reliability of each in vitro culture procedure before its practical application. Moreover, analyzing the effect on somaclonal variation of different factors under specific in vitro culture conditions may help to control its occurrence, for either increasing its frequency or limiting its appearance, depending on the objectives of the tissue culture system. Investigations carried out in olive revealed a significant influence of the genotype, the in vitro propagation method and the length of culture period. The genotype played a main role in most works in which somaclonal variation was observed.

Keywords Micropropagation · *Olea europaea* · Olive · Organogenesis · Somaclonal variation · Somatic embryogenesis · In vitro tissue culture

1 Introduction

The olive is an evergreen tree with an average height of about 4–8 m, although it can grow up to 15 m at maturity. The trunk is thick and the bark is gray to greyish-green in color. The crown is rounded, although slightly lobulate. Tree features, such as density of the crown, growth habit, wood color and internode length, vary depending on the cultivar (Rapoport 1997). The olive fruit is a drupe and, in contrast to other oil

C. Sánchez-Romero (✉)
Departamento de Botánica y Fisiología Vegetal, Universidad de Málaga, Campus de Teatinos s/n, 29071 Málaga, Spain
e-mail: c.sanchez@uma.es

C. Sánchez-Romero (ed.), *Somaclonal Variation: Basic and Practical Aspects*,
https://doi.org/10.1007/978-3-031-51626-9_10

crops, the oil accumulates in the mesocarp (Baldoni and Belaj 2009). The olive tree has great longevity. Its life span is typically longer than 500 years, although trees older than 2000 years are under cultivation in numerous regions (Baldoni and Belaj 2009).

The olive (*Olea europaea* subsp. *europaea*) belongs to the Oleaceae family, subfamily Oleideae, genus *Olea*. According to Green and Wickens (1989) and Green (2002), six sub-species can be distinguished based on their morphology and geographical distribution: subsp. *europaea*, subsp. *cuspidata*, subsp. *laperrini*, subsp. *maroccana*, subsp. *cerasiformis*, and subsp. *guanchica*. The sub-species *europaea* included two botanical varieties: var. *sativa*, corresponding to the cultivated olive, and var. *sylvestris*, corresponding to wild olive (Green and Wickens 1989; Green 2002). Cultivated olive (*Olea europaea* subsp. *europaea* var. *sativa*) consists of a great variety of cultivars, which have been empirically selected and vegetatively propagated in the traditional olive-producing countries (Rallo et al. 2018).

Olive is a diploid species with 46 chromosomes ($2n = 2x = 46$) (Baldoni and Belaj 2009). Its genome size has been determined in different investigations. According to Besnard et al. (2008), the mean 2C-value in Mediterranean samples of *Olea europaea* subsp. *europaea* was 3.39 ± 0.13 pg, including olive cultivars and wild olive. This estimation was intermediate between those reported for Italian cultivars (2C-value between 3.90 pg and 4.66 pg) (Rugini et al. 1996; Bitonti et al. 1999) and Portuguese trees (2C-value between 2.90 and 3.19 pg) (Loureiro et al. 2007). Contrarily to some Italian cultivars, where high genome size heterogeneity was found, the results obtained by Loureiro et al. (2007) suggested a low intraspecific variation, at least among the studied cultivars and between them and wild olive. As pointed out by Besnard et al. (2008), such different results could be due to comparisons of different genotypes and methodological differences in flow cytometry.

The olive is indigenous to the Mediterranean Basin (Díez et al. 2016). The earliest evidences of olive domestication is from approximately 6800–5800 BP, in the Jordan river valley (Zohary and Spiegel-Roy 1975). The domesticated olive originated from the Mediterranean wild olive populations (Weiss 2015). Results of molecular analysis of sets of wild olives and cultivars from different Mediterranean regions suggest that olive trees have undergone different domestications processes in diverse regions (Baldoni and Belaj 2009). Empiric local selection within wild olives and crosses between previously selected or introduced cultivars and other native cultivars or wild olives in all growing areas have yielded a high number of local cultivars (Rallo et al. 2018). Thus, although domestication probably began in the Eastern Mediterranean, selection for cultivars took place at several independent locations (Lavee and Zohary 2011; Besnard et al. 2013).

From its origin place, domesticated olive spread by the shores of the Mediterranean Sea and since 3000 BC the olive tree has been extensively cultivated along the Mediterranean basin (Baldoni and Belaj 2009). In this region, the olive is one of the most important crops, with an enormous impact on the economy, history, culture and environment (Baldoni and Belaj 2009). Although olive products have been valued since ancient times (Fabbri et al. 2009), in the last decades, olive oil has acquired greater importance due to its high nutritional quality and healthy effects compared with other plant oils. This has led to worldwide expansion of olive cultivation to new countries such as Argentina, Chile, Mexico, USA, Japan, China, New Zealand, Australia, and South Africa (Sánchez-Romero 2021). Currently, the olive is cultivated over 10,338,179 ha, with a production of 23,054,310.6 tonnes, with Spain, Italy and Turkey, as the main producing countries (FAOSTAT 2021). At present, the olive is the most economically important oil tree crop in temperate areas (Díez et al. 2016) and the second most important worldwide, after oil palm (Baldoni and Belaj 2009).

2 Somaclonal Variation

The growth of plant cells in vitro and their regeneration into plants is an asexual process, which only involves mitotic division of cells. Therefore, the expected result is clonal multiplication of genetically uniform plants (Karp 1994). However, during in vitro culture, uncontrolled variations can sometimes be observed. According to Larkin and Scowcroft (1981), plant tissue culture itself generates variability, and they coined the term somaclonal variation in order to refer to variation arising in the cell cultures, regenerated plants and their progenies.

Somaclonal variation occurs randomly and it is unpredictable in nature, provoking both positive and negative changes. Therefore, it supposes a problem in tissue cultures systems in which the objective is the production of true-to-type plants, such as clonal propagation, genetic transformation or germplasm conservation (Leva and Petruccelli 2007). This inconvenience is especially important in plant species with a long-life cycle, as olive, in which some negative changes may become evident after a number of years, in their adult stage. However, somaclonal variation can also constitute a source of variability, which can be used in plant breeding programs. This application is especially useful in vegetatively propagated crops and species with a narrow genetic basis, which are difficult to improve through conventional breeding (Jain et al. 1998; Remotti 1998). Moreover, it is considered an interesting alternative compared to other more controversial biotechnological tools, such as genetic transformation (Bradaï et al. 2016a).

Different factors influence somaclonal variation, such as variation preexisting in explants, the in vitro culture method, culture medium and conditions, length of the culture period, and the genotype and type of tissue used as explant source. Investigating their influence in specific in vitro culture techniques and genotypes used in olive can help to control its occurrence, for either increasing its frequency in culture systems focused on generation of genetic variability, or limiting its appearance, when the objective is the development of genetically stable plants.

Evaluation of somaclonal variation is, in consequence, an important requirement in order to determine the reliability of each in vitro culture protocol. Analysis of somaclonal variation can be carried out by using different approaches, such as phenotypic, cytological, biochemical or molecular analysis.

3 Somaclonal Variation in Olive

The olive tree has been in vitro cultured by using different techniques, such as micropropagation, organogenesis and somatic embryogenesis (Sánchez-Romero 2019; Rugini et al. 2020). Different works have investigated somaclonal variation when utilizing these tissue culture techniques (Table 1).

3.1 Micropropagation

In vitro propagation of plants from axillary buds or shoots has been applied in olive since the 1980s (Leva 2009). Although some drawbacks have conditioned micropropagation in olive, different optimization works have allowed the development of sufficiently effective protocols (Leva et al. 2012). Currently, in vitro propagation by axillary bud stimulation is available for many olive cultivars (Lambardi et al. 2013; Rugini et al. 2020) and it is used for large-scale plant production by commercial laboratories (Rugini and De Pace 2016). In parallel, multiple investigations on somaclonal variation in olive micropropagated plants have been carried out (Table 1).

García-Férriz et al. (2002) optimized a protocol for micropropagation of adult olive trees. Rooted in vitro-propagated plants of the cultivars Arbequina, Picual and Empeltre showed a normal development; and three years later, plants in field produced normally. In addition to phenotypic analysis, one month after acclimatization, micropropagated plants were molecularly compared with mature plants provided by the Germplasm Bank of Córdoba (Spain) by using four random amplified polymorphic DNA (RAPD) primers. In vitro cultured plants were considered true to type, as they produced the same amplification banding pattern than control plants of the same cultivar.

Subsequently, Briccoli-Bati et al. (2006) assessed true-to-typeness of olive micropropagated plants of cultivars Carolea and Nocellara Etnea. In vitro propagated plants of each cultivar were obtained by introducing in vitro uninodal cuttings. After multiplication for 4 years, micropropagated shoots were rooted and acclimatized to ex vitro conditions. Eight years after planting, percentage of flower bud differentiation, fruit weight, yield, pruning material and trunk diameter were measured. The vegetative and productive performance of micropropagated olive plants was compared with those of grafted and own-rooted plants in the case of cv. Carolea and with grafted plants in the case of cv. Nocellara Etnea, due to its very poor rooting ability. In each variety, all the plant material was obtained from one single mother plant. Although a major role of the varietal component was highlighted, differences in field performance were observed in plants obtained through different propagation techniques. Micropropagated Carolea plants presented a lower cumulated yield over the trial period than grafted and own-rooted plants, due to poor fruit set, as a similar percentage of flower buds was measured. In Nocellara Etnea micropropagated plants, yield was slightly higher with respect to grafted plants, but fruit weight was significantly lower.

Table 1 Works on somaclonal variation in olive

Cultivar	Explant	In vitro method	In vitro protocol	Evaluation method	Source of variation	Reference
Tanche and Picual	Cotyledon segments	Organogenesis	Cañas and Benbadis (1988)	Morphometric analysis	-	Cañas and Benbadis (1988)
Nabali	Roots of germinated seeds	Somatic embryogenesis	Shibli et al. (2001)	Phenotypic analysis	–	Shibli et al. (2001)
Arbequina, Picual, Empeltre	Uninodal explants	Micropropagation	García Férriz et al. (2002)	RAPD and phenotypic analysis	–	García Férriz et al. (2002)
Maurino	–	Micropropagation	–	Growth parameters, RAPD analysis	–	Leva et al. (2002)
Carolea and Nocellara Etnea	Uninodal cuttings	Micropropagation	Briccoli Bati et al. (2006)	Phenotypic analysis: growth and reproductive characters	–	Briccoli Bati et al. (2006)
Frangivento	Immature cotyledon portions	Somatic embryogenesis	Leva et al. (1995b)	Phenotypic analysis	–	Leva and Petruccelli (2007)
Maurino	Nodal explants	Micropropagation	Leva et al. (1995a)	Morphological analysis: vegetative growth and reproductive behaviour	–	Leva (2009)
Frangivento	Immature cotyledon portions	Somatic embryogenesis	Leva et al. (1995b)	Morphological analysis	–	Leva (2009)

(continued)

Table 1 (continued)

Cultivar	Explant	In vitro method	In vitro protocol	Evaluation method	Source of variation	Reference
Kroneiki	Radicle and proximal parts of cotyledon	Somatic embryogenesis	–	RAPD analysis	Callus phase	Peyvandi et al. (2010)
Moraiolo	Axillary shoots explants micropropagated in vitro	Indirect somatic embryogenesis	Dell'Orco and Mencuccini (2000)	Agronomic characteristics: Morphological and biometric analysis	–	Mencuccini (2011)
Dezful Baghe-Malek and Dezful Safiabad	Internode cuttings	Micropropagation	–	Morphological analysis, RAPD analysis, flow cytometry	Number of subcultures	Farahani et al. (2011)
Dezful Baghmalek and Dezful Safiabad	Internode cuttings	Micropropagation	–	Morphological analysis	Number of subcultures	Yari et al. (2011a)
Zard, Roughani and unkown Iks	Internode cuttings	Micropropagation	–	Morphological analysis	Number of subcultures	Yari et al. (2011b)
Maurino	Nodal explants	Micropropagation	Leva et al. (1995a)	RAPD and ISSR analysis	–	Leva and Petruccelli (2012)
Frangivento	Immature cotyledon portions	Somatic embryogenesis	Leva et al. (1995b)	RAPD analysis	–	Leva et al. (2012)
Koroneiki	Single node segments	Micropropagation	Noormohammadi et al. (2014)	Chloroplast rpl16 intron sequences and RAPD markers	–	Kangarloo et al. (2015)

(continued)

Table 1 (continued)

Cultivar	Explant	In vitro method	In vitro protocol	Evaluation method	Source of variation	Reference
Picual	Radicle of mature zygotic embryos	Somatic embryogenesis	Bradaï et al. (2016b)	Phenotypic evaluation: morphological and biometric analysis	Genotype and culture age	Bradaï et al. (2016a)
Dezful Baghmalek, Dezful Safiabad and Roughani	Internode cuttings	Micropropagation	–	Morphological analysis	Number of subcultures	Farahani et al. (2018)
Picual	Radicle of mature zygotic embryos	Somatic embryogenesis	Bradaï et al. (2016b)	SSR and RAPD analysis	Genotype and culture age	Bradaï et al. (2019)
Amin, Meshkat and Arbequina	Single node cuttings	Micropropagation-Organogenesis	Mirzaei et al. (2021)	AFLP analysis and flow citometry	–	Mirzaei et al. (2021)

AFLP amplified fragment length polymorphism; *ISRR* inter simple sequence repeat; *RAPD* random amplified polymorphic DNA; *SSR* simple sequence repeat; – not indicated

The effect of the number of subcultures has been investigated in different investigations in different cultivars. In a first study, Farahani et al. (2011) examined the occurrence of somaclonal variation during micropropagation of two Iranian olive cultivars, Dezful Baghe-Malek and Dezful Safiabad. Shoot cultures were established from internode samples and monitored during seven subcultures by morphological, RAPD and flow cytometry analysis. Analysis of variance showed significant differences in the morphological traits of plants obtained from different subcultures. Moreover, neighbor joining (NJ) clustering and principle coordinate analysis (PCO) based on morphological data separated regenerated plants of the latter subcultures from the earlier ones, thus indicating that latter subcultures bring about a higher degree of morphological variations in both cultivars. In both cases, larger plants, with higher values of different growth parameters, were obtained at the seventh subculture. RAPD analysis also revealed a significant influence of the number of subcultures on somaclonal variation of micropropagated plants. Similar results were obtained in the two cultivars, with the parental genotype and plants from first subculture placed close to each other, plants developed from the intermediate subcultures located together at some distance, and plants from the sixth and seventh subcultures situated far from the other ones. Nevertheless, both genotypes differed in the level of somaclonal variation, as indicated their different Nei's genetic diversity and Shanon index values. Mean C-value obtained by flow cytometry analysis did not differ significantly among mother plants and those regenerated after different subcultures. Therefore, authors concluded that the genetic changes due to somaclonal variation were mainly caused by mutations (insertions/deletions) and not to by quantitative changes. In a later investigation, Farahani et al. (2018) examined changes in different morphological parameters over seven subcultures in three Iranian cultivars, those previously studied, Dezful BaghMalek and Dezful Safiabad, and Roughani. The results obtained revealed different levels of somaclonal variation depending on the cultivar. Greater somaclonal variation was detected in Dezful BaghMalek, compared to Dezful Safiabad and Roughani, especially in rooting percentage and root length. Roughani was the most stable genotype, specifically in leaf length and width, root length, leaf number and rooting percentage. In each cultivar, important fluctuations in the morphological characters assessed were observed throughout the subcultures. Number of branches was the most significant trait, with almost similar variation in the three cultivars tested.

A major effect of the genotype on somaclonal variation assessed by morphological parameters and important fluctuations throughout subcultures in the characters assessed were also reported by Yari et al. (2011a) in a similar investigation carried out in the cultivars Dezful BaghMalek and Dezful Safiabad. Yari et al. (2011b) also investigated somaclonal variation during five subcultures in three Iranian cultivars, Zard, Roughani and Iks, reporting analogous results to those described in previous studies (Yari et al. 2011a; Farahani et al. 2018). Different changes throughout time in culture were observed depending on the genotype.

Kangarloo et al. (2015) investigated somaclonal variation in micropropagated plants of cv. Koroneiki growing in the field. Leaf samples of 49 tissue culture-derived plants were genetically compared to those of the mother plant by using the sequence of the chloroplast *rpl16* intron and ten RAPD primers. A NJ dendrogram based on the RAPD data placed the 50 selected plants into four distinct clusters. Thirty-seven percent of the total genetic variation was due to "among-cluster" differences, while 63% was due to "within-cluster" variability. New Technology (TNT) analysis, a phylogenic tree based on the sequences of chloroplast *rpl16* gene introns, also produced four clusters. Therefore, both types of plants could be obtained during in vitro propagation of this cultivar: true-to-type plants and somaclonal variants, which differed from the mother plant in their RAPD marker pattern and rpl16 sequences. According to authors, the first group can be used by growers as a true cultivar, while the second group of plants should be tested for altered morphological and agronomic characteristics in the field and those with desirable features could be used in breeding programmes and for further studies.

Somaclonal variation was also analyzed in olive plants of cv. Maurino micropropagated by axillary buds (Leva et al. 2012). Analyzed plants were obtained after multiplication of the initial olive explants for 12 repeated subcultures (each one lasting 45 days), and subsequent transferrence to field conditions. For comparison of micropropagated plants with the mother plant, control plants were in parallel produced from semi hardwood cuttings obtained from the same donor plant. The same training system and agronomic operations were implemented for both types of plants. Plant fidelity was examined in different works by assessing their morphological and molecular performance in field. In a first morphological evaluation one year after field transplantation, no differences were detected in different vegetative parameters, although statistically significant differences in some of them could be observed at the end of the second year (Leva et al. 2002). When plants were six years old, they were newly analyzed. For determining growth and developmental behavior in field of micropropagated and control plants, 32 vegetative and reproductive traits were measured for two consecutive years (Leva 2009). No differences in most of the vegetative parameters were found, with no variations in growth habit, vegetative growth, canopy and trunk area. Although the leaves and drupes of microplants were slightly wider than those of the control plants, they still retained the characteristic shapes of the cultivar. The productivity of the fruiting shoots was also similar, although a higher number of flowers per inflorescence was found in micropropagated plants. No changes in flowering time were observed. Hence, uniformity with the control plants was inferred in terms of vegetative, reproductive and oil traits (Leva 2009).

Genetic stability of these plants was also screened over several years. In a first report, two years after transference to field, micropropagated plants were analyzed by RAPD markers. No differences in the amplification pattern were found among micropropagated plants, control plants and the donor plant (Leva et al. 2002). Seven years later, plants were newly compared with control plants (Leva and Petruccelli 2012). For detecting genetic changes, 40 RAPD and 10 inter simple sequence repeat (ISSR) primers were used. Homogeneous RAPD and ISSR patterns were obtained, thus revealing genetic stability in nine-year-old mature micropropagated olive plants cultured in field. The results obtained in separate investigations and using different molecular markers corroborated previous studies on phenotypic stability of Maurino micropropagated plants (Leva et al. 2003; Leva 2009).

According to these data, authors concluded that micropropagation does not affect either the field performance or the genetic fidelity of the olive plants and, therefore, axillary multiplication can be considered a safe mode for multiplication of true-to-type plants (Leva et al. 2012).

No agronomic and morphological variations have also been reported in micropropagated plants of cultivars Moraiolo, Canino, Frantoio, Leccino and Dolce Agogia (Rugini and Baldoni 2005).

3.2 Organogenesis

Although organogenesis has been reported in olive from both zygotic and mature tissues, it is not efficient enough for its application as regeneration method (Rugini et al. 2016). Consequently, it has been less used in olive (Table 1).

Cañas and Benbadis (1988) regenerated olive plants of cultivars Tanche and Picual by indirect organogenesis. Shoot development mainly occurred from calli induced from cotyledon fragments proximal to the embryo axes. After in vitro rooting, plantlets were transferred to soil in a greenhouse and six months later their variability was analyzed through a simple morphometric test in which the presence of some rare characters, such as double leaves, shoot dichotomy and triple whorl, was determined. Seedlings from both cultivars were used as control. Although morphological changes frequently occur in seedlings and shoots from trunk and big branch buds of plants in the field, probably associated with juvenility (Cañas and Benbadis 1988), the results obtained revealed a certain variability in the regenerated plants compared with seedlings. Greater variability was found in plants of cv. Tanche, than in those of cv. Picual.

Even when using optimized in vitro propagation protocols, the production of large calli is normal during stem cutting in vitro culture. Under these circumstances, shoots can be obtained from meristems, but also from callus, due to indirect organogenesis (Mirzaei et al. 2021). Mirzaei et al. (2021) examined the genetic stability of callus-induced shoots and meristem-induced shoots in three olive cultivars, Amin, Meshkat and Arbequina. Microshoots originated from calli were compared with those developed from meristems, which were considered as donor plants, since in previous investigations in vitro propagation of olive shoots in OM and zeatin-enriched media was proved to be stable. Both types of shoots were separately incubated on their optimized culture media for 6 months, with repetitive subcultures every 6 weeks, before analysis by flow cytometry and amplified fragment length polymorphism (AFLP) markers. Flow cytometry assessment found some discrepancy between callus- and meristem-induced shoots, which varied depending on the cultivar. AFLP assay also revealed different stability depending on the genotype, with Arbequina showing the highest dispersal (7–38%) between CIS and MIS, followed by the Iranian cultivars Amin (2–32%) and Meshkat (5–9%), which exhibited the highest stability.

3.3 Somatic Embryogenesis

As in other tree species, somatic embryogenesis is the main method for adventitious plant regeneration in olive (Sánchez-Romero 2021). In fact, it has been repeatedly used in this species and different protocols are available for plant regeneration from both juvenile and mature tissues (Sánchez-Romero 2018; 2019). In olive plants regenerated via somatic embryogenesis, both true-to-typeness and somaclonal variation have been reported (Sánchez-Romero 2019) (Table 1).

Shibli et al. (2001) regenerated plants of cv. Nabali from embryogenic callus induced from roots of germinated seeds. Acclimatized plants were transferred to a greenhouse and monitored for phenotypic variations. No morphological differences where observed among the regenerated plants and between them and those produced by micropropagation of axillary buds.

Mencuccini (2011) analyzed plantlets of cv. Moraiolo regenerated through indirect somatic embryogenesis. Somatic embryos were obtained from embryogenic callus induced from micropropagated axillary shoots. Regenerated plants were established in vivo in a greenhouse and subsequently transferred to the field, where they were analyzed by assessing different agronomic features, such as height and diameter of the trunk, height of the first stage of branches, canopy volume, presence and duration of the juvenile phase and flowering parameters. After three years in field, authors reported differences in the developmental behavior, with variations in plant height, number of flowers and number of fruits.

Plants of cv. Frangivento were obtained from embryogenic tissue induced from immature cotyledon portions and maintained by bimonthly subculture for three years (Leva et al. 1995a, b). All somatic embryos from which plants were obtained were produced in the same experiment, using the same in vitro protocol and derived from the same cotyledon culture. The somatic embryogenesis-derived plants were monitored during several years. In an initial analysis, six-month-old plants grown under greenhouse conditions did not exhibited phenotypic abnormalities (Leva et al. 1995a, b). However, during their subsequent cultivation in pots in the open air, they began to show a differential developmental behaviour. Three groups were phenotypically detected on the basis of several morphological measurements: reduced mean height (BOS), tall variant (COS) and plants with disorder in the canopy structure (Leva et al. 2000). Leva and Petruccelli (2007) considered the first two variant phenotypes, BOS (bush-olive somaclone) and COS (columnar-olive somaclone), to investigate the occurrence of somaclonal variation in olive somatic plants. Four representative trees were then selected per phenotype on the basis of similar tree height, development of the canopy and leaf area. Frangivento micropropagated plants (4 representative trees) developed from explants derived from the same donor plant used for induction of somatic embryogenesis and cultured under the same conditions were used as a putative control. Mature plants, eight-year-old, growing in the field were characterized by monitoring 32 morphological traits, including vegetative, inflorescence and fruit features (Leva 2009). Under similar growth conditions multiple traits showed significant differences among BOS and COS and putative-control, with a wide range of plant traits differently involved in somaclonal variation respect to the putative control plants. Compared to the putative controls, BOS plants showed differences in 19 traits (mainly, reduction in plant height and in the dimensions of leaves, inflorescences and fruits, and increase in feather shoot number, which jointly contributed to a more compact growth habit than that of the putative control plants), while 18 traits were different between COS and controls (mainly plant height and canopy volume and projection). Morphological differences between BOS and COS were also evident and both phenotypes were separated by cluster analysis, mainly related to growth habit and leaf, fruit and pit traits. BOS and COS were morphologically stable phenotypes, not reversing to normal appearance following a few generations of vegetative propagation (Leva and Petruccelli 2007).

BOS and COS somaclones and Frangivento donor plants grown in the field were also analyzed by using RAPD molecular markers (Leva et al. 2012). The results obtained revealed that both somaclone types and the control micropropagated plants shared 86% of RAPD markers. Some primers showed polymorphism between the somaclones and the donor plant, whereas others were polymorphic between the BOS and COS groups. In a second molecular analysis, by using the same 20 RAPD primers in 12-year-old somatic plants, the same patterns of differences between somaclones and donor plant, and bands specific to BOS and COS were observed as in the first genetic study (Leva et al. 2012). Leva et al. (2012) suggested that genetic changes occurred during the long-term maintenance of embryogenic cultures in vitro. Moreover, they reported the identification of DNA fragments putatively associated with growth habit.

The heritability of the variation in growth habit was also investigated in the progeny of somaclonal plants, obtained by vegetative propagation (rooted cuttings) and cultivated in the field under different environmental conditions. Preliminary assessment of the performance of somaclonal daughter plants indicated that morphological characteristics were comparable to those of the somaclonal mother plants. According to Leva et al. (2012), growth of the somaclonal daughter plants only depended on their inherent growth habit, not being affected by the environmental conditions at the planting site.

The results obtained in the different evaluations carried out in Frangivento somatic embryogenesis-derived plants revealed that the detected somaclonal variations were stable. As the associated morphological variations were positive and not accompanied by deleterious changes in other agronomically or horticulturally important traits, Leva et al. (2012) proposed a possible future utilization of the somaclones in breeding programs as rootstocks or new genotypes.

In cv. Kroneiki, Peyvandi et al. (2010) analyzed the genetic changes during the process of somatic embryogenesis in order to predict the steps at which genetic instability could be initiated. For this purpose, mother plants, plantlets regenerated from somatic embryos, seedlings (derived from zygotic embryo germination), embryogenic calli, non-embryogenic calli and zygotic embryos were analyzed by using 20 RAPD primers. The genetic similarity coefficient showed that regenerated plants were less than 75% similar to mother plants. In relation to the somatic embryogenesis process, authors pointed out that genetic variation may take place during the callus step.

Bradaï et al. (2016a, 2019) investigated the effect of the genotype and culture age on the occurrence of somaclonal variation in olive plants cv. Picual. Plants derived from eight embryogenic lines initiated from radicle segments of zygotic embryos and maintained by repetitive subculture for two and eight years were phenotypically and genetically evaluated, four and a half years after acclimatization.

Phenotypic evaluation was carried out by morphological and biometric analysis of vegetative and reproductive characters (Bradaï et al. 2016a). Variant phenotypes were identified by comparing somatic plants with those obtained from germinated seeds. Germination of mature zygotic embryos cv. Picual was performed in parallel to plant regeneration via somatic embryogenesis and seedlings subsequently followed the same steps than those derived from somatic embryos. Fourteen variant phenotypes were identified affecting plant vegetative and reproductive development. Vegetative abnormalities were related to growth habit, stem structure, arrangement of axillary shoots, phyllotaxy and leaf morphology. Variation in reproductive features included both morphological and physiological traits. Thus, different alterations were found in flower structure and an early flowering phenotype was observed in some plants from different embryogenic lines. Nevertheless, despite flower abnormalities, normal fruit development was observed in one plant of each of these lines. Different indicators of somaclonal variation were determined, such as frequency of appearance of the different variant phenotypes, frequency of variant phenotypes (determined, according to Tremblay et al. (1999), as the number of variant phenotypes/total number of plants), percentage of regenerated plants showing morphological alterations, and accumulation of variant phenotypes per plant. A significant effect of the genotype was observed in all cases; although, in general, higher variations were found in aged lines.

The genetic analysis was executed by using RAPD and simple sequence repeat (SSR) markers (Bradaï et al. 2019). Both molecular markers detected somaclonal variation, although RAPD markers were more informative than SSR markers. Intraclonal variation was detected in both age groups. Although somaclonal variation was slightly higher in plants derived from lines maintained in vitro for a longer time, genetic instability seems to be mainly genotype dependent. This investigation confirmed the influence of the genotype and time in culture on somaclonal variation in olive plants regenerated via somatic embryogenesis, previously found when assessing the same plants by phenotypic analysis (Bradaï et al 2016a). The genotype had a stronger effect in the appearance of somaclonal variation, influencing most of the parameters tested. Thus, although higher variation was in general observed in plants derived from aged lines compared to those maintained in culture for shorter periods, this effect was strongly influenced by the embryogenic line, with some lines especially susceptible to change (Bradaï et al. 2016a). Contrarily to Leva and Petruccelli (2012), who identified DNA fragments putatively associated with growth habit, a correlation among the samples affected by phenotypic changes and specific molecular variations was not found in this investigation.

In cv. Canino, Rugini et al. (2020) reported no phenotypic variation in sixteen-year-old plants growing in the field, although SSR analysis demonstrated that one of the somatic plants showed polymorphism for one marker. However, plants derived from more than 3 years cyclic somatic embryogenesis displayed narrow leaves and reduced growth compared to the mother plant and those derived from somatic embryos recovered during the first culture cycle (Rugini and De Pace 2016).

4 Conclusions

Both plant fidelity and occurrence of somaclonal variation have been observed in olive tissue culture. Few investigations have been especially designed to evaluate the effect of specific factors on somaclonal variation in this species. Nevertheless, a significant influence of the genotype, the in vitro propagation method and the length of culture period could be inferred. The in vitro culture technique showed a significant influence on somaclonal variation, with micropropagation giving rise to more stable plants than organogenesis and somatic embryogenesis. These results are in agreement with previous statements considering micropropagation as a reliable technique that can be used for commercial clonal propagation (Leva et al. 2002). Culture age also affected the frequency of somaclonal variation in olive. In general, higher variability was found in plants derived from cultures maintained for prolonged periods compared with those produced after few culture cycles. Nevertheless, the most relevant role was played by the genotype, whose effect was evident in most of investigations in which somaclonal variation was detected. As stated by Bradaï et al. (2016a) distinct olive genotypes responded differently to in vitro culture conditions, being some of them especially susceptible to show somaclonal variation. As proposed by Tremblay et al. (1999), highly unstable genotypes could constitute a valuable material to produce somaclones displaying interesting phenotypes for agronomic traits.

An important aspect in somaclonal variation investigations is the evaluation of the possible changes, which can be performed by phenotypic, cytological, biochemical and molecular techniques.

Genetic stability of tissue culture-derived plants is normally assessed by using molecular markers, with multiple cases in which more than one DNA-based technique is utilized. However, these markers only analyze a small fraction of the genome and, consequently, the abscence of variations does not ensure the genetic fidelity of the plants obtained (Fourré et al. 1997). Moreover, variations observed in DNA fragment profiles could not be significant, as mutations can occur in non-coding portions of the genome and not be expressed in olive appearance or performance (Mirzaei et al. 2021). Therefore, it is advisable to do complementary phenotypic analyses, although in some cases the variations observed cannot be necessarily the result of genome changes (Harding 2004). Phenotypic analyses are recommended to be executed for long periods of time, evaluating plants in the field after they have reached full maturity; in order to discriminate unstable epigenetic variations, often due to juvenility acquired in vitro (Rugini and De Pace 2016). In general, the joint use of different techniques is considered the most appropriate way of estimating plant stability after in vitro culture.

References

Baldoni L, Belaj A (2009) Olive. In: Vollmann J, Rajcan I (eds) Oil crops. Handbook of plant breeding 4. Springer, New York, pp 397–421

Besnard G, Garcia-Verdugo C, Rubio de Casas R et al (2008) Polyploidy in the olive complex (*Olea europaea*): Evidence from flow cytometry and nuclear microsatellite analyses. Ann Bot 101:25–30. https://doi.org/10.1093/aob/mcm275

Besnard G, Khadari B, Navascués M et al (2013) The complex history of the olive tree: from Late Quaternary diversification of Mediterranean lineages to primary domestication in the northern Levant. Proc R Soc B Biol Sci 280.https://doi.org/10.1098/rspb.2012.2833

Bitonti M, Cozza R, Chiapetta A et al (1999) Amount and organization of the heterochromatin in *Olea europaea* and related species. Heredity (edinb) 83:188–195. https://doi.org/10.1038/sj.hdy.6885640

Bradaï F, Pliego-Alfaro F, Sánchez-Romero C (2016a) Somaclonal variation in olive (*Olea europaea* L.) plants regenerated via somatic embryogenesis: Influence of genotype and culture age on phenotypic stability. Sci Hortic (Amsterdam) 213:208–215. https://doi.org/10.1016/j.scienta.2016.10.031

Bradaï F, Pliego-Alfaro F, Sánchez-Romero C (2016b) Long-term somatic embryogenesis in olive (*Olea europaea* L.): Influence on regeneration capability and quality of regenerated plants. Sci Hortic (Amsterdam) 199:23–31. https://doi.org/10.1016/j.scienta.2015.12.010

Bradaï F, Sánchez-Romero C, Martín C (2019) Somaclonal variation in olive (*Olea europaea* L.) plants regenerated via somatic embryogenesis: influence of genotype and culture age on genetic stability. Sci Hortic (Amsterdam) 251:260–266. https://doi.org/10.1016/j.scienta.2019.03.010

Briccoli Bati C, Godino G, Monardo D, Nuzzo V (2006) Influence of propagation techniques on growth and yield of olive trees cultivars "Carolea" and "Nocellara Etnea." Sci Hortic (Amsterdam) 109:173–182. https://doi.org/10.1016/j.scienta.2006.03.013

Cañas LA, Benbadis A (1988) In vitro plant regeneration from cotyledon fragments of the olive tree (*Olea europaea* L). Plant Sci 54:65–74. https://doi.org/10.1016/0168-9452(88)90056-8

Dell´Orco P, Mencuccini M (2000) Embriogenesi somatica e primi risultati nella trasformazione genetica dell'olivo. In: Proceedings of the Convegno Annuale SIGA, Bologna (Italy), p 142

Díez C, Moral J, Diego B, Rallo L (2016) Genetic diversity and conservation of olive genetic resources. In: Ahuja M, Jain S (eds) Genetic diversity and erosion in plants. Springer International Publishing

Fabbri A, Lambardi M, Ozden-Tokatli Y (2009) Olive breeding. In: Mohan Jain S, Priyadarshan P (eds) Breeding plantation tree crops. Tropical species. Springer, New York, NY, pp 423–465

FAOSTAT (2021) Online database available at link. https://www.fao.org/faostat/es/#data/QCL

Farahani F, Yari R, Masoud S (2011) Somaclonal variation in Dezful cultivar of olive (*Olea europaea* subsp. *europaea*). Gene Conserv 10

Farahani F, Yari R, Noormohammadi Z, Sheidai M (2018) Study of morphological trait changes in in-vitro plantlets of three olive cultivars ("Dezful Baghmalek", "Dezful Safiabad" and 'Roughani') domesticated in Iran. Acta Hortic 1199:87–90. https://doi.org/10.17660/ActaHortic.2018.1199.14

Fourré J, Berger P, Niquet L, André P (1997) Somatic embryogenesis and somaclonal variation in Norway spruce: morphogenetic, cytogenetic and molecular approaches. Theor Appl Genet 94:159–169

García-Férriz L, Ghorbel R, Mari A et al (2002) Micropropagation from adult olive trees. Acta Hortic 586:879–882. https://doi.org/10.17660/ActaHortic.2002.586.191
Green P (2002) A revision of *Olea* L. Kew Bull 57:91–140
Green P, Wickens G (1989) The *Olea europaea* complex. In: Tan K, Mill R (eds) The Davis and Hedge Festschrift. Edinburgh University Press, Edinburgh, pp 287–299
Harding K (2004) Genetic integrity of cryopreserved plant cells: a review. Cryo-Letters 25:3–22
Jain SM, Ahloowalia BS, Veilleux RE (1998) Somaclonal variation in crop improvement. In: Jain SM, Brar DS, Ahloowalia BS (eds) Somaclonal variation and induced mutations in crop improvement. Springer, Netherlands, Dordrecht, pp 203–218
Kangarloo B, Noormohammadi Z, Farahani F et al (2015) An assessment of tissue culture-induced genetic variability in olive (*Olea europaea* L.) using chloroplast rpl16 intron sequences and single primer amplification reaction (SPAR) markers. J Hortic Sci Biotechnol 90:719–723. https://doi.org/10.1080/14620316.2015.11668737
Karp A (1994) Origins, causes and uses of variation in plant tissue cultures. In: Vasil I, Thorpe T (eds) Plant cell and tissue culture. Springer, Dordrecht, pp 139–151
Lambardi M, Ozudogru EA, Roncasaglia R (2013) In vitro propagation of olive (*Olea europaea* L.) by nodal segmentation of elongated shoots. In: Lambardi M, Ozudogru EA, Jain SM (eds) Protocols for micropropagation of selected economically-important horticultural plants. Humana Press, Totowa, NJ, pp 33–44
Larkin PJ, Scowcroft WR (1981) Somaclonal variation—a novel source of variability from cell cultures for plant improvement. Theor Appl Genet 60:197–214. https://doi.org/10.1007/BF02342540
Lavee S, Zohary D (2011) The potential of genetic diversity and the effect of geographically isolated resources in olive breeding. Isr J Plant Sci 59:3–13. https://doi.org/10.1560/IJPS.59.1.3
Leva A (2009) Morphological evaluation of olive plants propagated in vitro culture through axillary buds and somatic embryogenesis methods. Plant Sci 3:37–43
Leva AR, Petruccelli R (2007) Field performance of olive somaclones derived from somatic embryos. Acta Hortic 748:181–189. https://doi.org/10.17660/ActaHortic.2007.748.22
Leva AR, Petruccelli R (2012) Monitoring of cultivar identity in micropropagated olive plants using RAPD and ISSR markers. Biol Plant 56:373–376. https://doi.org/10.1007/s10535-012-0102-6
Leva AR, Petruccelli R, Muleo R, Goretti R, Bartolini G (1995a) Influenza di fattori trofici, regolativi e condizioni di coltura in vitro di diverse cultivar di olivo. In: Proceedings of the National Symposium "L'Olivicoltura Mediterranea: Stato e Prospettive della Coltura e della Ricerca", Rende (Italy), pp 239-248
Leva A, Muleo R, Petruccelli R (1995b) Long-term somatic embryogenesis from immature olive cotyledons. J Hortic Sci 70:417–421
Leva A, Muleo R, Polsinelli L et al (2000) Diversi habitus di crescita e profili di amplificazione RAPD in somacloni di olivo cv Frangivento ottenuti da embriogenesi somatica. In: Atti "V Giornate Scientifiche" S.O.I- Sirmione 28–30 Marzo, pp 403–404
Leva AR, Petruccelli R, Montagni G, Muleo R (2002) Field performance of micropropagated olive plants (cv. Maurino): Morphological and molecular features. Acta Hortic 586:891–894. https://doi.org/10.17660/ActaHortic.2002.586.194
Leva A, Muleo R, Petruccelli R (2003) Reproductive, vegetative and agronomic behavior of the micropropagated olive cv. Maurino. Acta Hortic 616:309–312. https://doi.org/10.17660/ActaHortic.2003.616.43
Leva AR, Petruccelli R, Rinaldi LMR (2012) Somaclonal variation in tissue culture: A case study with olive. In: Leva A, Rinaldi LMR (eds). Recent advances in plant in vitro culture. IntechOpen, pp 123-150
Loureiro J, Rodriguez E, Costa A, Santos C (2007) Nuclear DNA content estimations in wild olive (*Olea europaea* L. ssp. *europaea* var. *sylvestris* Brot.) and Portuguese cultivars of *O. europaea* using flow cytometry. Genet Resour Crop Evol 54:21–25. https://doi.org/10.1007/s10722-006-9115-3

Mencuccini M (2011) Produzione di somacloni da cultivar di olivo e prime osservazioni in campo. Acta Italus Hortus, pp 113–116

Mirzaei L, Yadollahi A, Kermani MJ et al (2021) Evaluation of genetic stability in olive callus-induced and meristem-induced shoots using flow cytometry and amplified fragment length polymorphism techniques. Plant Methods 17:1–15. https://doi.org/10.1186/s13007-021-00724-7

Noormohammadi Z, Kangarloo-Haghighi B, Sheidai M, Farahani F, Ghasemzadeh-Baraki S (2014) Genetic stability versus somaclonal variation in tissue culture-regenerated olive plants (Olea europaea cv. Kroneiki). Eur J Exp Biol 4:135–140

Peyvandi M, Nemat Farahzadi H, Arbabian S et al (2010) Somaclonal variation among somatic-embryo derived plants of *Olea europaea* L "cv. Kroneiki." J Sci Islam Repub Iran 21:7–14

Rallo L, Barranco D, Díez C et al (2018) Strategies for olive (*Olea europaea* L.) breeding: cultivated genetic resources and crossbreeding. In: Al-Khayri JM, Jain SM, Johnson DV (eds) Advances in plant breeding strategies: fruits. Springer International Publishing, pp 535–600

Rapoport H (1997) Botánica y morfología. In: Barranco D, Fernández-Escobar D, Rallo L (eds) El cultivo del olivo. Ediciones Mundi-Prensa, Madrid, pp 35–58

Remotti PC (1998) Somaclonal variation and in-vitro selection for crop improvement. In: Jain SM, Brar DS, Ahloowalia BS (eds) Somaclonal variation and induced mutations in crop improvement. Springer, Netherlands, Dordrecht, pp 169–201

Rugini E, Baldoni L (2005) *Olea europaea* Olive. In: Litz R (ed) Biotechnology of fruit and nut crops. CABI Publishing, Wallingford, Oxfordshire, UK, pp 404–428

Rugini E, De Pace C (2016) Olive breeding with classical and modern approaches. In: Rugini E, Baldoni L, Muleo R, Sebastiani L (eds) The olive tree genome. Springer, Cham, pp 163–193

Rugini E, Pannelli G, Ceccarelli M, Muganu M (1996) Isolation of triploid and tetraploid olive (*Olea europaea* L.) plants from mixoploid cv. 'Frantoio' and 'Leccino' mutants by in vivo and in vitro selection. Plant Breed 115:23–27. https://doi.org/10.1111/j.1439-0523.1996.tb00865.x

Rugini E, Cristofori V, Silvestri C (2016) Genetic improvement of olive (*Olea europaea* L.) by conventional and in vitro biotechnology methods. Biotechnol Adv 34:687–696. https://doi.org/10.1016/j.biotechadv.2016.03.004

Rugini E, Baldoni L, Silvestri C et al (2020) *Olea europaea* Olive. In: Litz RE, Pliego-Alfaro F, Hormaza J (eds) Biotechnology of fruit and nut crops, 2nd edn. CAB International, Wallingford, pp 343–376

Sánchez-Romero C (2018) Olive *Olea europaea* L. In: Jain SM, Gupta P (eds) Step wise protocols for somatic embryogenesis of important woody plants, vol II. Springer International Publishing, Cham, pp 25–38

Sánchez-Romero C (2019) Somatic embryogenesis in *Olea* spp. Plant Cell Tissue Organ Cult 138:403–426. https://doi.org/10.1007/s11240-019-01652-0

Sánchez-Romero C (2021) Somatic embryogenesis in Olive. Plants 10:1–6. https://doi.org/10.3390/plants10030433

Shibli RA, Shatnawi M, Abu-Ein A-J (2001) Somatic embryogenesis and plant recovery from callus of "Nabali" olive (*Olea europea* L.). Sci Hortic (Amsterdam) 88:243–256. https://doi.org/10.1016/S0304-4238(00)00241-7

Tremblay L, Levasseur C, Tremblay F (1999) Frequency of somaclonal variation in plants of black spruce (*Picea mariana*, Pinaceae) and white spruce (*P. glauca*, Pinaceae) derived from somatic embryogenesis and identification of some factors involved in genetic instability. Am J Bot 86:1373–1381

Weiss E (2015) "Beginnings of fruit growing in the old world"—two generations later. Isr J Plant Sci 62:75–85. https://doi.org/10.1080/07929978.2015.1007718

Yari R, Farahani F, Sheidai M et al (2011a) The effects of prolonged vegetative reproduction of the two Iranian olive cv. tree cultivars (Dezful Baghmalek and Dezful Safiabad) on morphological traits. Afr J Biotechnol 10:9076–9081. https://doi.org/10.5897/ajb10.836

Yari R, Farahani F, Sheidai M (2011b) Study of morphological traits changes in prolonged vegetative reproduction of three olive tree cultivars domesticated (Zard, Roughani and X) in Iran. Afr J Agric Res 6:6320–6325. https://doi.org/10.5897/AJAR10.448

Zohary D, Spiegel-Roy P (1975) Beginnings of fruit growing in the old world. Science 187:319–327. https://doi.org/10.1126/science.187.4174.319

Somaclonal Variation in Rice: Retrotransposon *Tos17* and Structural Variation with NGS Analysis

Akio Miyao

Abstract Phenotypic changes in regenerated plants observed from early times of the development of tissue culture methods in 1960s and 1970s, was termed somaclonal variation in 1981. The causes for these changes have been a long-standing interest among plant biologists. In early 1990s, transposition events of endogenous retrotransposon *Tos17* were found in regenerated plants from cultured cells of *japonica* rice cultivar Nipponbare. The transposition of *Tos17* was recognized as one cause for somaclonal variation. *Tos17* prefers genic regions for their transposition. However, contribution of *Tos17* transposition for somaclonal variation is estimated to be less than 5% of total phenotypic changes. Next generation sequencing (NGS) technology enables whole genome sequence analysis of rice individuals. Analysis of genome sequence from regenerated plants by the recently developed software PED (2019) now enables analysis of somaclonal variations appearing as genomic changes of large duplications and deletions and frequent single base mutations, as well as transpositions of *Tos17*. Because we now are able to know the genomic changes at the nucleotide sequence level using the new genomic analysis tools, detection of high frequency changes of genome in somaclones will contribute significantly to breeding and functional genomics of rice.

Keywords Next generation sequencing (NGS) · *Oryza sativa* · Retrotransposon · Somaclonal variation · Tissue culture · *Tos17*

1 Introduction

After the report of tissue culture method of tobacco in 1939, a lot of efforts for development of plant tissue culture, especially crops, were made (White 1939). Continuous growth and propagation of identical plants in an artificial environment are very attractive for plant growers. This might enable growing agricultural products in a factory-like environments even at winter season. However, there were initially difficulties

A. Miyao (✉)
Institute of Crop Science, NARO, 2-1-2, Kannondai, Tsukuba, Ibaraki 305-8518, Japan
e-mail: miyao@affrc.go.jp

C. Sánchez-Romero (ed.), *Somaclonal Variation: Basic and Practical Aspects*,
https://doi.org/10.1007/978-3-031-51626-9_11

in quality control of such products. The first report of successful redifferentiation of callus from rice tissue culture pointed out unstable phenotypes in regenerated individuals (Nishi et al. 1968). Morphologic, cytogenetic and enzymatic variations in sugarcane regenerated form callus culture were also reported (Heinze and Mee 1971). Often chromosome mosaics and loss and addition of chromosomes in cultured cells were found in the early cell culture studies. The instability of cultured cells was thus recognized as an important issue to be solved.

During decade of 1980s, the idea to use the instability as a genetic resource of variability for plant improvement, so called 'somaclonal variation' was proposed (Larkin and Scowcroft 1981). They stated that somaclonal variation is a universal feature among plants and useful to introduce new phenotypes into crops. Subsequently, somaclonal variation has been widely used as a breeding tool for crop improvement (Evans 1989).

The machinery of somaclonal variation has been keenly studied among scientists. Change of chromosomal number, translocations, and large deletions were observed by cytogenetic analysis in early cell culture studies. In her Nobel prize lecture, Barbara McClintock said 'It may be safe to state that no two of the callus derived plants are exactly alike, and none is just like the plant that donated the cell or cells for the tissue culture.' (McClintock 1984). Transposable element (TE) activity was found first in regenerated maize plant from tissue culture (Peschke et al. 1987). Activation and transposition of endogenous retrotransposons *Tto1* and *Tos17* were found from tobacco and rice tissue culture (Hirochika 1993; Hirochika et al. 1996). *Tos17* was shown to prefer coding regions for target of transposition. During the period from the late 1990s to early 2000s, a mutant panel containing about 50,000 lines by *Tos17* transposition was laboriously constructed (Miyao et al. 2003). Although the transposition of TEs is one of the key factors of somaclonal variation, the effect of TE transposition for somaclonal variation was relatively small, indicating the existence of other factors. History and overview for somaclonal variation of rice is already described in an excellent review (Neelakandan and Wang 2012). Epigenetic change on somaclonal variation has also been reported (Wang et al. 2013; Bednarek and Orłowska 2020; Ghosh et al. 2021; Liu et al. 2023). However, there are still unclear issues about the correlation between epigenetic change and somaclonal variation.

2 Activation of Transposable Elements in Cultured Cells of Rice

Sanger method for nucleotide sequence analysis and polymerase chain reaction (PCR) method brought new perspectives for somaclonal variation in the beginning of 1990s (Sanger et al. 1977; Saiki et al. 1985). Dr. H. Hirochika suspected that activation and transposition of TEs is involved in the cause of somaclonal variation. However, there were some difficulties to detect TE transposition in cultured cells. Because TEs are multicopy and occupy a large portion of the genome, TEs were easily cloned by plaque hybridization. However, it was difficult to detect TE transpositions themselves, due to their multicopy nature. He predicted that the activated TEs in cell culture would be transcribed, and their RNA could be detected. Two primers corresponding to two conserved motifs (QMDVKT and YVDDM) in reverse transcriptase domain of retrotransposon were successfully used to amplify the cDNA from active retrotransposons in cell culture (Hirochika et al. 1992).

Among the 20 TEs detected by these experiments, *Tos10*, *Tos17* and *Tos19* are active under tissue culture conditions, *Tos17* being the most active in cultured cells (Hirochika et al. 1996). Retrotransposon *Lullaby* reported by a French group as active retrotransposons in cultured cells turned out to be the same as *Tos19* (Picault et al. 2009). Transcript of *Tos17* was dramatically increased in cultured cells, although *Tos17* transcripts were detected only in small amounts in leaves. Transpositions of *Tos17* were detected by Southern hybridization of enzyme digested DNA of regenerated rice plant using *Tos17* coding region as a probe. The number of transposed *Tos17* was from 5 to 30 copies by the hybridization experiment. Because four out of eight target sites were coding regions and two out of four inserted loci were transcribed, *Tos17* was thought to be a useful in obtaining somaclonal mutants in rice.

3 Mutant Panel of Rice

The Rice Genome Research Program in Japan was launched at 1991 when I joined to NIAR (formerly NIAS, currently NARO). The first goal was to create a genetic map of DNA markers. Parents of F_2 individual for RFLP mapping were *japonica* rice Nipponbare and *indica* rice Kasalath (Kurata et al. 1994). After the creation of the genetic map, the second goal was to determine the whole genome sequence of Nipponbare. Full-length cDNA sequence of Nipponbare leaf, root and callus also determined in this project (Rice Full-Length cDNA Consortium 2003). Fortunately, Nipponbare is grows well in N6 medium and *Tos17* activity in cultured cells of Nipponbare is high. A mutant panel containing nearly 50,000 regenerated rice was created during five years (Miyao et al. 2003). Flanking regions of *Tos17* insertion were selected by suppression PCR and thermal asymmetric interlaced (TAIL) PCR method (Miyao et al. 1998) and sequenced by Sanger method. To determine the genomic positions of *Tos17* insertions, flanking sequences were searched against rice reference genome by BLAST search (Altschul et al. 1997). A total of 1,252,536 sites were mapped on the reference Nipponbare genome IRGSP1.0 (IRGSP 2005). Because *Tos17* is an endogenous retrotransposon in Nipponbare, insertion lines are not the genetically modified (GM) rice like as T-DNA insertion mutant lines. Mutant lines of *Tos17* are able to grow in non-restricted rice field, but T-DNA insertion lines are required to grow in restricted areas regulated for GM plants. Phenotypes of F_2 individuals of about 50,000 *Tos17* insertion lines in rice field were observed by a collaboration of 7 laboratories (Miyao et al. 2007). The non-GM feature was a great advantage for the experimental field work.

Half of the created lines showed distinguishable phenotypes. Indeed, almost all of M_2 individuals seemed different, including subtle phenotypic changes, as B. McClintock told. Photographs of segregated phenotypes with observation comments were stored into a relational database. The database was published in 2003, together with the website of 'Mutant Panel Database' https://tos.nias.affrc.go.jp. The website provides an insertion map of *Tos17* on the rice reference genome. If insertions of *Tos17* were detected different positions in coding regions of same gene and the disrupted plants showed same phenotype, the gene may involve the phenotype. Phenotypes were categorized into 53 visible classes. In total, 123,840 phenotypic evidence items were listed. Figure 1 shows the number of lines belonging to each categorized phenotype. Various phenotypes from germination to seed products have been observed. Most frequent phenotype is a sterile or partially sterile plant.

Many TEs create a target site duplication (TSD) at the integration site. From determined data of flanking sequence of *Tos17*, it was revealed that *Tos17* makes 5 base TSD and has a palindromic consensus sequence, ANGTT-TSD-AACNT (Miyao et al. 2003). Although there is base preference around TSD, there were no base preferences within TSD themself. In addition to the existence of 5 base preference for the flanking TSD, large scale preferences of *Tos17* insertions to specific loci were also detected. However, sequence specificity for the large-scale preference locations was not detected. On the other hand, protein kinase, especially disease resistance protein kinase genes are preferred by *Tos17* insertion. *Tos17* tends to prefer specific gene classes for the insertion. *Tos17* also prefers the RNA polymerase largest subunit gene for the insertion. Three largest subunit genes are present in the rice genome. Similarity of nucleotide sequence between chromosome 5 and 8 of the gene is 68%. *Tos17s* were specifically inserted into the gene at chromosome 5 (Fig. 2). Even

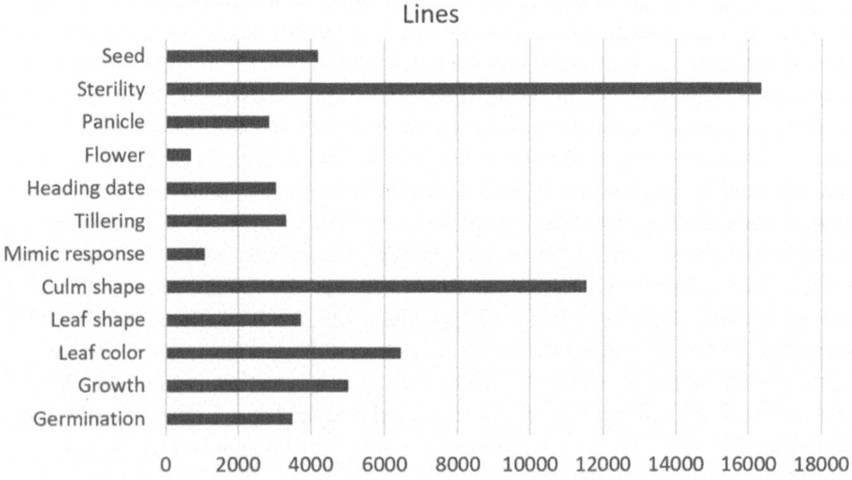

Fig. 1 Number of lines among phenotypic classes

though this region has almost completely the same nucleotide sequence between RNA polymerase genes on chromosome 5 and 8, *Tos17* transposed into only chromosome 5.

This preference may depend on protein–protein interaction between *Tos17* protein and a DNA binding protein on the preferred genome regions, rather than nucleotide sequence level association. There was no correlation between transcriptional activity on the target locus and the preference of *Tos17* insertions. Thus, the interacting protein(s) might not be transcription factors. The *Tos17* insertion preference tends to be on rapidly evolving genes, such as repetitive disease resistant kinase genes; this is interesting for understanding the functional segment of chromosome.

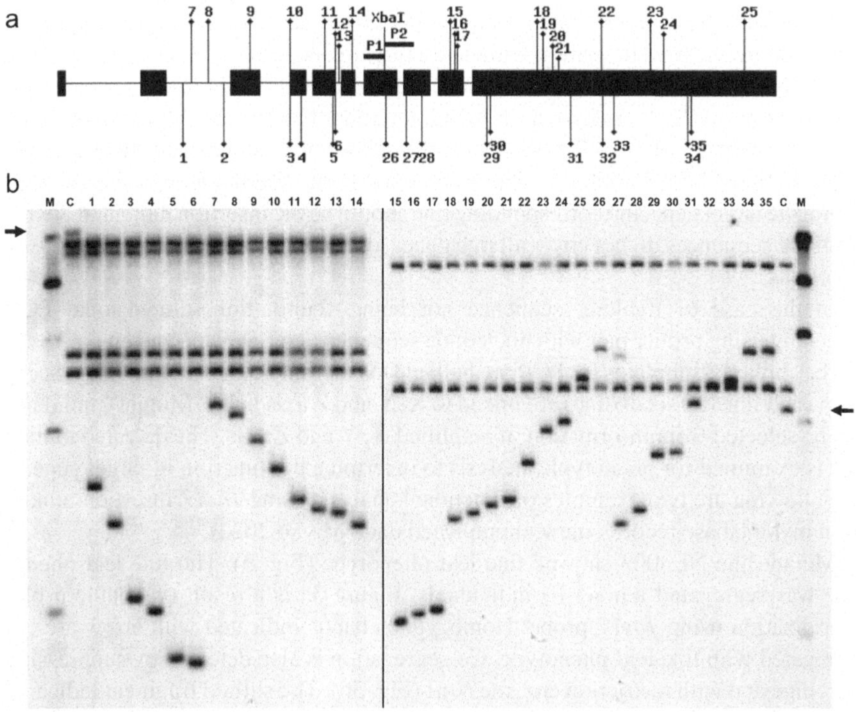

Fig. 2 *Tos17* insertions in RNA polymerase largest subunit gene. **a** Structure of RNA polymerase gene, *Xba*I recognition site and positions of *Tos17* insertion. Insertions with forward direction are shown above the structure figure. Reverse insertions are shown below the structure. P1 and P2 are used as probes for Southern blot hybridization. **b** Autoradiograms of hybridization using *Xba*I digested genome DNA from *Tos17* insertion lines. Lane numbers correspond to lines which have *Tos17* at the insertion positions shown with same number in panel a. Left and right panels are results using P1 and P2 probes, respectively. M is λ/*Hin*dIII marker. C is a control of Nipponbare. Bands indicated arrows on control lanes are shifted due to the *Xba*I site within *Tos17*. Only one band from the control is shifted. This indicates that *Tos17s* were inserted into specific RNA polymerase largest subunit gene among the largest subunit genes (Miyao and Hirochika 2002)

4 Practical Aspects for Functional Analysis of Genes by Transposition of *Tos17* Due to Somaclonal Variation in Rice

Using the rice mutant panel, many gene functions have been studied in different laboratories. The functional analysis follows either the forward and the reverse genetics approach. An exemplary flow of the forward genetics approach consists of observation of segregated phenotype in F_2 plants, searching co-segregated *Tos17* insertions by Southern blot hybridization and/or PCR, selecting inserted genes and then doing complementation experiments with the targeted gene. The DNA fragment of the targeted gene of *Tos17* is transformed into the mutant plant by agrobacterium infection method. If the transformed gene is functional, mutant phenotype should be recovered by the transformation with entire target gene.

On the other hand, the reverse genetics approach can be as follows, starting at the detection of *Tos17* insertion on the target gene. Nucleotide sequence of target gene is searched by BLAST program against dataset of flanking sequence for *Tos17* insertion from about 50,000 *Tos17* insertion lines. If the flanking sequence is detected within the target gene, the corresponding line should be the insertion mutant of *Tos17*. Flanking sequences do not cover all rice genes, due to the heavy preference of *Tos17* insertions.

In the case of flanking sequence not being found, three-dimensional PCR screening using primer pair with nucleotide sequence of target gene and end of *Tos17* can be applied. Templates for PCR are bulked DNAs from 50,000 mutant lines. Seeds from each line (10 seeds/line) are mixed to X, Y and Z axis bulks. Mutant candidates can be selected from information of amplified X, Y and Z axis. The detected mutant line is examined for phenotypic analysis to determine the function of target gene.

Following are two examples of functional analysis using *Tos17* insertion mutant from my database records (new unpublished data, Miyao 2022).

Mutant line NE3003 showed fine leaf phenotype (Fig. 3). The fine leaf phenotype was segregated among F_2 individuals. Figure 3a is a result of Southern blot hybridization using *Tos17* probe. Homozygous bands indicated with arrow are co-segregated with fine leaf phenotype. Co-segregation is also detected by suppression PCR digested with restriction enzyme *Nru*I (Fig. 3b). The shifted fragment indicated with arrow was sliced out from the agarose gel. Extracted DNA was sequenced by dye-deoxy termination method. The flanking sequence of *Tos17* insertion is a part of full-length cDNA clone (accession number AK108569). The hybridization pattern also shows same co-segregation of fine leaf phenotypes (bottom of Fig. 3a).

Figure 3c is the complementation of fine leaf phenotype with AK108569 gene transformation. Transformant of NE3003 (−/−) line with empty vector (pPZP2) shows the fine leaf phenotype (left side). The transformant of pPZP2 with AK108569 gene fragment recovers to the wildtype phenotype. The fragment from AK108569 cDNA clone is complemented by gene disruption of *Tos17* which causes the fine leaf phenotype. The nucleotide sequence of AK108569 has similarity of GDSLG lipase gene. Figure 3d shows sequences of the reaction center of GDSLG lipase and

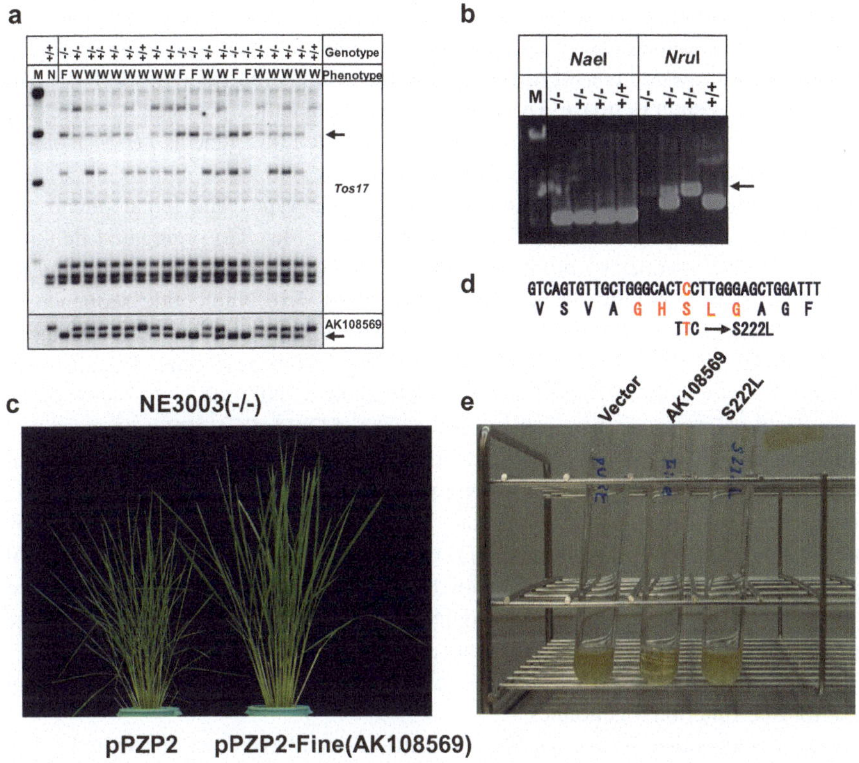

Fig. 3 Identification of causative gene for fine leaf mutation. **a** Co-segregation of fine leaf phenotype and genotype of *Tos17* insertion among F_2 individuals of NE3003 line. F is Fine leaf, W is Wild-type, N is Nipponbare control, M is λ/*Hin*dIII marker. *Tos17* genotypes are shown above phenotype description. +/+ Wild-type, +/− Heterozygous, −/− Homozygous insertion of *Tos17*. Upper side is autoradiogram of *Xba*I digested genome DNA detected by *Tos17* probe. Lower side is AK108569 probe. **b** Isolation of co-segregate flanking sequence of *Tos17* by suppression PCR method. Fragments on electropherogram in *Nru*I digest were co-segregated with *Tos17* insertions shown in panel a. Nucleotide sequence of the fragment pointed with left allow was similar with full-length cDNA of accession number AK108569. **c** Complementation of fine gene mutant by AK108569. AK108569 DNA was transformed into Fine leaf mutant (NE3003) by Agrobacterium infection method. The plant transformed the empty vector (pPZP2) showed the fine leaf phenotype. The transformant of pPZP2-FINE with cDNA sequence of AK108569 showed reverted wild-type phenotype. **d** Sequence of reaction centre of GHSLG lipase and site directed mutagenesis to serine at position 222 to Leucine. **e** Overexpression of GHSLG lipase/Esterase (AK108569) and S222L mutant in *Escherichia coli.* Cell lysis was observed by the lipase/esterase activity. Modification of the reaction centre (S222L) did not cause the cell lysis (new unpublished data, Miyao 2022)

site directed mutagenesis to make S222L mutation. The over expression of wildtype lipase in *E. coli* causes cell lysis. On the other hand, the over expression of S222L lipase did not show cell lysis (Fig. 3e). The serine at 222 amino acid position is a part of the reaction center of lipase. These results indicate that the lipase gene involves the leaf shape development. A Chinese group also reported this gene as *THIS1* isolated by map-based cloning (Liu et al. 2013).

Mutant line NF7008 shows an early death phenotype. The lethal phenotype of F_2 plants co-segregated with *Tos17* insertion (Fig. 4a). Co-segregated flanking sequences were obtained by suppression PCR. Reactions from *Fsp*I and *Pvu*II digests contained specific fragments in mutant DNA (Fig. 4b). Nucleotide sequence of specific fragments showed the *Tos17* target was the Os11g0621300 gene. This gene has a glycine-rich domain and similarity to polygalacturonase gene. Transformant of empty vector (pPZP2) into homozygous mutant of NF7008 (−/−) showed an early death phenotype, whereas the transformant of Os11g0621300 gene recovered the lethal phenotype (Fig. 4c). Although the polygalacturonase-like gene expressed with low level during the lifetime, its expression was increased at the time of heading (Fig. 4d). A Chinese group has also isolated this gene by T-DNA tagging method; they named the gene *GRDP1* (Zhao et al. 2021). They reported it as a lesion mimic mutant *spl-D*, showing higher accumulation of H_2O_2, spontaneous cell death, decreased chlorophyll content, upregulation of defence-related genes, and enhanced disease resistance. Both results are almost identical.

These examples show that the flow of functional analysis has become a useful tool for rice functional genomics.

So far (December 2022), our *Tos17* insertion lines have been distributed 3,179 times to 867 scientists around the world. Based on work using the distributed lines, a large number of gene functions have been reported. For example, homeobox gene *OSH15* (Sato et al. 1999), *phytochrome A* gene (Takano et al. 2001), *zeaxanthin epoxidase* gene from a viviparous mutant (Agrawal et al. 2001), *cellulose synthase* gene (Tanaka et al. 2003), *starch synthase IIIa* (Fujita et al. 2007, 2011), cell cycle regulation gene *RSS1* (Ogawa et al. 2011), seed dormancy regulator gene *Sdr4* (Sugimoto et al. 2010), *arbuscular receptor-like kinase 1* gene regulating arbuscular mycorrhizal symbiosis (Roth et al. 2018) and so on. Thus, the transposon tagging system involved in somaclonal variation has proven to be powerful tool for rice functional genomics when combined with the developed panel of rice lines. In addition, the mutation of *starch synthase IIIa* has been used for development of new rice cultivars, 'Akitasarari' and 'Akitaparari' with high amylose content (Nakamura et al. 2018).

5 TE Transposition Detected by NGS Analysis

During the 2000s, Next Generation Sequencing (NGS) technologies have been developed. Currently, we are able to obtain whole genome sequence data from individuals. Obtaining whole genome sequence data from individuals is helpful for understanding the somaclonal variation. However, because NGS data comprise of a huge number

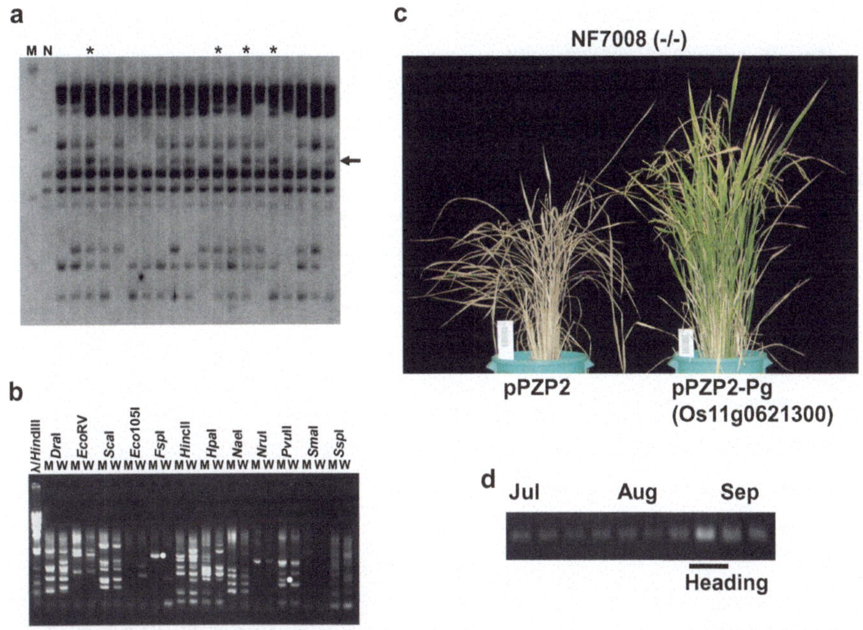

Fig. 4 Functional characterization of lethal phenotype of NF7008 line. **a** autoradiogram of Southern blot hybridization of F_2 individuals detected by *Tos17* probe. Genomic DNAs were digested the restriction enzyme *Xba*I. M is λ/*Hin*dIII marker. N is control (Nipponbare). Individuals showing lethal phenotype are marked with asterisks. Hybridization signals indicated by left arrow are co-segregated with the phenotype. **b** Isolation of flanking DNA fragment of *Tos17* insertion by the suppression PCR method. M is DNA from mutant (lethal) individual. W is wild type. Mutant-specific amplifications marked with white circles at the right side of band in *Fsp*I and *Pvu*II digests are detected. Nucleotide sequence from the band have a similarity of polygalacturonase gene (Os11g0621300). **c** Lethal mutant of NF7008 was complemented by transformation of Os11g0621300 gene. Transformant of empty vector pPZP2 showed a lethal phenotype. Transformant of the polygalacturonase gene pPZP2-Pg was recovered from the lethal phenotype. **d** Expression of the polygalacturonase gene detected by RT-PCR (new unpublished data, Miyao 2022)

of short-read sequences, the majority of short pieces from repetitive sequences in the genome interfere in the analysis of genetic change of somaclonal variation. Indeed, transposable element (TE) itself is a repetitive sequence. It is difficult to detect transposition of TE using standard mapping program for NGS analysis.

At the position of TE transposition, the sequence of some nucleotides is duplicated at the target site, called 'Target Site Duplication'. TE usually makes a fixed size of TSD. *Tos17* makes a 5 bp TSD for insertion (Fig. 5a). This short-read sequence with *Tos17* target site should contain head or tail sequence of *Tos17* at the junction of genome sequence.

The Transposon Insertion Finder (TIF) method for TE analysis proceeds as follows. Short-read sequences with the target are easily selected by 'grep' command

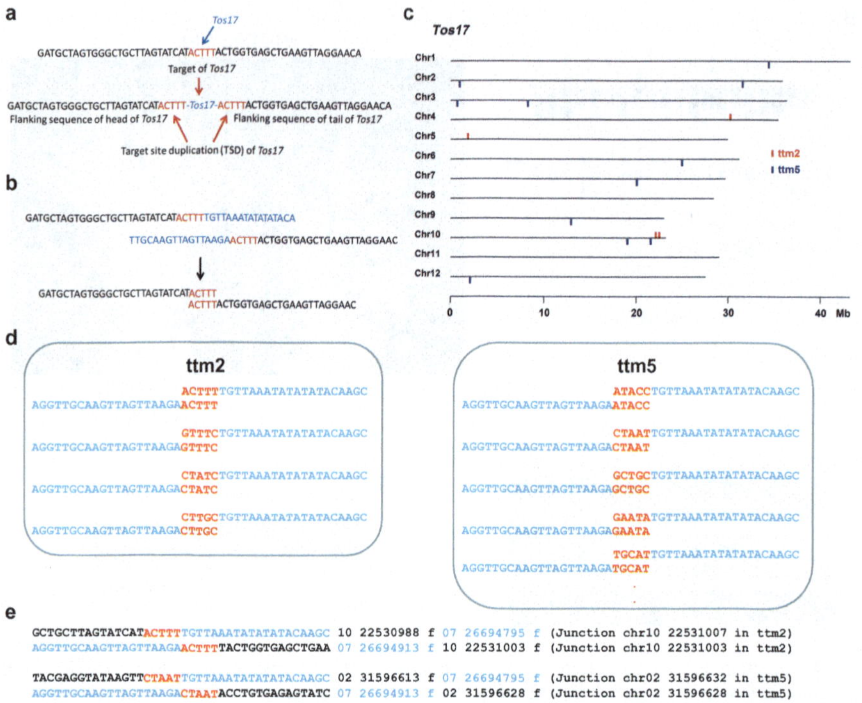

Fig. 5 Detection of *Tos17* insertion using NGS data from regenerated plants. **a** *Tos17* (blue) is inserted the target site. TSD by *Tos17* transposition is indicated with red. **b** TIF algorithm: NGS short reads containing *Tos17* sequence are selected. After removing *Tos17* sequences, flanking sequences of *Tos17* are joined by TSD sequence and then joined sequences are mapped the sequence against reference sequence. **c** Map of insertion of *Tos17* in rice reference genome (IRGSP1.0). **d** TEF algorithm (TSD method): NGS short reads are cut out to small fragments with 20 base plus TSD length in each position. Pieces are grouped by TSD sequence and then head and tail sequences of TE candidates are selected. Active TEs should show different TSD sets between ttm2 and ttm5. **e** TEF algorithm (Junction method): NGS short reads are cut out to 40 base fragments in each position. Junction of TE insertion should2014 show different map positions for the first and last 20 base sequences on the fragment and the junctions should have same TSD sequence (red) (Nakagome et al. 2014; Miyao and Yamanouchi 2022)

of Unix with head or tail sequence of *Tos17*. After removing head and tail sequence from selected sequences, upstream and downstream sequences of *Tos17* are grouped by TSD sequences (Fig. 5b). Inserted position of *Tos17* can be detected by searching with upstream and downstream sequences with TSD. The position of TSD is the target of *Tos17* insertion (Algorithm 1). Otherwise, upstream and downstream sequences are searched against the reference genome sequence and obtain mapped positions of junction. Target positions will be detected as mapped positions of upstream and downstream junctions of *Tos17* with TSD size (5 bp) apart (Algorithm 2). These algorithms were implemented by the perl script, tif.pl (Nakagome et al. 2014).

Figure 5c shows inserted map position of *Tos17* in ttm2 (NC7756) and ttm5 (ND6834). The TIF script is for detection of TE transposition, not only transpositions in somaclonal variation, but also transpositions due to other causes, e.g. due to a change of DNA methylation. In the analysis, the numbers of *Tos17* transpositions were 4 and 11 in rice lines ttm2 and ttm5, respectively. Thus, the determination of TE transpositions from NGS short reads was successful by the developed TIF algorithm.

TIF is the detection tool of inserted position of desired TE by grouping TSDs in flanking sequences of *Tos17*. If candidate of head and tail sequences for unknown TEs can be grouped by their TSDs, head and tail sequences of new TEs will be detected. Because insertion points on the genome are different between two samples, grouped TSD sequence with same head and tail sequence must be different between two samples (Fig. 5d, TSD method). On the other hand, first 20 base and last 20 base sequences from 40 base fragments on the junction of TE insertion must map to different positions on the reference genome and a TSD must exist at the junctions in the case of TSD forming TEs (Fig. 5e, Junction method). Two algorithms have been implemented in the perl script, tef.pl. This Transposable Element Finder (TEF) tool is for detection of active (transposed) TEs by comparing NGS sequences between two samples (Miyao and Yamanouchi 2022). TEF detected *Tos17* transposition between rice lines ttm2 and ttm5. In addition, *Tos17* transpositions were detected in a core collection of *japonica* rice: jrc18 (Hassokuho), jrc19 (Wataribune), jrc21 (Akamai), jrc26 (Aikoku), jrc32 (Omachi), jrc35 (Kabashiko), jrc37 (Shinyamadaho 2), jrc42 (Touboshi), jrc43 (Akamai), jrc44 (Karahoushi) and jrc53 (Raiden) by TEF. Because these lines are cultivars and are not from cell culture, *Tos17* might have been activated by other environmental factors during breeding process.

6 Point Mutations Detected in NGS of Regenerated Plants

Insertion of *Tos17* is one of the causes for somaclonal variation in cultured cells of rice. However, results of co-segregation analysis between *Tor17* insertions and change of phenotypes indicated that the contribution of *Tos17* insertion is only 5% or less. We attempted to search for other TEs activated by cell culture of rice. *Tos10* and *Tos19* have activity of transposition, but their activities were less than *Tos17*. Other unknown factors are thus involved for the somaclonal variation.

In 2010, we obtained the first NGS data sets from regenerated rice. Short reads were mapped by BWA program and single nucleotide polymorphisms (SNPs) were detected (Li 2011). In addition, one or two base insertions and deletions were also detected (Miyao et al. 2012a). The overall mutation rate of rice regenerated after 5-month cell culture is 1.74×10^{-6} base substitutions per site per generation. The spontaneous mutation rate of *Arabidopsis thaliana* is 7×10^{-9} (Ossowski et al. 2010). These values indicate that the mutation rate of somaclonal variation of rice is about 250-fold higher than spontaneous mutation of *Arabidopsis thaliana*. The most

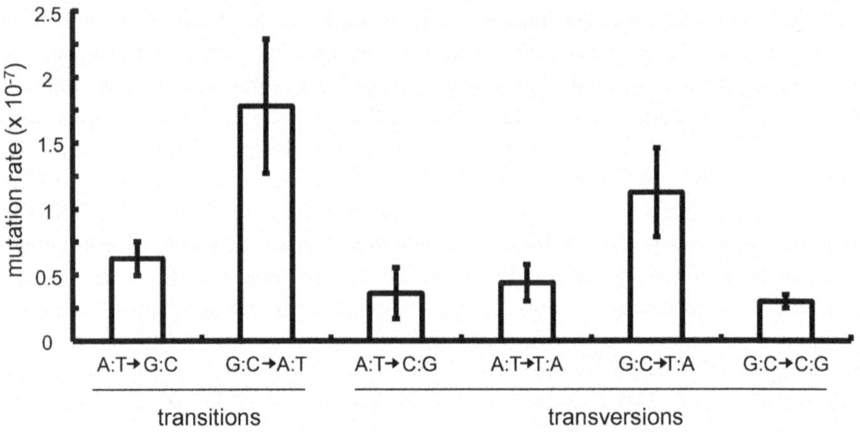

Fig. 6 Spectrum of SNPs of somaclonal variation. Cytosine to thymine transition is the most frequent (Miyao et al. 2012a)

frequent mutation type is G:C to A:T transition for both plants (Fig. 6). C to T transition is caused by deamination of methyl-cytosine. Spectra of somaclonal base substitutions and spontaneous mutations in *Arabidopsis* show almost the same pattern, except for a low level of G:C to A:T transitions in rice compared to *Arabidopsis*. The mutation rate of C to T transition of *Arabidopsis* is 8.7-fold greater than T to C transition. On the other hand, the rate of somaclonal mutation of rice is 2.7-fold. In addition, the rate of *Arabidopsis* is 1.1-fold (Jiang et al. 2011). In the paper of molecular mutation spectrum of *Arabidopsis*, Ossowski described that the C to T transition is introduced by deamination of methylated cytosines and ultraviolet light-induced mutagenesis. Because the deamination of methylated cytosines is a simple chemical reaction, the speed of the deamination between rice and *Arabidopsis* are almost same.

If the fixed C to T transition rate is true, the relatively low level of C to T transition in regenerated rice indicates that almost all types of mutation rate are increased in cultured cells (Fig. 6).

7 Detection of Structural Changes by Polymorphic Edge Detection

At the time of our first analysis of mutation (substitution) spectrum in rice, the length of NGS short reads was 33 or 35 bases. This read length was too short to detect structural variations. As the NGS technology improved, 100 or more bases in a read could be obtained. For detection of mutations from newer NGS data, combination of bwa for mapping and GATK for following detection of mutations is nowadays a standard analysis flow (Li and Durbin 2009; Poplin et al. 2017). Unfortunately, GATK is not able to detect deletions, in the case that the length of deletion is longer than short read length. Due to this limitation, we could not recognize most of structural changes in somaclonal variation.

I have found that bidirectional alignment of the reference sequence to the short read sequence solves the problem to detection of structural variants. The new Polymorphic Edge Detection (PED) method, including two algorithms was developed to detect the edges of DNA polymorphisms (Miyao et al. 2019). One is based on counting *k*-mers which contain a polymorphic base. The other is based on bidirectional alignment. Although the *k*-mer method is not suitable to detect structural variations, it can detect SNPs with direct comparison of NGS data from two samples without a reference genome sequence. On the other hand, the bidirectional alignment method can detect structural variations in addition to SNPs.

Figure 7a shows the bidirectional alignment of a 280 kb deletion in 3m1 regenerated rice line. Plants 1m1, 3m1 and 5m1 were regenerated after 1, 3 and 5 months in liquid cell culture of Nipponbare. Short read sequence is placed at middle. Upper side is reference sequence aligned from 5'-end and lower side is from 3'-end. Mismatch between upper side sequence and short read appears at chromosome 8 position 23,388,732. On the other hand, a mismatch in the lower side appears at chromosome 8 position 23,668,838. The mismatches are the polymorphic edges. The deletion is confirmed by PCR amplified fragment (120 bp) with a specific primer set across the junction which is only amplified by the 280 kb deletion. The deletion was found to have occurred in the regenerated plant from the 3 months cell culture (Fig. 7b). The deletion was a specific event in the 3m1 plant.

Figure 7c shows that indel events depend on the duration of cell culture. Regenerated plant from 1 month cell culture has almost the same level of indels to 10 times selfed Nipponbare (n1s10) which is the control for spontaneous mutation. The number of indels increase linearly with the duration of cell culture. Most of the indels are specific events in 1m1, 3m1 or 5m1. Only one insertion was detected as a common mutation from 3m1 and 5m1. This result indicates that most of indels occur independently or randomly in each cell in liquid culture.

The PED algorithm with bidirectional alignment is also suitable to detect insertions, inversions, translocations, and base substitutions as well as large deletions. Although the bidirectional alignment detects large deletions, the algorithm cannot detect insertions larger than the read length. Because most of large insertions are transpositions of TEs, TIF and TEF programs are better than PED program to detect TE insertions.

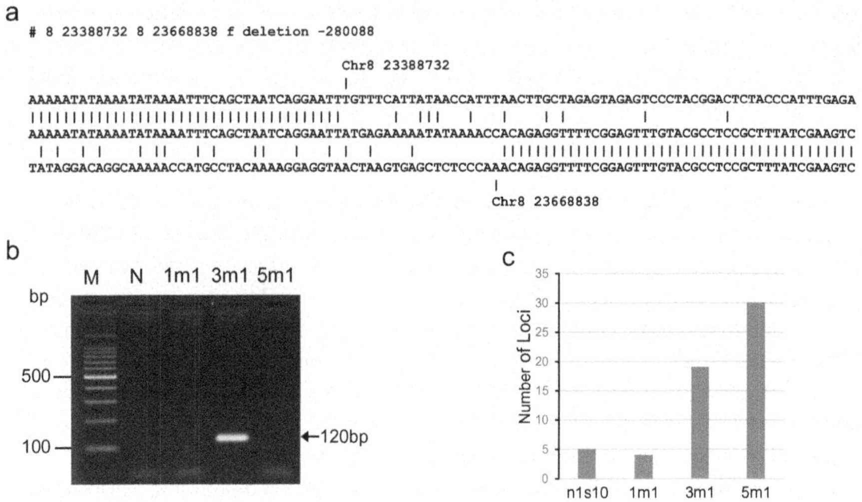

Fig. 7 Detection of large deletion by PED. a Bidirectional alignment detected a 280 kb deletion. **b** Detection of PCR amplified fragment in 3m1 line. M is 100 base ladder marker. Sizes of 100 and 500 base are indicated. N is Nipponbare. 1m1, 3m1 and 5m1 are DNA of regenerated plants from 1-, 3- and 5-months cultured cells of Nipponbare. **c** Time course of accumulation of indels. Control for spontaneous mutations of indel is 10 times selfed Nipponbare (n1s10) (Miyao et al. 2019, Unpublished data, Miyao 2022)

8 Copy Number Variations in Regenerated Rice

Segmental duplication is another type of insertions which cannot be detected by bidirectional alignment. Counting *k*-mers (e.g. $k = 20$) of short read sequences enables us to detect such segmental duplications. All short read sequences are split to *k*-mers in each position, and then frequencies of each *k*-mer sequence are counted. Repetitive *k*-mer sequences in the reference genome are removed and pairs of unique *k*-mer sequence and frequency are selected. The *k*-mer count method for copy number variation (CNV) detection has been implemented in the PED program as the extension of the *k*-mer method. Figure 8a is the comparison of *k*-mer frequencies between rice lines ttm2 and ttm5. The upper side bars mean more frequent in ttm2, and the lower side bars are in ttm5. The bias indicates the CNV. Bars at chromosome 5 indicate the 1.5 Mbp segmental duplications. From the information of junctions in short read sequences, it was determined that the sequence starts at 709,319 and is repeated after the position 2,200,548 (Fig. 8c). Figure 8b shows the restriction map before the duplication and fragment sizes. Figure 8c shows the estimated restriction map after the duplication and fragment sizes. Southern hybridization using the probe flanking the junction shows the new fragment in the lane of ttm2 (Fig. 8d). Because sizes of new fragments are matching the estimated sizes, the bias of chromosome 5 in Fig. 8a detected by *k*-mer frequency analysis is demonstrated to be the result of segmental duplication.

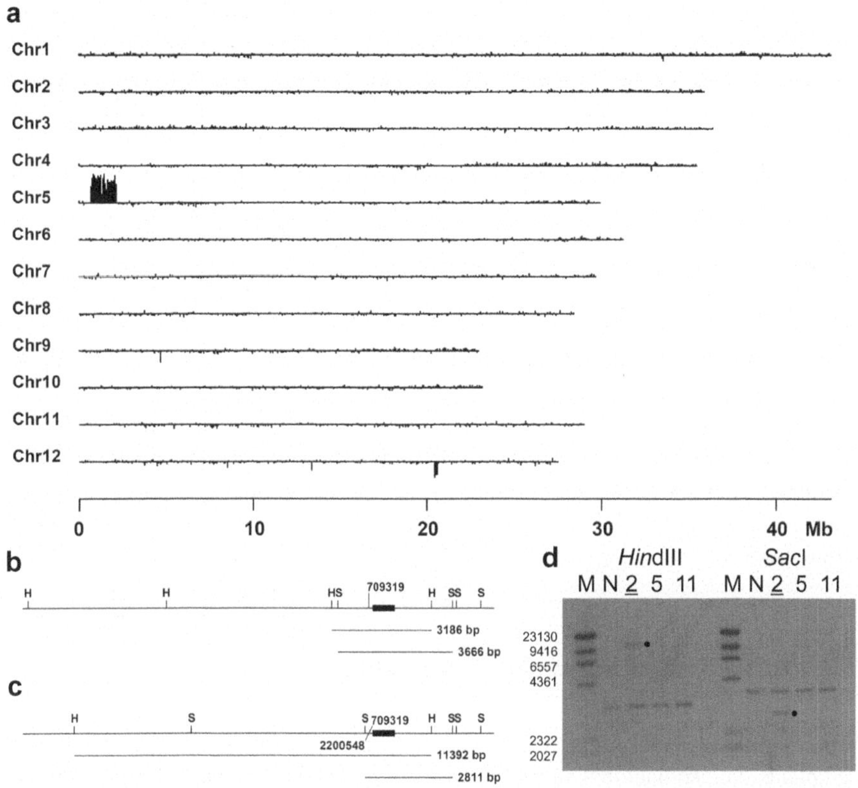

Fig. 8 Segmental duplication in ttm2. **a** Difference of mapped *k*-mer between ttm2 (upper side) and ttm5 (lower side of chromosome lines). Chromosome 5 in ttm2 has 1.5 Mb duplication. **b** Restriction map before the duplication. H is *Hin*dIII. S is *Sac*I. Junction position is 709,319 in chromosome 5. Thick line is the position of hybridization probe. Lines and numbers under the map are estimated fragment size digested by the enzymes. **c** Restriction map after the duplication. Segment from 709,319 to 2,200,548 are duplicated. Diagonal line on the map shows the junction. Fragment sizes are changed by the duplication. **d** Detection of fragments on the junction by Southern blot hybridization. M is λ/*Hin*dIII marker. N is Nipponbare. 2, 5 and 11 are ttm2, ttm5 and ttm11, respectively. *Hin*dIII and *Sac*I digests of genomic DNA were electrophoresed and then hybridized with the probe indicated by the thick line on the restriction enzyme map. Size of fragments at lane of ttm2 marked by black circle are matched to estimated fragment sizes in panel c (Miyao et al. 2012b)

9 Epigenetic Changes of Genome

Using regenerated rice from cultured cells, genome wide demethylation was observed by NGS analysis with bisulfate sequencing technique (Stroud et al. 2013). Change of methylation state in regenerated rice may affect their phenotypes. Indeed, a rice mutant, *Epi-d1*, shows a dwarf phenotype. Although there is no sequence change in the *DWARF1* gene, the phenotype was correlated to the DNA methylation in the promoter region (Miura et al. 2009). DNA methylation may contribute to the suppression of TE activity (Slotkin and Martienssen 2007). In *Arabidopsis thaliana*, DNA methylation and histone H1 jointly silence TEs (Choi et al. 2020). The change of methylation state in cultured cells may affect activation of suppressed TE activity. It was reported the mutation of chromatin remodeling protein gene *DDM1* causes decrease the methylation level of DNA and induces TE transposition (Miura et al. 2001). DDM1 protein interacts with histone variant H2A.W and the change of H2A.W stat prevents TE mobility (Osakabe et al. 2021). Thus, the changes in histone variants and DNA methylation affects TE transposition.

In rice studies, mutation of chromomethylase gene *OsCMT3a* causes TE transpositions (Cheng et al. 2015). Mutants of *OsCMT3a* were isolated from *Tos17* insertion lines. The mutant showed a decrease of methylation content of DNA. Whole genome resequencing revealed TE transpositions. *Dasheng, Osr4, Osr14, mPing, DaiZ, Line1-6_OS* were detected in the mutant. As in the case of *Arabidopsis*, low methylation of genome activates some TEs in rice. However, *Tos17* and *Tos19* transpositions in low methylation condition of DNA are at quite low level compared to cultured cells. There might exist some other factors for transposition of *Tos17* in cultured cells.

10 Future Use of Somaclonal Variation

At the beginning of cell culture technology for plants, phenotypic changes of regenerated plants, i.e., somaclonal variation was recognized. During past 30 years, molecular mechanisms of somaclonal variations have been revealed with development of molecular biology, genome sequencing and analysis techniques. In the cultured cells of rice, retrotransposon *Tos17* is activated and transposed. In addition, NGS technology enables whole genome analysis of regenerated rice and brings new opportunities for understanding and using somaclonal variation.

As *Tos17* prefers genic regions of the rice genome for insertion, *Tos17* becomes a useful tool to determine gene functions. Hundreds of gene functions have been examined by the mutant panel of *Tos17* insertion lines developed in Japan. Although some preference of nucleotide sequence in target sites of *Tos17* was detected, the genome wide preference could not explain the base preference around target sites. Research showed that *Tos17* inserted only one gene out of three RNA polymerase largest subunit genes with high frequency, although the preferred regions have almost same nucleotide sequences among three genes. The preference of *Tos17* insertions may be due to protein–protein interactions rather than recognition of nucleotide sequences. The unknown interactions may relate to the protein complex of *Tos17* integration and some nucleosome proteins, such histones or other modification protein complexes for regulation of transcription or building of chromatin structure. The sequence-independent insertion preferences of *Tos17* indicates the appearance of specific structural change of chromatin in cultured cells. After the target protein of *Tos17* is known, this knowledge of *Tos17* transpositions may help to reveal the unknown chromatin structure involved.

In cultured cells of rice, transcription of *Tos17* is activated. Because TE transposition disrupts gene structure and damages cell activity, almost all TEs in the genome are suppressed during transcription to avoid such effects and further TE invasions. In cultured rice cells, DNAs are demethylated (Stroud et al. 2013). The demethylation may activate the *Tos17* transcription. Study of *Tos17* disruption mutant of chromomethylase gene *OsCMT3a* showed activations of several TEs. However, transposition of *Tos17* was not dramatically increased in the chromomethylase mutant (Cheng et al. 2015). Highly methylated and inactive *Tos17* in cultivar Moritawase was demethylated by 5-azacytidine treatment of the cultured cell and showed the activation of *Tos17* transcription (Cheng et al. 2006). However, treatment of 5-azacytidine which cause low methylation of genomic DNA did not affect the *Tos17* activation in cultivar Matsumae (Liu et al. 2004). DNA demethylation in cultured rice cells may not be the major determinant of activation of *Tos17*.

On the other hand, a mutation of chromatin remodeling factor *ddm1* of *Arabidopsis* activates some TEs transposition (Miura et al. 2001). RNAi mutant of histone H3K9 methyltransferase of rice showed DNA demethylation as well as histone H3K9 demethylation. Transpositions of *Tos17* were observed in the RNAi mutant (Ding et al. 2007). TE activation in cultured cells may be due to structural change of chromatin. If DNA methylation is the major factor of TE suppression, demethylation of DNA will cause activation of all TEs. However, the low methylation environment in cultured cells activates only *Tos17* retrotransposon. In addition, *Tos17* was not transposed in *ddm1* mutant of rice which causes low methylation of chromatin (Higo et al. 2012; Numa et al. 2015). DNA methylation may be the secondary factor for repression of *Tos17* by condensation of heterochromatin.

TE detection software, TEF, revealed that *Tos17* was transposed previously during the breeding history of *japonica* cultivars (Miyao and Yamanouchi 2022). TEF also detected 328 kinds of TE among 50 *japonica* cultivars. In inter- and intraspecies recombinant inbred populations of *Lotus japonicus* activation of retrotransposon has been demonstrated (Fukai et al. 2022). TEs may activate in environments with mixed chromatin states by plant crossing for breeding.

NGS techniques have enabled analysis of mutations of somaclonal variation from whole genome sequence of regenerated plant. TE transpositions, point mutations, deletions and duplications can all be detected from genomic sequence of regenerated rice plants. The mutation rate of SNPs is hundreds-fold increased, compared to spontaneous mutation. However, somatic mutations of leaf DNAs which are not inherited to next generation, is almost at the same rate. The mutation rate of SNPs in cultured cells may be controlled by the same mechanisms as in somatic leaf cells. On the other hand, large deletions and duplications are frequently observed from regenerated plants. They may be introduced by chromosome breaks due to disturbances in cell cycle, such as asynchronous mitosis during cell culture (Heinze and Mee 1971).

For analysis of NGS data, it was difficult to detect large structural changes from NGS short reads using the standard analysis flow with bwa, Samtools and GATK (Li and Durbin 2009; Li 2011; Poplin et al. 2017). I devised a new TIF, TEF and PED methods to analyze mutations on somaclonal variation (Nakagome et al. 2014; Miyao et al. 2019, Miyao and Yamanouchi 2022). The bidirectional alignment method of PED clearly detects structural changes of somaclonal variation, even large, several hundred kb long deletions which cannot be detected by the previous standard analysis methods. With the new algorithms, it was revealed that number of deletions is increased depending on the culture time, in addition to large duplications which are also introduced by cell culture. The structural changes, such as deletions and duplications may be a major determinant of somaclonal variation, as well as transposition of TEs. The mutations introduced by cell culture have already proved by be very useful for functional genomics. Thus, partial duplications or translocations of chromosomes reported in the early times of cell culture study (Heinze and Mee 1971) should still be considered and reinvestigated to verify and study them further.

At present, the development of NGS technology is fast and whole genome sequences of individuals can be obtained easily and relatively cheaply. Almost all mutations on unique sequence regions in regenerated plants can be identified using the PED software. In addition, TEF detects transpositions of active TEs. Combined use of these analysis tools for detection of TE transpositions and structural changes, including large deletions and duplications, will greatly promote the efficiency of gene tagging. NGS analysis for regenerated plants and creation of mutation databases will help building an even more efficient next generation mutant panel for functional genomics of rice and for understanding the molecular basis of somaclonal variation. This will make rice breeding easier and more efficient, to meet the demand for accelerated production of new varieties, e.g. for coping with climate change.

Acknowledgements I am grateful to Dr. Ilkka Havukkala for critical reading of the manuscript and discussion. I would like to thank Dr. Hirohiko Hirochika. He gave me the opportunity to make the mutant panel using *Tos17* retrotransposon found by him. I also thank a lot of people involved in making the mutant panel and giving assistance for analysis. This work was mainly supported by the Ministry of Agriculture, Forestry, and Fisheries of Japan, the Enhancement of Center-of-Excellence, Special Coordination Funds for Promoting Science and Technology in Japan, and the Program for the Promotion of Basic Research Activities for Innovative Biosciences.

References

Agrawal GK, Yamazaki M, Kobayashi M, Hirochika R, Miyao A, Hirochika H (2001) Screening of the rice viviparous mutants generated by endogenous retrotransposon *Tos17* insertion. Tagging of a zeaxanthin epoxidase gene and a novel *ostatc* gene. Plant Physiol 125(3):1248–1257. https://doi.org/10.1104/pp.125.3.1248

Altschul SF, Madden TL, Schäffer AA, Zhang J, Zhang Z, Miller W, Lipman DJ (1997) Gapped BLAST and PSI-BLAST: a new generation of protein data-base search programs. Nucleic Acids Res 25(17):3389–3402. https://doi.org/10.1093/nar/25.17.3389

Bednarek PT, Orłowska R (2020) Plant tissue culture environment as a switch-key of (epi)genetic changes. Plant Cell Tiss Organ Cult 140:245–257. https://doi.org/10.1007/s11240-019-01724-1

Cheng C, Daigen M, Hirochika H (2006) Epigenetic regulation of the rice retrotransposon *Tos17*. Mol Genet Genomics 276:378–390. https://doi.org/10.1007/s00438-006-0141-9

Cheng C, Tarutani Y, Miyao A, Ito T, Yamazaki M, Sakai H, Fukai E, Hirochika H (2015) Loss of function mutations in the rice chromomethylase OsCMT3a cause a burst of transposition. Plant J 83(6):1069–1081. https://doi.org/10.1111/tpj.12952

Choi J, Lyons DB, Kim MY, Moore JD, Zilberman D (2020) DNA methylation and histone H1 jointly repress transposable elements and aberrant intragenic transcripts. Mol Cell 77(2):310–323. https://doi.org/10.1016/j.molcel.2019.10.011

Ding Y, Wang X, Su L, Zhai J, Cao S, Zhang D, Liu C, Bi Y, Qian Q, Cheng Z, Chu C et al (2007) SDG714, a histone H3K9 methyltransferase, is involved in *Tos17* DNA methylation and transposition in rice. Plant Cell 19(1):9–22. https://doi.org/10.1105/tpc.106.048124

Evans DA (1989) Somaclonal variation–genetic basis and breeding applications. Trends Genet 5(2):46–50. https://doi.org/10.1016/0168-9525(89)90021-8

Fujita N, Yoshida M, Kondo T, Saito K, Utsumi Y, Tokunaga T, Nishi A, Satoh H, Park JH, Jane JL et al (2007) Characterization of SSIIIa-deficient mutants of rice: the function of SSIIIa and pleiotropic effects by SSIIIa deficiency in the rice endosperm. Plant Physiol 144(4):2009–2023. https://doi.org/10.1104/pp.107.102533

Fujita N, Nakamura Y, Yoshida M, Saito K, Hirochika H, Miyao A (2011) JP Patent 4711762

Fukai E, Yoshikawa M, Shah N, Sandal N, Miyao A, Ono S, Hirakawa H, Akyol TY, Umehara Y, Nonomura KI et al (2022) Widespread and transgenerational retrotransposon activation in inter- and intraspecies recombinant inbred populations of *Lotus japonicus*. Plant J 111(5):1397–1410. https://doi.org/10.1111/tpj.15896

Ghosh A, Igamberdiev AU, Debnath SC (2021) Tissue culture-induced DNA methylation in crop plants: a review. Mol Biol Rep 48:823–841. https://doi.org/10.1007/s11033-020-06062-6

Heinze DJ, Mee GWP (1971) Morphologic, cytogenetic and enzymatic variation in *Sac-charum* species hybrid clones derived from callus culture. Am J Bot 58:257–262. https://doi.org/10.1002/j.1537-2197.1971.tb09971.x

Higo H, Tahir M, Takashima K, Miura A, Watanabe K, Tagiri A, Ugaki M, Ishikawa R, Eiguchi M, Kurata N et al (2012) *DDM1* (*Decrease in DNA Methylation*) genes in rice (*Oryza sativa*). Mol Genet Genomics 287:785–792. https://doi.org/10.1007/s00438-012-0717-5

Hirochika H (1993) Activation of tobacco retrotransposons during tissue culture. EMBO J 12(6):2521–2528. https://doi.org/10.1002/j.1460-2075.1993.tb05907.x

Hirochika H, Fukuchi A, Kikuchi F (1992) Retrotransposon families in rice. Mol Gen Genet 233(1–2):209–216. https://doi.org/10.1007/BF00587581

Hirochika H, Sugimoto K, Otsuki Y, Tsugawa H, Kanda M (1996) Retrotransposons of rice involved in mutations induced by tissue culture. Proc Natl Acad Sci USA 93(15):7783–7788. https://doi.org/10.1073/pnas.93.15.7783

International Rice Genome Sequencing Project (2005) The map-based sequence of the rice genome. Nature 11:436(7052):793–800. https://doi.org/10.1038/nature03895

Jiang C, Mithani A, Gan X, Belfield EJ, Klingler JP, Zhu JK, Ragoussis J, Mott R, Harberd NP (2011) Regenerant *Arabidopsis* lineages display a distinct genome-wide spectrum of mutations conferring variant phenotypes. Curr Biol 21:1385–1390. https://doi.org/10.1016/j.cub.2011.07.002

Kurata N, Nagamura Y, Yamamoto K, Harushima Y, Sue N, Wu J, Antonio BA, Shomura A, Shimizu T, Lin SY (1994) A 300 kilobase interval genetic map of rice including 883 expressed sequences. Nat Genet 8(4):365–372. https://doi.org/10.1038/ng1294-365

Larkin P, Scowcroft W (1981) Somaclonal variation—a novel source of variability from cell cultures for plant improvement. Theor Appl Genet 60:197–214. https://doi.org/10.1007/BF02342540

Li H (2011) A statistical framework for SNP calling, mutation discovery, association mapping and population genetical parameter estimation from sequencing data. Bioinformatics 27(21):2987–2993. https://doi.org/10.1093/bioinformatics/btr509

Li H, Durbin R (2009) Fast and accurate short read alignment with Burrows-Wheeler transform. Bioinformatics 25(14):1754–1760. https://doi.org/10.1093/bioinformatics/btp324

Liu W, Zhang D, Tang M, Li D, Zhu Y, Zhu L, Chen C (2013) THIS1 is a putative lipase that regulates tillering, plant height, and spikelet fertility in rice. J Exp Bot 64(14):4389–4402. https://doi.org/10.1093/jxb/ert256

Liu X, Zhu K, Xiao J (2023) Recent advances in understanding of the epigenetic regulation of plant regeneration. aBIOTECH. https://doi.org/10.1007/s42994-022-00093-2

Liu Z, Han FP, Tan M, Shan XH, Dong YZ, Wang XZ, Fedak G, Hao S, Liu B (2004) Activation of a rice endogenous retrotransposon *Tos17* in tissue culture is accompanied by cytosine demethylation and causes heritable alteration in methylation pattern of flanking genomic regions. Theor Appl Genet 109(1):200–209. https://doi.org/10.1007/s00122-004-1618-8

McClintock B (1984) The significance of responses of the genome to challenge. Science 226:792–801. https://doi.org/10.1126/science.15739260

Miura A, Yonebayashi S, Watanabe K, Toyama T, Shimada H, Kakutani T (2001) Mobilization of transposons by a mutation abolishing full DNA methylation in *Arabidopsis*. Nature 411(6834):212–214. https://doi.org/10.1038/35075612

Miura K, Agetsuma M, Kitano H, Yoshimura A, Matsuoka M, Jacobsen SE, Ashikari M (2009) A metastable *DWARF1* epigenetic mutant affecting plant stature in rice. Proc Natl Acad Sci U S A 106(27):11218–11223. https://doi.org/10.1073/pnas.0901942106

Miyao A (2022) Unpublished data from *Tos17* mutant panel experiments

Miyao A and Hirochika H (2002) Large scale sequence analysis of rice mutant panel: characteristics of *Tos17* target loci. Paper presented at the 25th conference on the molecular biology society of Japan, Yokohama, 11–14 Dec 2002

Miyao A, Yamanouchi U (2022) Transposable element finder (TEF): finding active transposable elements from next generation sequencing data. BMC Bioinf 23:500. https://doi.org/10.1186/s12859-022-05011-3

Miyao A, Yamazaki M, Hirochika H (1998) Systematic screening of mutants of rice by sequencing retrotransposon insertion sites. Plant Biotech 15(4):253–256. https://doi.org/10.5511/plantbiotechnology.15.253

Miyao A, Tanaka K, Murata K, Sawaki H, Takeda S, Abe K, Shinozuka Y, Onosato K, Hirochika H (2003) Target site specificity of the *Tos17* retrotransposon shows a preference for insertion within genes and against insertion in retrotransposon-rich regions of the ge-nome. Plant Cell 15(8):1771–1780. https://doi.org/10.1105/tpc.012559

Miyao A, Iwasaki Y, Kitano H, Itoh J, Maekawa M, Murata K, Yatou O, Nagato Y, Hirochika H (2007) A large-scale collection of phenotypic data describing an insertional mutant population to facilitate functional analysis of rice genes. Plant Mol Biol 63(5):625–635. https://doi.org/10.1007/s11103-006-9118-7

Miyao A, Nakagome M, Ohnuma T, Yamagata H, Kanamori H, Katayose Y, Takahashi A, Matsumoto T, Hirochika H (2012) Molecular spectrum of somaclonal variation in regenerated rice revealed by whole-genome sequencing. Plant Cell Physiol 53(1):256–264. https://doi.org/10.1093/pcp/pcr172

Miyao A, Nakagome M, Ohnuma T, Sasaki H, Kanamori H, Ikawa H, Katayose Y, Takahashi A, Matsumoto T, Hirochika H (2012b) Somaclonal variation detected in rice by next-generation sequencing. Paper presented at the 20th plant and animal genome conference, San Diego, 14–18 Jan 2012

Miyao A, Kiyomiya JS, Iida K, Doi K, Yasue H (2019) Polymorphic edge detection (PED): two efficient methods of polymorphism detection from next-generation sequencing data. BMC Bioinf 20(1):362. https://doi.org/10.1186/s12859-019-2955-6

Nakagome M, Solovieva E, Takahashi A, Yasue H, Hirochika H, Miyao A (2014) Transposon insertion finder (TIF): a novel program for detection of de novo transpositions of transposable elements. BMC Bioinf 15:71. https://doi.org/10.1186/1471-2105-15-71

Nakamura N et al (2018) Akitasarari. https://www.hinshu2.maff.go.jp/vips/cmm/apCMM112.aspx?TOUROKU_NO=29393&LANGUAGE=English

Neelakandan AK, Wang K (2012) Recent progress in the understanding of tissue culture-induced genome level changes in plants and potential applications. Plant Cell Rep 31:597–620. https://doi.org/10.1007/s00299-011-1202-z

Nishi T, Yamada Y, Takahashi E (1968) Organ redifferentiation and plant restoration in rice callus. Nature 219:508–509. https://doi.org/10.1038/219508a0

Numa H, Yamaguchi K, Shigenobu S, Habu Y (2015) Gene body CG and CHG methylation and suppression of centromeric CHH methylation are mediated by DECREASE IN DNA METHYLATION1 in Rice. Mol Plant 8(10):1560–1562. https://doi.org/10.1016/j.molp.2015.08.002

Ogawa D, Abe K, Miyao A, Kojima M, Sakakibara H, Mizutani M, Morita H, Toda Y, Hobo T, Sato Y et al (2011) RSS1 regulates the cell cycle and maintains meristematic activity under stress conditions in rice. Nat Commun 2:278. https://doi.org/10.1038/ncomms1279

Osakabe, Jamge B, Axelsson E, Montgomery SA, Akimcheva S, Kuehn AL, Pisupati R, Lorković ZJ, Yelagandula R, Kakutani T (2021) The chromatin remodeler DDM1 prevents transposon mobility through deposition of histone variant H2A.W. Nat Cell Biol 23(4):391–400. https://doi.org/10.1038/s41556-021-00658-1

Ossowski S, Schneeberger K, Lucas-Lledó JI, Warthmann N, Clark RM, Shaw RG, Weigel D, Lynch M (2010) The rate and molecular spectrum of spontaneous mutations in *Arabidopsis thaliana*. Science 327(5961):92–94. https://doi.org/10.1126/science.1180677

Peschke VM, Phillips RL, Gengenbach BG (1987) Discovery of transposable element activity among progeny of tissue culture-derived maize plants. Science 238(4828):804–807. https://doi.org/10.1126/science.238.4828.804

Picault N, Chaparro C, Piegu B, Stenger W, Formey D, Llauro C, Descombin J, Sabot F, Lasserre E, Meynard D et al (2009) Identification of an active LTR retrotransposon in rice. Plant J 58(5):754–765. https://doi.org/10.1111/j.1365-313X.2009.03813.x

Poplin R, Ruano-Rubio V, DePristo MA, Fennell TJ, Carneiro MO, Van der Auwera GA, Kling DE, Gauthier LD, Levy-Moonshine A, Roazen D et al (2017) Scaling accurate genetic variant discovery to tens of thousands of samples. bioRxiv 201178. https://doi.org/10.1101/201178

Rice Full-Length cDNA Consortium (2003) Collection, mapping, and annotation of over 28,000 cDNA clones from *japonica* rice. Science 301(5631):376–379. https://doi.org/10.1126/science.1081288

Roth R, Chiapello M, Montero H, Gehrig P, Grossmann J, O'Holleran K, Hartken D, Walters F, Yang SY, Hillmer S et al (2018) A rice serine/threonine receptor-like kinase regulates arbuscular mycorrhizal symbiosis at the peri-arbuscular membrane. Nat Commun 9(1):4677. https://doi.org/10.1038/s41467-018-06865-z

Sanger F, Nicklen S, Coulson AR (1977) DNA sequencing with chain-terminating inhibitors. Proc Natl Acad Sci USA 74(12):5463–5467. https://doi.org/10.1073/pnas.74.12.5463

Sato Y, Sentoku N, Miura Y, Hirochika H, Kitano H, Matsuoka M (1999) Loss-of-function mutations in the rice homeobox gene *OSH15* affect the architecture of internodes resulting in dwarf plants. EMBO J 18(4):992–1002. https://doi.org/10.1093/emboj/18.4.992

Saiki RK, Scharf S, Faloona F, Mullis KB, Horn GT, Erlich HA, Arnheim N (1985) Enzymatic amplification of beta-globin genomic sequences and restriction site analysis for diagnosis of sickle cell anemia. Science 230(4732):1350–1354. https://doi.org/10.1126/science.2999980

Slotkin RK, Martienssen R (2007) Transposable elements and the epigenetic regulation of the genome. Nat Rev Genet 8(4):272–785. https://doi.org/10.1038/nrg2072

Stroud H, Ding B, Simon SA, Feng S, Bellizzi M, Pellegrini M, Wang GL, Meyers BC, Jacobsen SE (2013) Plants regenerated from tissue culture contain stable epigenome changes in rice. Elife 2:e00354. https://doi.org/10.7554/eLife.00354

Sugimoto K, Takeuchi Y, Ebana K, Miyao A, Hirochika H, Hara N, Ishiyama K, Kobayashi M, Ban Y, Hattori T et al (2010) Molecular cloning of *Sdr4*, a regulator involved in seed dormancy and domestication of rice. Proc Natl Acad Sci U S A 107(13):5792–5797. https://doi.org/10.1073/pnas.0911965107

Takano M, Kanegae H, Shinomura T, Miyao A, Hirochika H, Furuya M (2001) Isolation and characterization of rice phytochrome A mutants. Plant Cell 13(3):521–534. https://doi.org/10.1105/tpc.13.3.521

Tanaka K, Murata K, Yamazaki M, Onosato K, Miyao A, Hirochika H (2003) Three distinct rice cellulose synthase catalytic subunit genes required for cellulose synthesis in the secondary wall. Plant Physiol 133(1):73–83. https://doi.org/10.1104/pp.103.022442

Wang X, Wu R, Lin X, Bai Y, Song C, Yu X, Xu C, Zhao N, Dong Y, Liu B (2013) Tissue culture-induced genetic and epigenetic alterations in rice pure-lines, F1 hybrids and polyploids. BMC Plant Biol 13:77. https://doi.org/10.1186/1471-2229-13-77

White PR (1939) Potentially unlimited growth of excised plant callus in an artificial nutrient. Am J Bot 26:59–64. https://doi.org/10.1002/j.1537-2197.1939.tb12867.x

Zhao X, Qiu T, Feng H, Yin C, Zheng X, Yang J, Peng YL, Zhao W (2021) A novel glycine-rich domain protein, GRDP1, functions as a critical feedback regulator for controlling cell death and disease resistance in rice. J Exp Bot 72(2):608–622. https://doi.org/10.1093/jxb/eraa450

The manufacturer's authorised representative in the EU is Springer Nature Customer Service Centre GmbH, Europaplatz 3, 69115 Heidelberg, Germany. If you have any concerns regarding our products, please contact ProductSafety@springernature.com

Printed and bound by CPI Group (UK) Ltd, Croydon, CR0 4YY
01/10/2026
02225896-0002